Behavioral Epidemiology
and Disease Prevention

NATO ASI Series

Advanced Science Institutes Series

A series presenting the results of activities sponsored by the NATO Science Committee, which aims at the dissemination of advanced scientific and technological knowledge, with a view to strengthening links between scientific communities.

The series is published by an international board of publishers in conjunction with the NATO Scientific Affairs Division

A	**Life Sciences**	Plenum Publishing Corporation
B	**Physics**	New York and London
C	**Mathematical and Physical Sciences**	D. Reidel Publishing Company Dordrecht, Boston, and Lancaster
D	**Behavioral and Social Sciences**	Martinus Nijhoff Publishers
E	**Engineering and Materials Sciences**	The Hague, Boston, and Lancaster
F	**Computer and Systems Sciences**	Springer-Verlag
G	**Ecological Sciences**	Berlin, Heidelberg, New York, and Tokyo

Recent Volumes in this Series

Volume 82—Receptor-Mediated Targeting of Drugs
Edited by G. Gregoriadis, G. Poste, J. Senior, and A. Trouet

Volume 83—Molecular Form and Function of the Plant Genome
Edited by Lous van Vloten-Doting, Gert S. P. Groot, and Timothy C. Hall

Volume 84—Behavioral Epidemiology and Disease Prevention
Edited by Robert M. Kaplan and Michael H. Criqui

Volume 85—Primary Photo-Processes in Biology and Medicine
Edited by R. V. Bensasson, G. Jori, E. J. Land, and T. G. Truscott

Volume 86—Wheat Growth and Modelling
Edited by W. Day and R. K. Atkin

Volume 87—Industrial Aspects of Biochemistry and Genetics
Edited by N. Gürdal Alaeddinöglu, Arnold L. Demain, and Giancarlo Lancini

Volume 88—Radiolabeled Cellular Blood Elements
Edited by M. L. Thakur

Series A: Life Sciences

Behavioral Epidemiology
and Disease Prevention

Edited by
Robert M. Kaplan
and
Michael H. Criqui

University of California, San Diego
La Jolla, California

Plenum Press
New York and London
Published in cooperation with NATO Scientific Affairs Division

Proceedings of a NATO Advanced Research Workshop on
Behavioral Epidemiology and Disease Prevention,
held April 14–20, 1983,
at the Bellagio Study and Conference Center, Lake Como, Italy

Library of Congress Cataloging in Publication Data

NATO Advanced Research Workshop on Behavioral Epidemiology and
Disease Prevention (1983: Bellagio Study and Conference Center)
  Behavioral epidemiology and disease prevention.

  (NATO ASI series. Series A, Life sciences; v. 84)
  "Proceedings of a NATO Advanved Research Workshop on Behavioral
Epidemiology and Disease Prevention, held April 14–20, 1983, at the Ballagio
Study and Conference Center, Lake Como, Italy"—T.p. verso.
  "Published in cooperation with NATO Scientific Affairs Division."
  Includes bibliographies and index.
  1. Medicine and psychology—Congresses. 2. Epidemiology—Congresses.
3. Health behavior—Congresses. 4. Medicine, Preventive—Congresses. I.
Kaplan, Robert M. II. Criqui, Michael H. III. North Atlantic Treaty Organiza-
tion. Scientific Affairs Division. IV. Title. V. Series. [DNLM: 1. Behavioral Medi-
cine—congresses. 2. Epidemiology—congresses. 3. Preventive Medicine—
congresses. WA 105 N279b 1983]
R726.5.N36   1983   616.07′1   85-3611
ISBN 0-306-41929-7

©1985 Plenum Press, New York
A Division of Plenum Publishing Corporation
233 Spring Street, New York, N.Y. 10013

Printed in the United States of America

PREFACE

In the United States and in Europe, there has been an increasing interest in the relationship between individual behavior and disease. The American National Academy of Sciences (Hamberg, Elliott, and Parron, 1982), through its Institutes of Medicine, has estimated that as many as 50 percent of chronic disease cases can be traced to individual behaviors such as smoking, diet, exercise, etc. Similar conclusions have been reached by a variety of European investigators and institutes. The World Health Organization has also expressed considerable interest in individual behavior in relation to the development of chronic disease. Thus, throughout the NATO countries there has been increased awareness of the relationship between behavior and disease. However, communication among investigators in different countries has been rather limited. Further, many different scientific disciplines including psychology, sociology, medicine, microbiology, statistics, and epidemiology have all developed new and different literatures in this field. One purpose of this book is to bring together contributions from scientists in each of these fields. Much of the variance in individual health behavior occurs across countries rather than within countries. Thus, we can learn much from comparing behavior-disease relationships across countries. To date, there have been few studies which have had an adequate international basis for these comparisons.

Interest in behavioral epidemiology is a relatively recent phenomena. Thus, many scientists are entering the field without uniform background, experience, or training. This book discusses approaches common in a variety of NATO countries. The chapters focus on behavior patterns such as diet, exercise, and cigarette smoking which vary across countries. For example, cigarette smoking is on the decline in the United States while it is on the increase in several European countries. Thus, we can learn much about behavior and disease relationships by examining cross-national data. On the other hand, some observed relationships within countries may not occur in other countries. In order to sort out these data, international studies are needed. Contributors were selected because of known records of productive collaborative research.

Although there are a variety of recent books on epidemiology and on behavioral interventions, we are not aware of a publication that attempts to systematically integrate these areas from an international perspective.

v

Preface

With the assistance of a NATO Advanced Research Workshop grant, the contributors were brought together at the Bellagio Study and Conference Center near Lake Como in Italy in April of 1983. Preliminary versions of the chapters were presented and critiqued at the meeting. Then, using feedback from the other participants, each author continued to develop their manuscript until they eventually resulted in the chapters for this book.

The contributors to this volume represent only a small sample of the distinguished investigators interested in behavioral epidemiology. Our funding only allowed us to bring together a small group and we had to select participants to represent specific points of view. Clearly there are many others who could have offered different and important insights.

This book would not have been possible without the help of several individuals and institutions. First, we are grateful to the North Atlantic Treaty Organization, Division of Scientific Affairs who provided the financial support for the meeting. Second, The Rockefeller foundation generously provided the Bellagio Conference and Study Center for the meeting. This excellent facility provided an ideal environment for the nurturing of many of the ideas represented in this volume. Susan Garfield of the Rockefeller Foundation helped with many of the arrangements while Angela Barmettler served as an exceptional hostess at the Bellagio Center.

Besides the authors, many others contributed to this volume. Nan Criqui helped with many of the arrangements at the Bellagio Center. Debbe O'Brien and Becky Caruso helped with the countless clerical chores. Connie Toevs and Therese Chaucon worked on many aspects of the conference organization and the subsequent manuscript preparation. Valerie Ruud did a superb job of editing the papers. Finally, we are most appreciative of Barbara Dupuie Boyer who typed the entire manuscript and supervised the typesetting.

RMK
MHC
La Jolla, California, USA

Reference

Hamberg, D. A., Elliott, G. R., and Parron, D. L. *Health and Behavior Frontiers of Research in the Biobehavioral Sciences.* Washington, D.C.: National Academy Press, 1982.

CONTENTS

PART I

Methods in Epidemiology

PART II

Epidemiology

Contents

PART III

Stress

PART IV

Behavioral Interventions

PART I

METHODS IN EPIDEMIOLOGY

This first section reviews selected methodologic issues in the interface between epidemiology and behavioral science. The first paper, by Trichopoulos and Sparos provides a general overview of epidemiologic research methodology. The second paper, by Criqui, reviews response bias which is a behaviorally based and major problem in epidemiologic surveys. Finally, R. Kaplan describes general health outcome measures and the interface between epidemiology and policy science.

1

REFLECTIONS ON EPIDEMIOLOGIC METHODOLOGY IN THE STUDY OF DISEASE ETIOLOGY AND HEALTH SERVICES

Dimitrios Trichopoulos, M.D.
Loucas Sparos, M.D.

Some epidemiologists have written about the aims and uses of epidemiology, its strengths and limitations, and its achievements and prospects; others have concentrated on the conceptual and methodological aspects of the new discipline. Our objective is, by necessity, more modest; we shall discuss the evolution of epidemiologic methodology and the current thinking on methodology.

Epidemiology has flourished in the study of disease causation and in the planning and, particularly, the evaluation of health services. The separation of etiologic epidemiology into that of "infectious" and that of "non-infectious" diseases does not appear justifiable on theoretical grounds, since noninfectious factors are almost always involved in the etiology of infectious diseases, and infectious agents are present in the causal web of several neoplastic, neurologic, and metabolic diseases. Thus, in the first part of this chapter selected methodological aspects of general etiologic epidemiology will be considered. In the second part, the epidemiologic methods commonly used in the evaluation of health services will be presented more systematically. We believe that in etiologic epidemiology methodological sophistication has already advanced beyond the point of limited returns; whereas in the study of health services, inadequate attention has been paid to the development of evaluation for the various health measures, procedures, and programs.

1. Etiologic Epidemiology

The elaboration of concepts and the development of methods in etiologic research took place after the Second World War and coincided with a switch in emphasis from acute to chronic diseases. The temporal association was not incidental. In inductive and deductive etiologic research, the most popular effect parameter is the relative risk, including both the rate ratio and the risk ratio (Breslow and Day, 1980; MacMahon and Pugh, 1970; Miettinen, 1976; Rothman and Boice, 1979). Since most acute diseases are defined using etiologic criteria (associated with infinite values of relative risk) and are characterized by relatively short latent periods (usually hours, days, or weeks), etiologic research would appear to be methodologically simple, and limited only by the technology for the identification of the respective etiologic factors. By contrast, most chronic diseases are defined with manifestational criteria and are characterized by etiologic heterogeneity and very long latent periods. Under these circumstances the establishment of causality becomes difficult, even when there are no technological obstacles to the identification of the suspected agents. Indeed, most epidemiologic studies of chronic diseases focus on factors that may generate no more than a fivefold risk differential, several years before the clinical or even the biological onset of the disease. It should be clear that, for effective planning and analysis of these studies, more sophisticated, powerful, and informative methods are required. These methods form the core of what is now considered the discipline of modern epidemiology (Breslow and Day, 1980; Kleinbaum, Kupper, and Morgenstern, 1982; MacMahon and Pugh, 1970; Rothman and Boice, 1979).

The realization that "informativeness" is of such critical importance in the epidemiologic investigation of most chronic diseases and the demonstration of the superiority, in this context, of the case-control approach (Breslow and Day, 1980; Schlesselman, 1982) (stemming from the fact that in most natural situations most diseases are very rare, whereas many causal factors may not be so) help to explain why this approach occupies such a central position in current epidemiologic practice[1] and methodologic thinking. They help also to explain why the papers by

---

[1] In the introductory chapter to the most comprehensive book in print on the analysis of case-control studies (Breslow and Day, 1980), Cole estimated that in the field of cancer epidemiology case-control studies outnumber cohort studies by 2.2 to 1. Impressive as this figure may be it is still an underestimate. Cole's calculation is based on the entries in the "Directory of On-going Research in Cancer Epidemiology;" however, these are "prevalence" entries and, as such, tend to over-represent the cohort studies, on account of their much longer average duration.

Cornfield (1951) and Mantel and Haenszel (1959), together with the classical textbook of MacMahon, Pugh, and Ipsen (1960), have been the basic methodological references for a whole generation of epidemiologists.

Several methodological issues have attracted the interest of epidemiologists in the last thirty years, but none as regularly and as vividly as confounding (Miettinen and Cook, 1981). Although neither is new or peculiar to epidemiology, the term and concept have been identified with this discipline almost as much as has relative risk. There are several reasons for this preference, over and beyond the essential role of confounding in the interpretation of the observed associations. Unlike other forms of bias, confounding is not a consequence of inadequate study design or implementation, but an accurate reflexion of nature and of the established equilibrium of causal forces. Furthermore, the implications and the procedures to control confounding are very different in the cohort studies and in the (uniquely epidemiological) case-control investigations. Lastly, it is frequently impossible to gain insight into the nature of confounding without a clear understanding of the substantive issues involved--a fact that points to the biomedical elements of this subtle and complex concept (Miettinen and Cook, 1981).

Control of confounding is one of the two most pressing issues in epidemiologic research (Breslow and Day, 1980; Kleinbaum et al., 1982; MacMahon and Pugh, 1970; Rothman and Boice, 1979; Schlesselman, 1982) (the other being control of bias). Control can be achieved in the design stage by restriction of the exposure matrix or in the analysis stage by stratification or mathematical modeling. Notwithstanding the large and increasing popularity of modeling, we believe that the stratified analysis is usually adequate and always more comprehensible (Rothman and Boice, 1979). Even when modeling appears to be the only effective approach for complete control of confounding, it is still useful to do a stratified analysis for the two or three most important confounders (jointly), and the results are frequently surprisingly similar. Finally, individual matching as a procedure to increase efficiency in the stratified analysis is justified only in small case-control studies with several strong confounding effects.

Control of bias is the second major problem in epidemiologic research. Given the multitude of the sources of biases in epidemiologic work, their variable nature, and the limitations of the existing procedures to identify and quantify them, the complete elimination of bias from epidemiologic studies is perhaps unattainable. However, as Cole (1979) and others have pointed out, it is not the absence of bias but the balance of bias between the compared groups (cases and controls in retrospective studies, exposed and non-exposed in cohort investigations) that is the prerequisite for validity. This, in turn, indicates that in epidemiologic work comparability of the studied groups is a critical issue, while their representatives is not--unless the selection factor (which is responsible for the unrepresentativeness of the compared groups) interacts with the other causal factors of the disease under investigation.

The preceding discussion brings into focus the issue of interaction, which is central to both the concept of multifactorial causation (Rothman, 1976) and its biological and statistical considerations. Recent studies by Rothman, Greenland, and Walker (1980) and others (Kleinbaum et al., 1982), elaborating on the additivity of causal factors when they have pathogenetically independent effects and on their multiplicativity when they have complementary pathogenetic processes, may facilitate the biological interpretation of statistical models of disease causation based on epidemiological investigations. However, the notorious lack of power of most statistical methods to discriminate between additivity and multiplicativity at the usual range of relative effect (i.e., relative risks of up to fivefold) and the methodological complications that accompany most changes in the scale of measurement hamper these efforts.

Two other related methodological developments have grown beyond the traditional methodological boundaries and have affected the current epidemiologic practice. The first is based on the realization that the principal effect parameter in epidemiologic research is the rate ratio (rather than the risk ratio) and this is directly estimable from both the cohort and the case-control studies (Miettinen, 1976). The second concerns the general (almost conceptual) suitability of the proportional hazards model (and, to a lesser extent, of the logistic regression model) to all types of epidemiologic analysis (Breslow and Day, 1980; Kleinbaum et al., 1982; Schlesselman, 1982), and the implications of this suitability for such practical problems as the setting of rules for control selection (Breslow and Day, 1980; Cole, 1979).

We would like to conclude this section by pointing out two other domains in which important changes took place during the last decade, even though none of them relates to analytic methodology. The first change, which concerns, the presentation of results is the gradual introduction of statistical testing and point estimation with interval estimation and confidence setting (Breslow and Day, 1980; Cole, 1979, Rothman and Boice, 1979). The second change, which concerns the interpretation of results, is the increasing influence of those who advocate the Bayesian approach. This influence is evident not only in the formal "discussion" sections of many research papers but also in the handling of the "multiple comparisons" and other related issues, and in the critical judgement (Peto et al., 1976, 1977) of "positive" findings when they arise from "small" studies (i.e., studies with limited power).

2. Epidemiology in Health Services Evaluation

Evaluation, and in particular outcome evaluation, has been thought of as a distinct chapter in health services research and as a separate and rather late activity in the implementation of these services. However, as Knox (1979), and others, have pointed out, evaluation is better thought of as a continuous activity and as an integral part of all health services research and implementation. Since behavioral epidemiology deals as much with the implementation of behavioral changes for disease prevention and health promotion as with the understanding of the role of behavioral factors in disease etiology, and since there are relatively few texts dealing with the epidemiologic methods used in health services evaluation, these methods will be summarized here briefly. Part of the discussion that follows is based on a forthcoming volume of the Commission of the EEC, to which one of the authors of this chapter has contributed as author and member of the editorial board.

Health care encompasses simple preventive, diagnostic, and therapeutic measures, as well as whole systems of health care delivery. Preventive measures may aim at primary or at secondary prevention. In primary prevention, the objective is the reduction of the incidence of the disease by reducing the risk of onset, whereas in secondary prevention the objective is the reduction of the mortality of the disease by advancing the diagnosis at a presymptomatic state characterized by better clinical prognosis. Therapeutic measures may be substantial or symptomatic. The objective of substantial therapeutic measures is to affect the natural history of the corresponding disease, whereas the objective of symptomatic therapeutic measures is to alleviate one or more symptoms. Finally, complex systems of health care are mixtures of preventive, diagnostic, and therapeutic measures, brought together by administrative structures operating under economic, social, and political constraints. Their objectives are almost always multi-dimensional and so by necessity, is, their evaluation.

Preventive and therapeutic measures and programs may be evaluated with respect to their safety, effectiveness, and efficiency. Safety measures the potential for adverse effects; effectiveness measures the beneficial impact on the outcome under study; and efficiency introduces cost aspects into the evaluation. Since effectiveness refers to a pre-specified outcome, the selection of an appropriate outcome criterion is paramount. Inappropriate outcome criteria may be classified into two groups: (i) *Irrelevant or insensitive criteria.* For example, many psychotropic drugs may have a substantial impact on the quality of life, which is impossible to evaluate with the use of mortality or fatality indices. (ii) *Biased criteria.* For example, it would be inappropriate to compare the survival of screening-detected cases of cancer with the survival of cases detected after the appearance of the first symptoms (Cole and Morrison 1980), even without any beneficial effect of screening, the survival in the first group would be better because of the additive effect of the lead-time bias and the length-time bias.

In the following sections we shall review briefly the epidemiologic methods for the outcome evaluation of health services. These methods may be experimental with randomization, experimental without randomization, observational analytic, and observational descriptive. Experiments with randomization (randomized trials) are not always possible, particularly when the evaluation concerns complex programs or systems of health care. However, they represent the best existing method (Anonymous, 1980; Peto et al., 1976, 1977), and the other available procedures are frequently judged by the degree of their similarity to the RCT.

7

Therefore, knowledge of the structure and function of the RCT is essential for a more general understanding of the methodology of evaluation.

2.1 The randomized controlled trial (RCT)

The conceptual basis of the RCT is that individuals are allocated randomly and evenly into the study and control measure (preventive or therapeutic) or into the study and control program (a program is a combination of preventive and/or therapeutic and/or administrative measures). As a consequence, all factors and characteristics that may affect outcome are equally distributed between the two (or more) groups; irregularities can only occur by chance with a measurable probability. Therefore, RCTs are in principle free from confounding. In addition to the classical simple design of RCT, other randomized designs have been used or proposed (Zelen, 1979), mostly in clinical research. They include the following: (i) *Restricted randomization.* The principle of this design (which can safeguard validity and increase efficiency) is that randomization and all subsequent comparisons are done "within strata," homogeneous with respect to one or more factors of prognostic importance (Feinstein and Landis, 1976). (ii) *Sequential analysis.* In this design the data are continuously analyzed as the outcomes become known, and the trial is legitimately interrupted as soon as statistical significance is reached (Armitage, 1971). (iii) *Factorial designs.* These designs permit the simultaneous evaluation of two or more treatments as well as of their potential interactions (Armitage, 1971). (iv) *Co-operative clinical trials.* In many clinical trials the required number of patients is much larger than the number of patients admitted in any individual center during any reasonable time period. To overcome this difficulty multi-center clinical trials have been undertaken, involving two or several centers in the same or several countries (Cancer Research Campaign Working Party, Trials and Tribulation, 1980; Peto et al., 1976, 1977). (v) *Trials on collective units.* They are characterized by random distribution, not of individuals, but of groups of individuals (villages, factories, schools, etc). Collective units must be used when the intervention programs can only be applied to whole communities (e.g., campaigns through the media), when individually applied intervention programs have collective effects (e.g., herd immunity), and when there is danger of extensive "contamination" effects (i.e., unplanned cross-over of individuals between the various study and comparison groups). Collective units consist of clusters of individuals, and the parameters (notably the variance) should be calculated by considering the cluster, not the individual, as the unit (Cornfield, 1978).

A statistically significant result, even in a well planned and carefully executed RCT, must be interpreted with caution. The result may be a consequence of multiple testing or may reflect an unrecognized bias. Whether a real result is of sufficient medical importance and whether a real and important result is generalizable to other population groups should also be explored (Commstock, 1978; Sackett, 1980). Still more caution is required for the interpretation of a negative RCT (results statistically insignificant) (Anonymous, 1978). A small RCT can be negative, even though the study program is considerably more effective than the control program. Even the fact that the statistically nonsignificant result emerged from a large trial with a placebo control group does not necessarily imply that the study program is ineffective. Alternative explanations are that the dosage was inadequate, the pattern of administration inappropriate, or the compliance unsatisfactory.

2.2 Experimental evaluation of complex systems of health care

Randomized trials are eminently suitable for the evaluation of simple preventive and therapeutic measures. The comparison of multi-dimensional systems of health care should be based on similar principles, but the application of the RCT for the evaluation of even moderately complex health services is beset by difficulties, for the following reasons (United States National Commission for the protection of human subjects of biomedical and behavioral research: The Belmont Report. 1976): (i) Randomization is not always possible either because the administrative and organization difficulties are insurmountable or because the ethical problems are considered prohibitive, and (ii) The objectives of health care systems are usually composite and complex; therefore the evaluation is inevitably multidimensional, making all but impossible the randomized comparison of the individual components.

These difficulties have led to the use of non-randomized trials or of observational studies. In non-randomized trials there is always a deliberate administration of the intervention program, but one that relies on a reasoned choice and not on chance. Several types of non-randomized trials have been proposed or actually applied, including the following (Cook and Campbell, 1979): (i) *Uncontrolled trials with a single examination after the intervention.* The lack of a control group makes the interpretation of results difficult and often impossible. Such trials should be reserved for situations where a reference situation is implicitly present. (ii) *Before-after comparison.* The evaluation criterion is measured before and after the intervention. The method is sensitive and convenient, but it is difficult to use it to evaluate long-term effects, because several factors may interfere with both the intervention program and the outcome criterion. (iii) *Controlled non-randomized trials.* In this design, the program under evaluation is administered to consenting volunteers, and the frequency and/or level of the outcome among them is compared with the corresponding values among non-volunteers who have not received the study program. The validity of this design depends on the comparability between volunteers and non-volunteers with respect to factors that may affect the outcome of the study, and such comparability is, unfortunately, the exception rather than the rule. Although several statistical methods aimed to control confounding and other forms of measurable bias are now available, including stratification and multivariate modeling, none of them is as effective as randomization as a means to insure study validity. There are several

9

variations of this basic design. In one of them a before-after comparison is added; this allows the comparability of the groups before the intervention program to be evaluated, and enables changes in parameters for each of the groups to be measured.

2.3   Observational evaluations of complex systems of health care

Observational studies may be analytical (cohort or case-control) or descriptive (population correlations in time or place). They have long been used in etiologic research, but their application for the evaluation of the effectiveness of health services is realtively recent (Holland and Karhausen, 1978). Analytical studies are usually designed *ad hoc* and based on individual associations, whereas descriptive studies are usually derived from routine statistics and based on population correlations.

Retrospective (case-control) studies. Individuals with and without the outcome under consideration are compared with respect to their exposure to the study program. The retrospective approach has been used in evaluative studies of primary prevention and, more recently, has been successfully tried for the evaluation of screening procedures (Clarke and Anderson, 1979).

Prospective (cohort) studies. Individuals exposed and not exposed to the study program are compared with respect to the frequency or level of outcome under consideration. It should be remembered that the two (rarely more) groups are formed spontaneously, contrary to what happens in an experiment. Prospective etiologic studies and prospective studies of health care have many similarities, but they differ in a number of points including the fact that the parameters that affect the choice of program are frequently determinants of disease outcome (confounding by indication) (Greenland and Neutra, 1980; Miettinen, 1979). Nevertheless, biases are usually few and weaker in prospective than in retrospective studies of health care. Prospective studies have been used much more frequently than retrospective studies in the evaluatiion of preventive and therapeutic programs as well as of more complex systems of health care (Goujard et al., in press; Lee, Morrison, and Morris, 1957; Lipworth, Lee, and Morris, 1963). In all these studies control of confounding in the analysis (preferably by multivariate techniques) is paramount.

Descriptive observational studies. These studies are based on population correlations. In a typical population correlation study the proportion of those exposed to a particular program is correlated with the proportion of those who show the target outcome. The correlation may be over different countries (international correlation), over different regions of the same country (intra-country correlation), over various socio-economic groups, or over consecutive time periods (time trends). Thus, the mortality from cervical cancer may be correlated with the extent of application of cytological screening or the incidence of an infectious disease may be examined in relation to the intensification of a relevant vaccination campaign. In both these instances there is no information concerning any single individual; all the available data refer to collective units (population groups). The interpretation of population correlation studies is not easy. Difficulties arise from the use of populations as sampling units, the varying latent interval of the various health measures, and the presence of many confounding factors. Nevertheless these studies have been successfully used in the evaluation of several and varied types of health care (Guzick, 1978; Knox, 1976; Rumeau-Rouquette, 1979).

Frequently, health services are evaluated on the basis of empirical standards, which are considered as "intermediate" objectives ("process" evaluation). The value of these standards (Brook, 1977) rests on the assumption that they represent necessary (although hardly sufficient) conditions for the achievement of the corresponding ultimate objectives. The routine monitoring of health services functions in terms of empirical procedural standards is the central activity of medical audit (Shaw, 1980). The process evaluation may be done by systems analysis, although this approach is more suitable for management rather than evaluation (WHO Expert Committee, 1976). Systems analysis is particularly appealing because it allows an integrated (rather than analytical) approach to the evaluation of complex systems of health care.

References

Anonymous. Interpreting clinical trials. *British Medical Journal* 2 (1978) 1318.

Anonymous. Defending the controlled trials. *Lancet* 2 (1980) 731.

Armitage, P. *Statistical methods in medical research.* Oxford: Blackwell Scientific, 1971. Pp. 226-239, 415-425.

Breslow, N. and Day, N. *Statistical methods in cancer research.* Volume 1. The analysis of case-control studies. IARC Scientific Publications No. 32. Lyon: World Health Organization. International Agency for Research on Cancer, 1980.

Brook, R. H. Quality--Can we measure it? *New England Journal of Medicine* 296 (1977) 170-172.

Cancer research campaign working party, trials and tribulations: Thoughts on the organisation of multicentre clinical studies. *British Medical Journal* 2 (1980) 918-920.

Clarke, E. A. and Anderson, T. W. Does screening by "Pap" smears help prevent cervical cancer? *Lancet* 2 (1979) 1-4.

Cole, P. The evolving case-control study. *Journal of Chronic Diseases* 32 (1979) 15-27.

Cole, P. and Morrison, A. S. Basic issues in population screening for cancer. *Journal of the National Cancer Institute* 64 (1980) 1263-1272.

Commstock, G. W. Uncontrolled ruminations on modern controlled trials. *American Journal of Epidemiology* 108 (1978) 81-84.

Cook, T. D. and Campbell, D. T. *Quasi experimentation: Design and analysis issues for field setting.* Chicago, Ill.: Rand McNally College Publishing Co., 1979.

Cornfield, J. A method of estimating comparative rates from clinical data. Applications to cancer of the lung, breast and cervix. *Journal of the National Cancer Institute* 11 (1951) 1269-1275.

Cornfield, J. Randomization by group: A formal analysis. *American Journal of Epidemiology* 108 (1978) 100-102.

Feinstein, A. R. and Landis, R. J. The role of prognostic stratification in preventing the bias permitted by random allocation of treatment. *Journal of Chronic Diseases* 29 (1976) 277-284.

Goujard, J., Crost, M., Delecour, M., Dubois, O., Du Mazaubrun, C., Ponte, C., and Rumeau-Rouquette, C. Transfert des nouveau-nés dans la région nord. Facteurs de risque de mortalité des enfants de moins de 2000 g. *Archives Françaises de Pédiatrie,* (in press).

Greenland, S. and Neutra, R. Control of confounding in the assessment of medical technology. *International Journal of Epidemiology* 9 (1980) 361-367.

Guzick, D. S. Efficacy of screening for cervical cancer: A review. *American Journal of Public Health* 68 (1978) 125-133.

Holland, W. W. and Karhausen, L. (eds.) *Health Care and Epidemiology.* London: Henry-Kimpton, 1978.

Kleinbaum, D., Kupper, L., and Morgenstern, H. *Epidemiologic research. Principles and quantitative methods.* London: Lifetime Learning Publications, 1982.

Knox, E. G. Control of haemolytic disease of the newborn. *British Journal of Preventive and Social Medicine* 30 (1976) 163-169.

Knox, E. G. (ed.) *Epidemiology in health care planning.* Oxford: Oxford University Press, 1979.

Lee, J. A., Morrison, S. L. and Morris, J. N. Fatality from three common surgical conditions in teaching and non-teaching hospitals. *Lancet* 2 (1957) 785-790.

Lipworth, L., Lee, J. A., and Morris, J. N. Case fatality in teaching and non-teaching hospitals, 1956-59. *Medical Care* (London) 1 (1973) 71-76.

MacMahon, B. and Pugh, T. *Epidemiology. Principles and methods.* Boston: Little, Brown, 1970.

MacMahon, B., Pugh, T., and Ipsen, J. *Epidemiologic methods.* Boston: Little, Brown, 1960.

Mantel, N. and Haenszel, W. Statistical aspects of the analysis of data from retrospective studies of disease. *Journal of the National Cancer Institute* 22 (1959) 719-748.

Miettinen, O. Estimability and estimation in case-referent studies. *American Journal of Epidemiology* 103 (1976) 226-235.

Miettinen, O. S. Public health policy on coronary heart disease. *Heart Bulletin* 10 (1979) 165-167.

Miettinen, O. and Cook, E. Confounding: Essence and detection. *American Journal of Epidemiology* 114 (1981) 593-603.

Peto, R., Pike, M. C., Armitage, P., Breslow, N. E., Cox, D. R., Howard, S. V., Mantel, N., McPherson, K., Peto, J., and Smith, P. G. Design and analysis of randomized clinical trials requiring prolonged observation of each patient. I. Introduction and design. *British Journal of Cancer* 34 (1976) 585-612. II. Analysis and examples. *British Journal of Cancer* 35 (1977) 1-39.

Rothman, K. Causes. *American Journal of Epidemiology* 104 (1976) 587-592.

Rothman, K. and Boice, J., Jr. *Epidemiologic analysis with a programmable calculator.* Washington, D.C.: U.S. Government Printing Office, 1979.

Rothman, K., Greenland, S., and Walker, A. Concepts of interaction. *American Journal of Epiedemiology* 112 (1980) 467-470.

Rumeau-Rouquette, C. *Naître en France.* 1 vol. INSERM, 1979.

Sackett, D. L. The competing objectives of randomized trials. *New England Journal of Medicine* 303 (1980) 1059-1060.

Schlesselman, J. *Case-control studies. Design, conduct, analysis.* New York: Oxford University Press, 1982.

Shaw, C. D. Aspects of audit: (1) The background. *British Medical Journal* 1 (1980) 1256-1258.

United States. National Commission for the protection of human subjects of biomedical and behavioral research. The Belmont Report. Ethical principles and guidelines for the protection of human subjects of research. DHEW, Washington, 1976 (Pub. No. (OS) 78-0013).

WHO Expert Committee. Application of systems analysis to health management. *World Health Organization, Technical Report Series* No. 596. Geneva, 1976.

Zelen, M. A new design for randomized clinical trials. *New England Journal of Medicine* 300 (1979) 1242-1245.

2

THE PROBLEM OF RESPONSE BIAS

Michael H. Criqui, M.D., M.P.H.

Historically, there has been considerable concern about the problem of biased results in epidemiologic studies of volunteers because the sample is limited to persons willing to be evaluated (Friedman, 1974; Lilienfeld, 1976; MacMahon and Pugh, 1970; Mausner and Bahn, 1974). The realization that non-respondents might be quite different from respondents in certain critical characteristics has generated a free-floating uncertainty, since, by definition data is usually not available on non-respondents. Occasionally, however, investigators have attempted to gather data on identified non-respondents, usually by indirect means, such as data already recorded for other reasons, or by obtaining available subsequent information, such as cause of death on death certificates (Cobb, King, Chew, 1957; Doll and Hill, 1964; Gordon et al., 1959; Hammond, 1969; Heilbrun, Nomura, Stemmermann, 1982; Horowitz and Wilbeck, 1971; Wilhelmsen et al., 1976). Between 1972 and 1974, in a population study of over 5,000 older Rancho Bernardo, California, U.S. residents for cardiovascular disease, we directly contacted by telephone about 77 percent of the 1,103 non-respondents to our survey, and asked them questions concerning both risk factors for and the prevalence of cardiovascular diseases. We did this simply to reassure ourselves that the non-respondents were not so markedly different as to invalidate the results of our subsequent analyses of the data. What we found, although of seemingly limited portent initially, subsequently led us to years of research and analysis of this question and to date five publications in the epidemiologic literature (Austin et al., 1981; Criqui, 1979; Criqui, Austin, and Barrett-Connor, 1979; Criqui, Barrett-Connor, and Austin, 1978; Greenland and Criqui, 1981). Our insight was not epiphanic but evolutionary and stepwise, over a period years.

This chapter summarizes our current best understanding of this complex problem. After defining response bias, we will address response bias distortions in estimates of prevalence, in estimates of incidence, and finally in risk ratio estimates. We will employ only dichotomous definitions of both risk factors and diseases, because this simplifies the presentation and allows for 2x2 contingency table illustrations.

2. What is Response Bias?

There has been some argument over the term response bias. The argument centers around the observation that whether or not dissimilarities between respondents and non-respondents produce bias depends on the measurements and parameters of interest. Pragmatically, we measure risk factors (or exposures) and diseases (or outcomes), whether retrospectively, cross-sectionally, or prospectively, and are interested in (1) risk factor and disease prevalence and/or incidence and (2) an estimate of the extent of the association between the risk factor and the disease, usually the relative risk in prospective studies and the odds ratio when the investigation is retrospective. However, in a given situation the prevalence (or incidence) estimates for a data set may be exactly representative of the target population while the risk ratios are greatly distorted. Conversely, the prevalence (or incidence) estimates might be markedly distorted, but the risk ratio exactly representative. This principle is illustrated in Tables 1 to 3. Table 1 shows a target population of 1,000 middle-aged men, 200 of whom have a risk factor (e.g., elevated cholesterol) and 100 of whom have a disease (e.g., coronary heart disease). The four inner cells are labeled A, B, C, and D.

The prevalence of the risk factor (A+B) is 20 percent, of disease (A+C), 10 percent. If the study were prospective, disease would be incidence, and we would calculate the relative risk

$$\frac{A/A+B}{C/C+D} = \left(\frac{40/200}{60/800}\right) = 2.67$$

Were the data retrospective, the odds ratio would be

$$\frac{AD}{BC} = \frac{(40)(740)}{(160)(60)} = 3.08.$$

Table 1

Target Population

	Dis (disease)	$\overline{Dis}$ (no disease)	T (total)
RF (risk factor)	40 (A)	160 (B)	200 (A+B)
$\overline{RF}$ (no risk factor)	60 (C)	740 (D)	800 (C+D)
T (total)	100 (A+C)	900 (B+D)	1000 (A+B+C+D)

Table 2

	Dis	$\overline{Dis}$	T
RF	39(A)	61(B)	100
$\overline{RF}$	11(C)	389(D)	400
T	50	450	500

Table 2 shows a situation where only 50 percent of the target population responded.

The prevalence of the risk factor is unbiased (20 percent), as is the prevalence (or incidence) of the disease (10 percent). However, the relative risk,

$$\frac{39/100}{11/400} = 14.18$$

is distorted by more than a factor of 5; the odds ratio

$$\frac{(39)(389)}{(11)(61)} = 22.6$$

by more than a factor of 7. Thus unbiased prevalence (or incidence) does not necessarily imply unbiased risk ratios.

Table 3

	Dis	*Dis*	T
RF	40(A)	160(B)	200
$\overline{RF}$	6(C)	74(D)	80
T	46	234	280

Table 3 shows a situation where only 28 percent of the target population responded.

The prevalence of the risk factor is now greatly exaggerated at 200/280 or 71.4 percent and the prevalence of disease is also artifactually increased at 46/280 or 16.4 percent. However, the relative risk

$$\frac{40/200}{6/80} = 2.67$$

and the odds ratio

$$\frac{(40)(74)}{(160)(6)} = 3.08$$

are identical to the target population. Thus, risk ratios can be unbiased despite marked prevalence distortions.

These examples demonstrate that the term "response bias" is parameter specific, and it is more precise to speak of "prevalence response bias," or "relative risk response bias," or "odds ratio response bias."

3. Can Response Bias Distort Prevalence Estimates?

The answers to this question is the easiest, and in fact obvious--yes. If the respondent population has a greater (or lesser) proportion of a risk factor(s) or disease(s), this deviation is by definition a bias in prevalence, as described in the discussion in the previous section. What do we know about the nature of prevalence response bias in epidemiologic studies? Table 4, taken from one of our earlier papers (Criqui et al., 1978), shows several comparisons between respondents and non-respondents in the Rancho Bernardo study of cardiovascular disease. The first four questions refer to disease prevalence, the remainder to risk factor prevalence. Though not all the differences were statistically significant, a consistent trend emerged.

Compared to non-respondents, respondents were less likely to have disease but more likely to have risk factors, and thus we categorized the respondents as a "worried well" group. The only statistically significant exception to the risk factor trend was the greater prevalence of cigarette smoking in the non-respondents However, this finding seemed consistent with the worried well paradigm. Cigarette smoking is a self-selected variable, and thus may reflect the suggested "unworried" posture of the non-respondents. Although we found this "worried well" response pattern fascinating, we demonstrated that with our 82 percent response rate, the prevalence biases overall would be quite small, and, although at that time we were unsure, we guessed risk ratio biases would be small as well.

Other studies have addressed similar questions, although they have usually focused on either risk factors or diseases, not both. A similar decrease in disease prevalence has been found in respondents (Wilhelmsen et al., 1976) and respondents in an arthritis study had higher levels of worry and more specific worries than non-respondents, (Cobb, King, and Chew, 1957). Some studies have reported small if any differences between respondents and non-respondents (Kannel et al., 1979; Remington, Taylor, Buskirk, 1978). In general, studies that find significant differences find the "worried," or "well," or both patterns among respondents. Again, smoking has been the exception for the risk factor group, having been reported previously to be less common in respondents (Doll and Hill, 1964).

Although these patterns are of considerable theoretical interest, studies with good response rates and mild response bias patterns are likely to give good estimates of the true prevalence in the target population.

We should caution that the worried well pattern is probably not generalizable to all epidemiologic studies. If there is a stigma associated with a risk factor, such as alcohol in the fetal alcohol syndrome, the study population may underrepresent the worried. Similarly, if a risk factor is intrinsically linked to low participation, such as "social isolation," it may well be underrepresented in the study group. It also seems likely that disease prevalence will be overrepresented rather than underrepresented if, for instance, a particular study offered a new diagnostic or theraputic modality for a particular condition.

Table 4

Age-adjusted interview comparisons of respondents and non-respondents
Rancho Bernardo, California 1972 - 1974

	Males % Yes			Females % Yes		
	Respondents	Non-respondents	ρ^* value	Respondents	Non-respondents	ρ^* value
Questions on illness						
Hospitalization for heart failure	1.1	4.2	< .01	0.8	3.1	< .01
Hospitalization for heart attack	8.0	9.0	.23	3.0	4.1	.13
History of diabetes	5.7	4.9	.31	3.0	6.0	< .01
History of stroke	3.0	3.7	.28	1.3	1.7	.27
Questions on family history						
History of heart attack in first degree relative	34.2	22.8	< .01	40.5	25.1	< .01
If previous question yes, was it at age 50 or under	15.1	12.9	.32	18.9	17.8	.4
History of stroke in first degree relative	22.3	20.3	.23	30.5	15.6	< .01
History of diabetes in first degree relative	15.7	12.7	.10	17.8	14.8	.09
Question of risk factors other than family history						
History of hyperlipidemia	17.2	13.1	.03	14.0	7.9	< .01
Current diet therapy for hyperlipidemia	7.0	7.2	.46	6.1	4.7	.15
Current drug therapy for hyperlipidemia	2.9	4.3	.09	3.1	3.3	.41
History of hypertension	23.8	23.4	.44	24.7	24.4	.46
(No. of eggs weekly)†	(4.2)	(4.0)	.25	(3.2)	(3.0)	.08
Present cigarette smoker	22.4	26.9	.05	26.6	31.6	.03

* Z-test for differences between proportions.

† Numbers in parentheses are number of eggs, not % yes. T-test of means used for comparison.

4. Can Response Bias Distort Incidence Estimates?

Although one can measure the incidence of risk factors in prospective studies, *disease* incidence is usually the focus of study. Again, disease incidence can be distorted by response bias, as described earlier in this chapter. The data from the Rancho Bernardo study predicts a lesser subsequent disease incidence and mortality in respondents despite their higher levels of some risk factors, because they smoked less and had a lesser prevalence of morbidity at baseline. Indeed, other population studies have reported lower death rates in participants compared to non-participants (Cobb et al., 1957; Doll and Hill, 1964; Gordon et al., 1959; Heilbrun, Nomura, and Stemmermann, 1982; Wilhelmsen et al., 1976). Thus, the trend for incidence response bias is consistent and congruent with the "worried well" pattern of response bias.

5. Can Response Bias Distort Risk Ratio Estimates?

5.1. Theoretical considerations

Again, the reader has already gleaned earlier in this chapter that the answer is yes. But further information is required. Under what circumstances can this happen? Is the bias dependent upon whether the relative risk or the odds ratio is used?

We began to address these issues with a publication in 1979 (Criqui, 1979). We described a situation called "pure risk factor response bias," defined as response rates being different (i.e., biased) for risk factor categories (rows), but each row being unbaised with respect to disease distribution. In our 2x2 table, this means the two row totals $(A+B, C+D)$ are unequally represented, but the two cells comprising each row are proportionally related to each row's total as in the target population. Thus, 80 percent of those with the risk factor, row $A+B$, might respond compared to 60 percent of these without the risk factor, row $C+D$, but pure risk factor response bias would also require the representation in cells A and B was each 80 percent and cells C and D each 60 percent. If the relative risk were the estimate of association, then

$$\frac{(.8)A}{(.8)(A+B)} \bigg/ \frac{(.6)C}{(.6)(C+D)} = \frac{A}{A+B} \bigg/ \frac{C}{C+D}$$

so the relative risk would be unbiased. If the odds ratio were the measure of association, then

$$\frac{(.8)A\,(.6)D}{(.8)B\,(.6)C} = \frac{AD}{BC}$$

so it would similarly be unbiased. If this kind of response pattern were to occur to our Table 1 target population, the numbers would appear as in Table 5.

Table 5

	Dis	$\overline{Dis}$	T
RF	32(A)	128(B)	160
$\overline{RF}$	36(C)	444(D)	480
T	68	572	640

The relative risk would be

$$\frac{32/160}{36/480} = 2.67$$

and the odds ratio

$$\frac{(32)(444)}{(128)(36)} = 3.08$$

confirming their identity with the target population.

A second pattern of response was called "pure disease response bias," defined as response rates being different for disease categories (columns), but each column being unbaised with respect to risk factor distributions. In our 2x2 tables, this means the two column totals (A+C, B+D) are unequally represented, but the two cells comprising each column are proportionally related to each column's total as in the target population. Thus, 70 percent of those with disease, column A+C, might respond vs. only 50 percent of those without disease, column B+D, but pure disease response bias would also require the representation in cells A and C was each 70 percent and cells B and D each 50 percent. For the relative risk

$$\frac{(.7)(A)}{(.7)A + B(.5)} \bigg/ \frac{(.7)C}{(.7)C + D(.5)}$$

the terms are no longer cancellable, so error will occur in the relative risk. However, for the odds ratio

$$\frac{(.7)AD(.5)}{(.5)BC(.7)} = \frac{AD}{BC}$$

the estimate is unbiased. Thus, pure disease response bias will distort the relative risk estimate but not the odds ratio estimate. If this kind of response patterns were to occur in our Table 1 target population, the numbers would appear as in Table 6.

Table 6

	Dis	$\overline{Dis}$	T
RF	28	80	108
$\overline{\text{RF}}$	42	370	412
T	70	450	520

The relative risk would be

$$\frac{28/108}{42/412} = 2.54$$

an underestimate of about 5 percent, while the odds ratio would be

$$\frac{(28)(370)}{(80)(42)} = 3.08$$

an unbiased estimate.

It is well to note that in "pure risk factor response bias" (Table 5), the column totals are not exactly representative of the target population, but they deviate only to the extent necessarily dictated by the risk factor response bias. Similarly, the row totals in "pure disease response bias" (Table 6) deviate from exact representativeness only to the extent necessitated by the disease response bias.

Since the odds ratio can be used as a risk ratio in both retrospective and prospective studies, the reader may conclude from the above considerations that the odds ratio's inherent greater stability in certain response patterns makes it more attractive than the relative risk. We would caution that a close inspection of the algebraic relative risk formula preceding Table 6 suggests a "canceling property" and relative stability of the relative risk as long as the pure disease response bias is not 'extreme. Only at the extremes of pure disease response bias does the relative risk deviate greatly.

What happens when risk factor and disease response bias occur simultaneously? In this situation, the key question is whether the risk factor and disease response patterns are independent of one another. Suppose the risk factor response bias from Table 5 and the disease response bias from Table 6 occurred simultaneously and independent of one another. We could no longer label either response pattern "pure" since they now would occur jointly. The relative risk would be

$$\frac{(.8)A\,(.7)}{(.7)(.8)A+B\,(.8)(.5)}\Big/\frac{(.6)C\,(.7)}{(.7)(.6)C+D\,(.6)(.5)}=\frac{(.8)A\,(.7)}{(.8)(.7A+.5B)}\Big/\frac{(.6)C\,(.7)}{(.6)(.7C+.5D)}$$

which is identical to the algebraic expression for the relative risk in Table 6. The odds ratio estimate

$$\frac{(.7)(.8)AD\,(.6)(.5)}{(.5)(.8)BC\,(.6)(.7)}=\frac{AD}{BC}$$

remains unbiased. If this kind of response pattern were to appear in our Table 1 target population, the numbers would appear as in Table 7. The underlying marginal response probabilities and the response rates for the four inner cells are given in parentheses. The latter are simply the products of the former.

Numbers are carried to one decimal place to retain the mathematical exactness of the argument. The relative risk would be

$$\frac{(22.4/\,86.4)}{(25.2/247.2)}=2.54$$

which, as predicted, is identical to the underestimate in Table 6. The odds ratio would be

$$\frac{(22.4)(222)}{(25.2)(\,64)}=3.08$$

which is the predicted, unbiased result.

Remember that in Table 7 the assumption of independent response bias was made. What happens if this is not the case? We have described (Greenland and Criqui, 1981) a condition we call "differential response bias," i.e., risk factor response bias which differs across disease categories or disease response bias differing across risk factor categories. This kind of response pattern can be conceptualized as a kind of effect modification where the risk factor response bias is dependent upon (i.e., differential for) the presence or absence of disease or the disease response bias is dependent upon the presence or absence of the risk factor. A theoretical example would be Table 2 where cell A (risk factor and disease

Table 7

		(.7) Dis	(.5) Dis	T
(.8)	RF	22.4(.56)	64(.40)	86.4
(.6)	$\overline{RF}$	25.2(.42)	222(.30)	247.2
	T	47.6	286	333.6

present) has a very high response rate of 39/40 or 97.5 percent, whereas the rate in the other three cells varied from 18.3 percent to 52.6 percent. This marked overrepresentation of cell A produced an inflation in the numerator of both the relative risk and the odds ratio and a fivefold to sevenfold overestimation of the risk ratios. It seems likely that "differential response bias" is the source of major response bias errors in risk ratio estimates in epidemiologic studies.

5.2. Pragmatic considerations

So far our discussion of response bias distortions in risk ratios had dealt with contrived examples. Have we any real data to suggest what problems in risk ratios might arise in epidemiologic studies?

Before addressing this issue, we would like to point out why we feel that the issue of response bias is so appropriate to a meeting focusing on behavioral epidemiology. Risk ratio distortions that can occur as the result of misclassification bias, ascertainment bias, selection bias (to be distinguished from "self-selection" bias which we conceptualize as response bias), or "overmatching" are largely investigator-linked errors. Conversely, although one could argue the degree of response bias in part reflects the aggressiveness of the data collection effort, the decision to respond is clearly a product of the subject's behavior. The "worried well" response pattern outlined previously appears to be a recurring behavioral characteristic of human populations. It seems logical that the subjects with a concern about their risk status ("worried") would be more likely to respond to an investigation offering risk factor screening. We have previously speculated that ill persons might be less likely to respond because of poor motivation or mobility due to their infirmity and/or because of a "saturation" with the hospital-medical experience (Criqui, Barrett-Connor, and Austin, 1978).

Before we had fully considered the implications of the above findings, we used data gathered on the non-respondents in the Rancho Bernardo Heart Disease Survey to look at the percentage error occurring in odds ratios due to response bias. Tables 8 through 11 are adapted from an earlier paper (Criqui, Austin, and Barrett-Connor, 1979), and show separately for men and women respondent and target odds ratios and the percentage error of the respondent odds ratios for six risk factors and four diseases. The six risk factors are: family history of heart attack, family history of stroke, family history of diabetes, history of hyperlipidemia, history of hypertension, and current smoking. The four diseases were hospitalization for heart attack, hospitalization for heart failure, history of stroke, and history of diabetes.

25

Table 8

Target and Respondents Odds Ratios, Men Aged 30-79

	DISEASES					
	HOSPITALIZED FOR HEART ATTACK					
RISK FACTORS	TARGET N=2854	RESPONDENT N=2322	% ERROR	TARGET N=2854	RESPONDENT N=2322	% ERROR
FAMILY HX OF HEART ATTACK	2.42	2.58	6.69	1.52	2.27	49.52
FAMILY HX OF STROKE	1.31	1.13	-13.73	1.58	1.49	-5.96
FAMILY HX OF DIABETES	1.26	1.39	10.40	1.50	1.73	15.32
HX OF HYPERLIPIDEMIA	3.53	3.80	7.83	2.26	2.82	24.85
HX OF HYPERTENSION	1.61	1.63	1.28	2.25	2.60	15.38
CURRENT SMOKING	1.27	1.19	-6.10	1.48	1.52	3.04

Table 9

Target and Respondents Odds Ratios, Men Aged 30-79

	DISEASES					
				HISTORY OF STROKE		
RISK FACTORS	TARGET N=2854	RESPONDENT N=2322	% ERROR	TARGET N=2854	RESPONDENT N=2322	% ERROR
FAMILY HX OF HEART ATTACK	0.88	1.10	25.68	1.77	1.68	-5.32
FAMILY HX OF STROKE	1.88	1.34	-28.58	0.97	1.00	3.21
FAMILY HX OF DIABETES	1.25	1.32	5.25	3.13	2.79	-10.94
HX OF HYPERLIPIDEMIA	2.02	2.47	22.29	1.16	1.13	-2.52
HX OF HYPERTENSION	5.12	3.56	-30.44	1.73	1.66	-3.89
CURRENT SMOKING	0.75	0.73	-2.49	0.77	0.74	-3.82

The Problem of Response Bias

Table 10

Target and Respondents Odds Ratios. Women Aged 30-79

	DISEASES					
	HOSPITALIZED FOR HEART ATTACK			HOSPITALIZED FOR HEART FAILURE		
RISK FACTORS	TARGET N=3301	RESPONDENT N=2730	% ERROR	TARGET N=3301	RESPONDENT N=3720	% ERROR
FAMILY HX OF HEART ATTACK	2.08	1.97	-5.49	1.14	0.99	-13.09
FAMILY HX OF STROKE	1.51	1.65	9.09	0.87	1.17	34.55
FAMILY HX OF DIABETES	1.25	1.31	5.13	1.46	1.26	-13.53
HX OF HYPERLIPIDEMIA	2.59	2.75	6.11	1.83	2.90	57.91
HX OF HYPERTENSION	2.51	2.70	7.43	1.28	1.48	15.44
CURRENT SMOKING	0.71	0.59	-15.67	0.73	0.79	8.25

Table 11

Target and Respondents Odds Ratios, Women Aged 30-79

	DISEASES					
	HISTORY OF STROKE TARGET			HISTORY OF DIABETES RESPONDENT		
RISK FACTORS	N=3301	N=2730	ERROR	N=3301	N=2730	ERROR
FAMILY HX OF HEART ATTACK	0.71	0.7	8.05	0.91	1.23	34.83
FAMILY HX OF STROKE	2.00	2.95	47.64	1.48	1.56	5.53
FAMILY IIX OF DIABETES	2.17	1.60	-26.25	2.83	2.82	-0.48
HX OF HYPERLIPIDEMIA	1.59	1.72	8.75	1.53	1.47	-3.64
HX OF HYPERTENSION	2.83	3.50	23.82	1.84	1.82	-0.81
CURRENT SMOKING	1.67	1.83	10.04	1.48	1.67	12.66

Despite some errors close to 50 percent. we were struck by how close the respondent populations' odds ratios were to those in the target population. Table 12. adapted from a later publication, (Austin et al.. 1981) expresses this information another way. The numbers in the table are odds ratio error terms, defined as (response rate cell A) (response rate cell D)/(response rate cell B)(response rate cell C). Thus, this term is greater than one when the numerator is "overweighted" and less than one when the denominator is overweighted. When the odds ratio is unbiased, the error term is equal to one. Multiplying the odds ratio error term by the true odds ratio in the target population gives the observed odds ratio in the respondent population.

Table 12

Odds Ratio Error Terms for Men and Women, Aged 30-79 Years

Risk factors	Diseases			
	Heart attack	Heart failure	Stroke	Diabetes
Family History of Heart attack				
Men	1.10	1.63	1.28	0.96
Women	0.95	0.83	1.06	1.33
Family History of Stroke				
Men	0.90	0.81	0.65	1.05
Women	1.12	1.32	1.59	1.06
Family History of Diabetes				
Men	1.14	1.01	1.04	0.90
Women	1.04	0.66	0.66	0.98
Hyperlipidemia				
Men	1.11	1.23	1.25	0.98
Women	1.10	1.47	1.09	0.90
Hypertension				
Men	1.02	1.12	0.68	0.95
Women	1.08	1.01	1.31	1.00
Smoking				
Men	0.95	0.96	0.84	0.95
Women	0.81	0.94	0.99	1.10

Again, we see that the error terms in general are consistently close to unity. Thus, the "worried well" response pattern appears to reflect response as a function of risk factor status and disease simultaneously but independently, rather than a differential response bias or interactive effect. It is noteworthy that the "worried well" response pattern is not a prerequisite for minimized risk ratio bias. If the "unworried sick" or the "unworried well" were the observed response the odds ratio error would still be minimal because all response patterns which are fundamentally linked to exposure or disease categories will mathematically cancel in the odds ratio formula as illustrated earlier in the Table 7 discussion. This is equivalent to saying that such response biases are limited to the marginal row and/or column totals in a 2x2 contingency table. It is only when response occurs as a function of the inner cells; i.e., A, B, C, or D, that the odds ratio is altered. The example in Table 2 was of a disproportionate overrepresentation in cell A. A different example would be that subjects without coronary heart disease might come into a study if they had high cholesterol (cell B), but stay home if they had a normal cholesterol (cell D). This differential response bias; i.e., a differential response rate for disease absence depending on presence of the risk factor, would inflate cell B and deflate cell D in the $\frac{AD}{BC}$ odds ratio formula, thereby inflating the denominator, deflating the numerator, and producing an underestimate of the true odds ratio.

As noted, however, differential response bias patterns were the exception rather than the rule in the Rancho Bernardo Study. Other epidemiologic studies have found respondents to be more "worried" and/or more "well" (Cobb, King, and Chew, 1957; Doll and Hill, 1964; Gordon et al., 1959; Hammond, 1969; Heilbrun, Nomura, and Stemmermann, 1982; Horowitz and Wilbeck, 1971; Wilhelmsen et al., 1976). The frequent recurrence of this response behavior in epidemiologic studies leads us to the suspicion that these reassuring findings with respect to the effects of non-response on risk ratios in population studies may be at least somewhat generalizable, given the cautions expressed earlier, and odds ratio response bias in other studies may be similarly conservative. Interestingly, recently reported findings from a prospective cancer study showed little odds ratio bias due to non-response (Heilbrun, Nomura, and Stemmermann, 1982).

We would like to caution the reader that this chapter address only response bias. Odds ratio errors due to such problems as "overmatching," or misclassification, ascertainment, or selection bias are conceptually and mathematically different issues and are of potentially much greater concern, particularly in case-control studies.

References

Austin, M. A., Criqui, M. H., Barrett-Connor, E., and Holdbrook, M. J. The effect of response bias on the odds ratios. *American Journal of Epidemiology* 114 (1981) 137-143.

Cobb, S., King, S., and Chew, E. Difference between respondents and non-respondents in a morbidity survey involving clinical examination. *Journal of Chronic Diseases* 6 (1957) 95-108.

Criqui, M. H. Response bias and risk ratios in epidemiologic studies. *American Journal of Epidemiology* 109 (1979) 394-399.

Criqui, M. H., Austin, M., and Barrett-Connor, E. The effect of non-response on risk ratios in a cardiovascular disease study. *Journal of Chronic Diseases* 32 (1979) 633-638.

Criqui, M. H., Barrett-Connor, E., and Austin, M. Differences between respondents and non-respondents in a population-based cardiovascular disease study. *American Journal of Epidemiology* 108 (1978) 367-372.

Doll, R. and Hill, A. B. Mortality in relation to smoking: Ten years' observations of British doctors. *British Medical Journal* 1 (1964) 1399-1410, 1460-1467.

Friedman, G. D. *Primer of Epidemiology.* New York: McGraw-Hill, 1974, pp. 108-109

Gordon, T., Moore, F. E., Shurtleff, D., and Dawber, T. R. Some methodological problems in the long-term study of cardiovascular disease: Observations on the Framingham Study. *Journal of Chronic Diseases* 10 (1959) 186-206.

Greenland, S. and Criqui, M. H. Are case-control studies more vulnerable to response bias? *American Journal of Epidemiology* 114 (1981) 175-177.

Hammond, E. C. Life expectancy of American men in relation to their smoking habits. *Journal of the National Cancer Institute* 43 (1969) 951-962.

Heilbrun, L. K., Nomura, A., and Stemmermann, G. N. The effects of nonresponse in a prospective study of cancer. *American Journal of Epidemiology* 116 (1982) 353-363.

Horowitz, O. and Wilbeck, F. Effect of tuberculous infection on mortality risk. *American Review of Respiratory Disease* 104 (1971) 643-655.

Kannel, W. B., Feinleib, M., McNamara, P. M., Garrison, R. J., and Castelli, W. P. An investigation of coronary heart disease in families. The Framingham Offspring Study. *American Journal of Epidemiology* 110 (1979) 281-290.

Lilienfeld, A. M. *Foundations of Epidemiology.* New York: Oxford University Press, 1976, pp. 209-210.

MacMahon, B. and Pugh, T. F. *Epidemiology: Principles and Methods.* Boston: Little, Brown, 1970, pp. 218-220.

Mausner, J. S. and Bahn, A. K. *Epidemiology, an Introductory Text.* Philadelphia: W. B. Saunders, 1974, pp. 140-142.

Remington, R. D., Taylor, H. L., and Buskirk, E. R. A method for assessing volunteer bias and its application to a cardiovascular disease prevention programme involving physical activity. *Journal of Epidemiology and Community Health* 32 (1978) 250-255.

Wilhelmsen, L., Ljungberg, S., Wedel, H., and Werkö, L. A comparison between participants and non-participants in a primary prevention trial. *Journal of Chronic Diseases* 29 (1976) 331-339.

3

QUANTIFICATION OF HEALTH OUTCOMES FOR POLICY STUDIES IN BEHAVIORAL EPIDEMIOLOGY

Robert M. Kaplan, Ph.D.[1]

Epidemiological research requires the quantification of health outcomes. Some outcomes are *hard* or directly observable. These include mortality and observable lesions. Other outcomes, such as symptomatic complaints, are *soft*. The objective of health care is to achieve positive outcomes. Through prevention or medical care we hope to avoid negative outcomes.

Outcomes should not be confused with process. Many scientists seem to have confused variables known to be relevant to outcomes with outcomes themselves.

1. Supported by Grant K04 HL 00809 from the National Heart, Lung, and Blood Institute of the National Institutes of Health.

For example, hypertension is not an outcome. However, high blood pressure may be associated with negative outcomes or may mediate undesirable events. Similarly, stress, lack of exercise, cigarette smoking, and a variety of other factors are important because they may affect outcomes. The importance of epidemiologic research is in its identification of processes that may control outcomes. Strong epidemiological research avoids linking process directly to process. For example, a process-process study might link dietary changes to changes in serum cholesterol. A strong epidemiologic study would link dietary changes to health status. Thus, a quantitative expression of health status is very important.

1. Health Status

In a variety of publications, we have argued that a single index of health status is both feasible and highly desirable (Bush, Kaplan, and Berry, in press; Kaplan, 1982; Kaplan and Bush, 1982; Kaplan, Bush, and Berry, 1976, 1978, 1979; Kaplan and Ernst. 1983). The traditional approach to health status assessment focuses on measures of mortality. A variety of mortality indexes, such as the crude mortality rate and the age-specific mortality rate, exist. There are also a variety of disease-specific mortality rates. Sometimes, global comparisons of health care in different nations are made using extremely crude indices such as the infant mortality rate or the number of infants to die at or under one year per 1,000 live births.

There are many difficulties with focusing only on mortality as a health outcome. The most obvious problem is that mortality ignores all those who are alive. Most of health care is oriented toward improving the quality of life in addition to extending the duration of life. Measures of quality of life typically consider morbidity. The other extreme from mortality alone is the breakdown of morbidity into numerous specific disease categories. Considering different specific disease indicators makes it impossible to make rational comparisons between programs or treatments that have very different specific objectives. For many treatments in medical care, it is possible to measure effectiveness using a single indicator such as diastolic blood pressure or serum cholesterol. These approaches are not suitable, however, for comparing the relative output of different interventions for different disease groups.

Very specific measures may overlook the consequences or side effects of treatment (Jette, 1980; Mosteller, 1981). A treatment for hypertension, for example, may cause gastric irritation, nausea, and bed disability. Although the activities of different health care providers are diverse, they share the common objective of improving health status. That objective is to extend the duration of life and to improve the quality of life. Our work has been directed toward quantifying progress toward this objective. Further, we hope to assess the relative contributions of different approaches designed to improve health status. These approaches might include medical and surgical treatments, behavioral interventions, prevention and screening programs, and legislative changes (such as auto infant restraint laws). The method can combine hard and soft indicators, if desirable, into a single unit.

A major trend in this field is the use of decision models such as cost-benefit analysis and cost-effectiveness analysis. Both of these models are used to weight positive and negative alternatives in order to reach a rational decision about the utilization of resources. The methods differ in the unit they use to express the efficacy of a program or a treatment. In cost-benefit analysis, both the health outcome and the costs of the program are expressed in monetary units. In cost-effectiveness analysis, the consequences of a program are expressed in some non-monetary unit. The most popular approaches to cost-effectiveness analysis express outcomes in terms of years of life produced by a program or equivalents of well-years of life making adjustment for diminished quality of life. In some of our previous work (Kaplan and Bush, 1982) we make the distinction between cost-effectiveness and cost-utility. Cost-utility is a special case of cost-effectiveness that takes expressed preference for health states into consideration. Although we prefer the cost-utility terminology, the distinction is subtle and the terms cost-effectiveness and cost-utility will be used interchangeably for the purpose of this paper. Within the last decade, the growth of published cost-benefit and cost-effectiveness studies has been exponential. Recent trends suggest that cost-effectiveness analysis is emerging as the preferred method (Warner and Hutton, 1980).

2. Well-Years or Quality Adjusted Life Years

Our approach is to express the benefits of medical care, behavioral intervention, or preventive programs in terms of well-years. Others have chosen to describe the same outcome as Quality Adjusted Life Years (Weinstein and Stason, 1977). Well-Years integrate mortality and morbidity to express health status in terms of equivalents of well-years of life. If a man dies of heart disease at age 50 and we would have expected him to live to age 75, it might be concluded that the disease caused him to lose 25 life-years. If 100 men died at age 50 (and also had life expectancy of 75 years), we might conclude that 2,500 (100 men X 25 years) life-years had been lost.

Yet, death is not the only outcome of concern in heart disease. Many adults suffer myocardial infarctions leaving them somewhat disabled over a long period of time. Although they are still alive, the quality of their lives has diminished. Our model permits various degrees of disability to be compared to one another. A disease that reduces the quality of life by one-half will take away .5 well-years over the course of one year. If it affects two people, it will take away 1.0 well-year (equal to 2 X .5) over a one-year period. A medical treatment that improves the quality of life by .2 for each of five individuals will result in the production of one well-year if the benefit is maintained over a one-year period. Using this system, it is possible to express the benefits of the various programs by showing how many equivalents of years of life they produce. Yet not all programs have equivalent costs. In periods of scarce resources, it is necessary to find the most efficient uses of limited funds. Our approach may provide a framework within which to make policy decisions that require selection between competing alternatives. In recent years, expenditures for health care services have increased relentlessly. Preventive services must compete with traditional medical services for the scarce health care dollar. We believe preventive services can be shown to be competitive in such analyses. Yet, performing such comparisons requires the use of a formal decision model.

In the next section, the general model of health status assessment and cost-effectiveness analysis will be presented. Then, case studies from a variety of areas will be offered. These case studies include traditional surgical interventions such as bypass surgery, preventive medical programs such as pneumococcal vaccination programs, and behavioral prevention programs such as compliance programs for hypertension and behavioral interventions in chronic lung disease.

3. Building a Health Decision Model

The health decision model grew out of substantive theory in economics, psychology, medicine, and public health. These theoretical linkages have been presented in several previous papers (Bush, Chen, and Patrick, 1973; Chen, Bush, and Patrick, 1975; Fanshel and Bush, 1970). Building a health decision model requires at least five distinct steps.

Step 1: Defining a Function Status Classification. During the early phases of the health index project, a set of mutually exclusive and collectively exhaustive levels of functioning were defined. After an extensive, specialty-by-specialty review of medical reference works, we listed all of the ways that disease and injuries can affect behavior and role performance. Without considering etiology, it was possible to match a finite number of conditions to items appearing on standard health surveys, such as the Health Interview Survey (National Center for Health Statistics), the Survey of the Disabled (Social Security Administration), and several rehabilitation scales and ongoing community surveys. These items fit conceptually into three scales representing related but distinct aspects of daily functioning: Mobility, Physical Activity, and Social Activity. Physical Activity can be viewed as being composed of four levels, while the other two scales are thought of as having five distinct levels. Table 1 shows the steps from the three scales. In some of our previous work, we have referred to unique combinations of the three scales as Function Levels and 43 such levels have been observed to date (Kaplan, Bush, and Berry, 1976). Several investigators have used this function

Table 1

Dimensions and Steps for Function Levels in the
Quality of Well-Being Scale

Mobility	Physical Activity	Social Activity
Drove car and used bus or train without help (5)	Walked without physical problems (4)	Did work, school, or housework and other activities (5)
Did not drive, or had help to use bus or train (4)	Walked with physical limitations (3)	Did work, school, or housework but other activities limited (4)
In house (3)	Moved own wheelchair without help (2)	Limited in amount or kind of work, school, or housework (3)
In hospital (2)		
	In bed or chair (1)	
In special care unit (1)		Performed self-care but not work, school or housework (2)
		Had help with self-care (1)

status classification (or a modified version of it) as an outcome measure for health program evaluation (Reynolds, Rushing, and Miles, 1974; Stewart et al., 1978). However, the development of a truly comprehensive health status indicator requires several more steps.

Step 2: Classifying Symptoms and Problems. There are many reasons a person may not be functioning at the optimum level. Subjective complaints are an important component of a general health measure because they relate dysfunction to a specific problem. Thus, in addition to Function Level classifications, an exhaustive list of symptoms and problems has been generated. Included in the list are 35 complexes of symptoms and problems representing all of the possible symptomatic complaints that might inhibit function. A few examples of these symptoms and problems are shown in Table 2.

Step 3: Weights for the Quality of Well-Being. Each combination of Function Level and Symptom/Problem complex might describe how a person

Table 2

Examples of Symptom-Problem Complexes and Linear Adjustments for Level of Well-Being Scores

Symptom-Problem Complex	Adjustment
1. Pain, stiffness, numbness, or discomfort of neck, hands, feet, arms, legs, or several joints	-.0343
2. One *hand or arm* missing, deformed (crooked), paralyzed (unable to move), or broken (includes wearing artificial limbs or braces).	-.0608
3. Burn over large areas of face, body, arms or legs.	-.1100

functioned on a particular day in his or her life. One function state might be described for example, as follows:

> In house
> Walked with physical limitations
> Performed self-care, but not work, school, or housework
> Pain, stiffness, numbness, or discomfort of neck,
>   hand, feet, arms, legs, or several joints

As we noted earlier, the health decision model includes the impact of health conditions upon the quality of life. This requires that the desirability of health situations be evaluated on a continuum from death to completely well. An evaluation such as this is a matter of utility or preference, and thus, function level-symptom/problem combinations are scaled to represent precise degrees of relative importance.

Human judgment studies are needed to determine weights for the different states. We have asked random samples of citizens from the community to evaluate the relative desirability of a good number of health conditions. In a series of studies, a mathematical model was developed to describe the consumer decision process. The validity of the model has been cross validated with an R^2 of .94 (Kaplan, Bush, and Berry, 1978). These weights, then, describe the relative desirability of all of the function states on a scale from zero (for death) to 1.0 (for optimum function). Thus, a state with a weight of .50 is viewed by the members of the community as being about half as desirable as optimum function or about halfway between optimum function and death.

Using these weights, one component of the general model of health is defined. This is the "Quality of Well-Being Scale," which is the point in time component of the Health Status Index (Fanshel and Bush, 1970; Kaplan, Bush, and Berry, 1976). The Quality of Well-Being score for any individual can be obtained from values associated with his/her function level, adjusted for symptom or problem.

The example above shows Function Level 19 for which a weight of .5824 has been obtained. This weight is adjusted by the value associated with the symptom/problem which in this case is -.0343 (associated with complex number 1 in Table 2). So, the Quality of Well-Being score is .5481 (=.5824 + (-) .0343). Using the symptom-problem adjustments, the Index becomes very sensitive to minor "top end" variations in health status. For example, there are symptom-problem complexes for wearing eyeglasses, having a runny nose, or breathing polluted air. These symptom adjustments apply even if a person is in the top step in the other three scales. For example, a person with a runny nose receives a score of .84 on the Quality of Well-Being Scale when he is at the highest Function Level (see Kaplan et al., 1976). Thus, the Index can make fine as well as gross distinctions.

Mathematically, the Quality of Well-Being Score may be expressed as

$$W = \frac{1}{N} \sum_{K=1}^{L} W_K N_K$$

where

W	is the symptom standardized time-specific Quality of Well-Being Score,
	K indexes the Function Levels $[K=1,....,L]$.
W_K	is the Quality of Well-Being (weight, utility, relative desirability, social preference) for each Function Level, standardized (adjusted) for all possible Symptom/Problem Complexes,
N_K	is the number of persons in each Function level, and
N	is the total number of persons in the group, cohort, or population.

Thus, Quality of Well-Being is simply an average of the relative desirability scores assigned to a group of persons for a particular day or a defined interval of time.

Several studies attest to the reliability (Kaplan, Bush, and Berry, 1978; Bush, Kaplan, and Berry, 1983) and validity (Kaplan, Bush, and Berry, 1976) of the Quality of Well-Being Scale. For example, convergent evidence for validity is given by significant positive correlations with self-rated health and negative correlations with age, number of chronic illnesses, symptoms, and physician visits. However, none of these other indicators were able to make the fine discrimination between health states which characterize the Quality of Well-Being Scale. These data support the convergent and discriminant validity of the Scale (see Kaplan, Bush, and Berry, 1976).

Step 4: The Well-Life Expectancy. Quality of Well-Being is only one of the two major components of the Health Decision Model. The other component requires consideration of transitions among the levels over time. Consider the health situation we described earlier (e.g., Function Level 19). Suppose that this condition described two different individuals; one who was in this condition because of participation in a marathon race and another because of arthritis. The fact that these individuals are in these conditions for different reasons is reflected by different expected transitions to other levels over the course of time. The marathon runner probably is sore from her ordeal, but is expected to be off and running again within a few days. However, the arthritis sufferer will probably continue to convalesce at a low level of function. A Health Decision Model must consider both current functioning and probability of transition to other Function Levels over the course of time. When transition is considered and documented in empirical studies, the consideration of a particular diagnosis is no longer needed. We fear diseases because they affect our current functioning or the probability that there will be a limitation in our functioning some time in the future. A person at high risk for heart disease may be functioning very well at present, but may have a high probability of transition to a lower level (or death) in the future. Cancer would not be a concern if the disease did not affect current functioning or the probability that functioning would be affected at some future time.

When weights have been properly determined, health status can be expressed precisely as the expected value (product) of the preferences associated with the states of function at a point in time and the probabilities of transition to other states over the remainder of the life-expectancy. Quality of Well-Being (W) is a static or time-specific measure of function, while the Well Life Expectancy (E) also includes the dynamic or prognostic dimension. The Weighted Life Expectancy is the product of Quality of Well-Being times the expected duration of stay in each Function Level over a standard life period. The equation for the Weighted Life Expectancy is

$$E = \sum_{K=1}^{L} W_K Y_K$$

where

> E is the symptom-standardized Weighted Life Expectancy in equivalents of completely well-years, and

> Y is the expected duration of stay in each Function Level or case type estimated with an appropriate statistical (preferably stochastic) model.

An example computation of the Weighted Life Expectancy is shown in Table 3. Suppose that a group of individuals was in a well state for 65.2 years, in a state of non-bed disability for 4.5 years, and in a state of bed disability for 1.9 years before their deaths at the average age of 71.6 calendar years. In order to make adjustments for the diminished quality of life they suffered in the disability states, the duration of stay in each state is multiplied by the preference associated with the state. Thus, the 4.5 years of non-bed disability become 2.7 equivalents of well years when we adjust for the preferences associated with inhabiting that state. Overall, the Weighted Life Expectancy for this group is 68.5 years. In other words, disability has reduced the quality of their lives by an estimated 3.1 years.

Table 3

Illustrative Computation of the Weighted Life Expectancy

State	Y_K	W_K	W_K	Y_K
Well	A	65.2	1.00	65.2
Non-bed disability	B	4.5	.59	2.7
Bed disability	C	1.9	.34	.6
Total		71.6		68.5

Source: Kaplan and Bush, 1982

$$Weighted\ Life\ Expectancy = \sum_{K=1}^{L} W_K\, Y_K = 68.5\ Well\ Years$$

$$Current\ Life\ Expectancy = \sum_{K=1}^{L} Y_K = 71.6\ Calendar\ Years$$

Step 5: Estimating the Cost/Effectiveness Ratio. The San Diego Health Index Group has shown in a variety of publications how the concept of a Well or Weighted Life Expectancy can be used to evaluate the effectiveness of programs and health interventions. The output of a program has been described in a variety of publications as Quality Adjusted Life Years (Bush, Chen, and Patrick, 1973; Bush, Fanshel, and Chen, 1972), Well Years, Equivalents of Well-Years, or Discounted Well-Years (Kaplan, Bush, and Berry, 1970, 1976; Patrick, Bush, and Chen, 1973a, 1973b). Weinstein (1980a, 1980b) has popularized the concept and calls the same output Quality-Adjusted Life Years (QALYS) and this has recently been adopted by the Congressional Office of Technology Assessment (1979). It is worth noting that the Quality Adjusted Life Years terminology was originally introduced by Bush, Patrick, and Chen in 1973, but later abandoned because it has surplus meaning. The term "wellness" or "well-years" implies a more direct linkage to health conditions. Whatever the term, the Index shows the output of a program in years of life adjusted by the quality of life which has been lost because of diseases or disability.

4. Utilization of a Well-Years Measure

By comparing experimental and control groups on a health status index, it is possible to estimate the output of a program in terms of the well-years it produces. This is shown as the area between curves representing the two groups in Figure 1. Dividing the cost of the program by the well years it yields gives the cost-effectiveness ratio.

5. Examples of Cost-Utility Studies

In this section, studies using the cost-utility model well be reviewed. A variety of studies with very different specific objectives have been chosen to highlight the general nature of the model.

Coronary Artery Bypass Surgery (CABS). Despite some controversy (Braunwald, 1977; DeBakey and Lawrie, 1979; Hultgren et al., 1978), coronary artery bypass surgery (CABS) has become a major treatment for systematic coronary artery disease. The number of procedures performed in the United States has steadily grown to an estimated 110,000 procedures in 1980 at an estimated cost of $15,220 per operation (Kolata, 1981). Because of the significance of the procedure and the expenses associated with it, Weinstein and Stason (1982) conducted a systematic evaluation of the literature on CABS using a cost-utility model. The data for the Weinstein and Stason evaluation were provided by clinical reports, systematic longitudinal data banks, and clinical trials including the major trials conducted by the European Coronary Surgery Study Group and the Veterans Administration Cooperative Study.

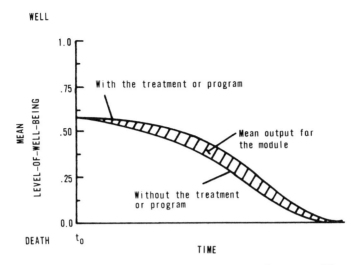

Figure 1. Theoretical comparison of treated and untreated groups. The area between the two curves is the output or benefit of a program in Well-Year units.

Source: Kaplan and Bush, 1982.

The analysis considered the benefit for a 55-year old male population, since 55 years is approximately the median age for receipt of CABS. The analysis considers only those men who would be deemed operable by cardiologists on the basis of clinical characteristics and angiography. The analysis was done separately for men with obstruction (defined as 50 percent or more) of 1, 2, or 3 coronary arteries or left main coronary artery disease. In each of these cases, ventricular function was good--with at least a 40 percent ejection fraction. The analysis for patients with poor ventiruclar function will not be considred here.

In order to calculate quality adjusted life years, Weinstein and Stason needed to integrate morbidity and mortality information. They used data about symptomatic relief from the European study (European Coronary Surgery Study Group, 1979, 1980) and from the Montreal Heart Institute (Campeau et al., 1979). They also simulated the benefit results using a variety of preferences for observed levels of functioning and symptomatic angina.

41

The approach used by Weinstein and Stason does not depend on specific data sets. Figure 2 shows differences in the life expectancy as obtained from different data sets. Notice that the VA data and the European data differ in their evaluation of the benefits of surgery for one-vessel and two-vessel disease. The VA data suggest that surgery may be detrimental in these cases, while the European data indicate there will be benefits. The figure also shows how these two trials and other data are merged to obtain central assumptions that are operative in the analysis. However, the analysis can also consider differing assumptions, and the impact these assumptions have upon quality adjusted life expectancy. Under the assumption that the preference for life with angina is .7 (on a scale from 0 to 1), Weinstein and Stason estimated the benefits of surgical treatment over medical treatment for the various conditions. They found that the benefits and well-years or quality adjusted life years would be .5, 1.1, 3.2, and 6.2 years for 1 vessel, 2 vessels, 3 vessels, and left main disease respectively. Figure 3 shows the cost-utility of bypass surgery under the central assumptions. As the figure shows, the cost-effectiveness ranges from $30,000/year for one vessel disease to $3,800/year for left main artery disease. Weinstein and Stason performed these analyses under a variety of assumptions. In doing so, they revealed the impact of considering quality of life. One assumption ignored quality of life and considered only life expectancy. The cost-effectiveness of bypass surgery for one vessel disease under this assumption cannot be estimated since surgery has no effect upon survival. However, many of the benefits of surgery are directed toward the quality of life rather than survival. A model that did not integrate mortality and morbidity would have missed these benefits.

In summary, the Weinstein and Stason (1983) analysis demonstrates that the cost-utility of CABS differs by characteristic of disease state. However, the cost-utility figures compare favorably with those from other widely-advocated medical procedures and screening programs.

Pneumococcal Vaccine. There is little question that vaccination programs have had a major beneficial impact. Diseases that once threatened the masses, such as polio and smallpox, have virtually been eliminated. The largest promoter and user of vaccines has been the United States government. Every year or two, Congress is asked to enact new legislation on vaccines. The legislation authorizes government purchase and distribution of vaccines for use by state and local government programs. Nevertheless, the production of vaccines in the United States has steadily declined since the 1940s (Riddough and Willems, 1980). Preventive health measures, like vaccination programs, have given way to the use of curative pharmaceuticals, including the sulfa drugs, the penicillins, and the tetracyclines (Office of Technology Assessment, 1979). A number of factors have contributed to the decrease in the number of product licenses issued for vaccines (see Riddough and Willems, 1980). Vaccines are not entirely without risk. Even very safe products may be associated with a small probability of side effects. For

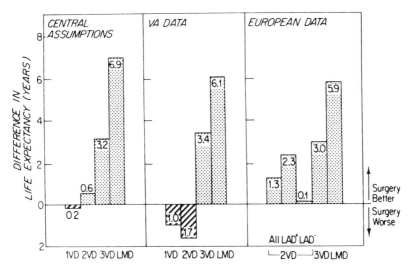

Figure 2. Effect of annual survival estimates on differences in life expectancy.

Source: Weinstein and Stason, 1982, reproduced with permission.

example, some recipients of the swine flu vaccine developed a temporary form of paralysis known as Guillian Barré Syndrome. Popular journalists, such as those from CBS's *60 Minutes*, (a popular magazine program in the U.S.) filmed these people and suggested that the promoters of the vaccine program were risking the lives of the public for personal and political benefit. These media presentations did not consider the potential overall benefit of the program, and public faith in vaccination programs deteriorated. In order to make an informed decision about the program, it is necessary to have some measure of effectiveness that maps benefits and side-effects on the same scale. The United States Congressional Office of Technology Assessment (OTA) (1979) performed such an analysis for pneumococcal vaccines using a health-decision model.

Using the data from the existing clinical studies, the OTA was able to assemble estimates of the effectiveness of the vaccine. However, data also suggested that about 5 percent of the recipients reacted with swelling and fever, and that severe reactions, such as Guillian Barré Syndrome are expected to occur in one case per 100,000.

About 15 percent of pneumonia cases are caused by pneumococci. The multivalent polysaccharide vaccine provides protection against 14 types of pneumococci. These 14 types account for 75 percent of the cases of pneumococcal infections. The vaccine is estimated to be 80 percent effective for these 14 types.

Using the health decision model it was estimated that the cost per well-year yielded by the vaccine is $4,800 (1979 dollars). This differs for different age

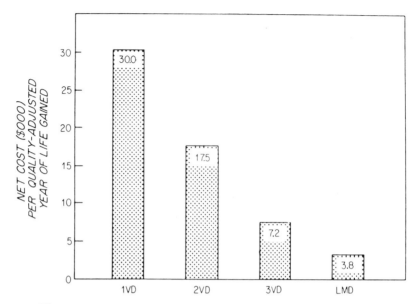

Figure 3. **Cost-effectiveness of coronary artery bypass surgery.**

Source: Weinstein and Stason, 1982, reproduced with permission.

groups. For example, the costs for young children, who rarely die of pneumococcal pneumonia, were as high as $77,200/year, while the costs for the aged (older than 65 years), who are the most frequent victims, are less than $1,000 per well-year.

And what about Guillain Barré Syndrome, which has caused so much concern. Considering the decreased effectiveness of reducing the well-year yield by adding the Guillian-Barré disabled (at the same rate as from Swine Flu program) or adding to the costs by the increase in insurance premiums inflates the cost/well-year to $4,900--a modest increase of $100. For the elderly it has no measurable effect. Stated another way, the large number of well-years produced by avoiding pneumonia through pneumococcal vaccine greatly outweigh the setbacks caused by the low probability of severe side effects. Further, the vaccine program produces well-years at a fraction of the cost of curative programs. Once a common unit of outcome is defined, the appropriate decision becomes obvious.

This section reviewed the cost-utility of two very different health care interventions: coronary artery bypass surgery and pneumococcal vaccine. The same system can be used to estimate the benefits of non-medical interventions such as passive restraints in automobiles and laws requiring reduced exposure to toxic substances in the work place. However, our focus here is on behavioral programs, and the remainder of this section will review cost-utility studies on the benefits of behavioral interventions.

Adherence to Antihypertensive Medication. Hypertension is a major public health problem because of its high prevalence and its association with heart disease and stroke. Many people are unaware that they have hypertension, or those who are aware are unwilling to take the necessary actions to control the condition.

Weinstein and Stason (1976) and other colleagues at the Center for the Analysis of Health Practices at the Harvard School of Public Health have analyzed the cost-utility of hypertension screening programs in great detail. They report the cost-utility for programs screening severe hypertension (diastolic > 105 mm Hg) to be \$4,850/well-year while the corresponding figure for mild hypertension screening programs (diastolic 95-104 mm Hg) to be \$9,800/year in 1976 dollars.

However, their analysis also considered a variety of factors that influence these cost-utility ratios. One of the most important factors is adherence to the prescribed medical regimen once cases have been detected. The figures given above assume full adherence to the regimen. Yet, substantial evidence reveals that full or 100 percent adherence is rare (see DiMatteo and DiNicola, 1982). Compliance with antihypertensive medications is of particular interest because taking the medication does not relieve symptoms. In fact, medication adherence can increase rather than decrease somatic complaints. More studies have been devoted to compliance among hypertensive patients than to compliance in any other disease category. Some studies suggest that behavioral intervention can be very useful in increasing adherence to prescribed regimens (Haynes et al., 1976).

In their analysis, Weinstein and Stason (1977) considered the value of programs designed to increase adherence to antihypertensive medication. Two separate problems were considered. First, there are drop-outs from treatment. Second, there is failure to adhere to treatments that have been prescribed. The two cases may differ in their cost. One extreme is the patient who fails to see a physician and purchase medication. Here the costs would be very low. The other extreme would be the patient who remains under medical care, purchases medications, but does not take them. In this case, the costs would be high.

45

Weinstein and Stason refer to these as the maximum cost assumption and the minimum cost assumption. Under the minimum cost assumption patients do not receive the full benefits of medication because of incomplete adherence. Yet they also do not spend money. According to Weinstein and Stason, the cost-effectiveness under this assumption is very similar to full adherence in which patients receive the benefits of medication but make full expenditures. Under the maximum cost assumption, the effect of incomplete adherence is substantial, particularly for those beginning therapy beyond the age of 50. Earlier, it was noted that the costs to produce a well-year for a national sample (U.S.) were $4,850 for those with pretreatment diastolic blood pressure greater than 105 mm Hg. With incomplete adherence, these value increase to $6,400 under the minimum cost assumption and $10,500 under the maximum cost assumption. For mild hypertensive screening (diastolic blood pressure 95-104 mm Hg.), the $9,880 per well-year under the full adherence assumption rose to $12,500 under the minimum cost assumption and $20,400 under the maximum cost assumption.

Since adherence under the maximum cost assumption appears to have a strong affect upon cost-utility, it is interesting to consider the value of behavioral interventions to improve adherence. Several studies have shown the value of behavioral interventions (Shapiro and Goldstein, 1982) and it is reasonable to assume that a successful behavioral intervention that will improve adherence rates by 50 percent (Haynes et al., 1976). Weinstein and Stason considered the cost-utility of interventions that would improve adherence by 50 percent under the maximum cost assumption. The results of this analysis are shown in Figure 4 for a hypothetical program that would reduce diastolic blood pressure from 110 mm Hg to 90 mm Hg. The figure also shows the differential expected cost-effectiveness for programs designed for males and for females. As the figure shows, the intervention would improve the cost-effectiveness for both males and females and at each age of program initiation. The interaction between gender and age of initiation of therapy reveals the finding from epidemiologic studies that blood pressure is better controlled in women than in men. In summary, the figure demonstrates that even an expensive program can improve cost-utility because it produces substantial improvements in outcome relative to its costs.

In the Weinstein and Stason monograph, a variety of other hypothetical conditions were considered. Under the assumption that the program improves adherence by 50 percent, a significant benefit of the program remained under the maximum cost assumption. However, under the minimum cost assumption the hypothetical adherence intervention would have a significant benefit if it increased adherence by 50 percent but no significant effect if it increased adherence by only 20 percent.

A variety of other assumptions in the Weinstein and Stason analysis need to be considered. For example, they make (and discuss) many other assumptions about the relationship between hypertension and outcome, the linear relationship

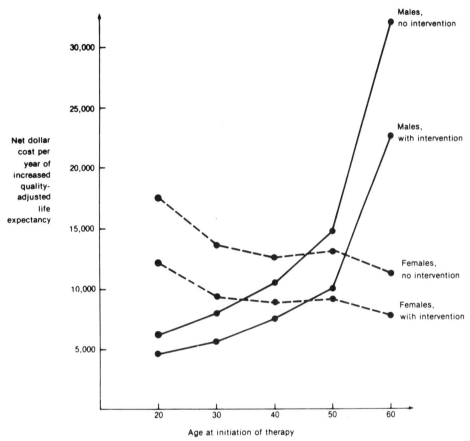

Figure 4. Effect of a hypothetical behavioral program that improves adherence by 50 percent under the maximum cost assumption.

Source: Weinstein and Stason, 1976, reproduced with permission.

between adherence and outcome, and the effect of adherence programs. However, some data support the reasonableness of each of these assumptions.

Community-wide Interventions to Prevent Heart Disease. Another approach to cost-effectiveness is provided in an exceptional monograph by Berwick, Cretin, and Keeler (1980). This analysis differs from the others considered in this paper because the unit of benefit is years of life saved unadjusted for quality of life. However, the Berwick and colleagues analysis is unusually thoughtful and helps pinpoint some of the major problems in this area. The purpose of the analysis was to compare the cost-effectiveness of different programs designed to lower cholesterol in childhood. The approaches included screening, early dietary intervention, and population-wide educational programs. Data on the relationship between cholesterol and heart disease were taken from the Framingham Heart study. Estimates of the benefits of community-wide behavioral interventions were extrapolated from the Stanford Heart Disease Prevention Program (Farquhar et al., 1977) and the North Karelia project (Salonen et al., 1979). The costs of a mass media campaign were estimated to be about $.50 per person per year. This is approximately the amount spent to advertise coffee, tea, and cocoa products. It was assumed that the mass media program would reduce cholesterol by 2 percent. This program was compared with a hypothetical school education program which may cost $5.00 to reduce cholesterol by 1 percent. However, the school education program may be argued to be more effective because it directs resources specifically toward children.

Discount rates are used by economists to estimate the future value of current money. Programs designed for children to prevent heart disease in later life are difficult to analyze. The money spent on them now could also be invested and by the time the benefits accrue, the money would have increased in value. Future value of money can be calculated as $(1+D)^Y$ where D is the discount rate and Y is the number of years. Benefits can also be discounted using the same system. Berwick et al. (1980) considered the cost-effectiveness of mass media versus school education interventions for males and females using three different discount rates (0 percent, 5 percent and 10 percent). They found that the discount rate had a very substantial effect upon the cost-effectiveness. This is because current money is invested to buy benefits which occur much later in the life cycle.

Comparing the cost-effectiveness of three different programs: targeted screening, school education, and mass media, Berwick and colleagues concluded that mass media campaigns may be the most cost-effective approach. One of the most important lessons to be learned from the Berwick, Cretin, and Keeler study is that the value of different approaches to the same problem can be readily compared using cost-effectiveness analysis. However, the analyses may be sensitive to certain assumptions such as the discount rate. At low discount rates, primary prevention programs are very cost-effective. However, at high discount rates, the value of these programs becomes questionable. A comparison of programs designed for children versus those designed for the elderly will be given in the next section.

Another issue raised by the Berwick, Cretin, and Keeler study is the data base upon which these cost-effectiveness studies are performed. As we have seen, these analyses are based on many assumptions. For example, Berwick and colleagues base their estimates on total cholesterol, yet current work suggests that HDL or apoproteins (see Avogaro, this volume) may be better predictors of mortality. Others (Stallones, 1983) have suggested that the relationship between diet and heart disease may be independent of cholesterol. These analyses also make the assumption that dietary changes will result in reductions in heart disease. This, too, is a matter of continuing debate (Laren et al., 1983; MRFIT, 1982). These authors are aware of the limitations, and the analysis can vary these assumptions and determine their impact upon any policy decision.

Behavioral Programs for Patients with Chronic Lung Disease. The analyses above all use secondary data for the cost-effectiveness calculations. Unfortunately, such analyses depend upon a variety of unverified assumptions. In this section, data are presented from a prospective study that attempted to combine observations of changes in health status with cost-effectiveness analysis.

Chronic Obstructive Pulmonary Disease (COPD) is one of the most rapidly growing health problems in most western countries. In the United States, deaths due to COPD are rising at a rate of 1.4 percent per year. COPD is now the most rapidly increasing of the top ten leading causes of death. COPD is the second leading cause of permanent disability for older adults. From 1970 to 1975, the mortality rate from Chronic Obstructive Pulmonary Disease increased from 16 per 100,000 of the population to 19 per 100,000 (Brashear, 1980). In addition to its association with premature mortality, COPD is also a major cause of permanent disability and decreased quality of life. Morbidity from COPD results in approximately 34 days of restricted activity per 100 person years. In the United States from January 1975 to December 1976, an estimated 163.4 million office-based physician visits were attributed to respiratory diseases. These visits comprised approximately 14 percent of all office visits for any condition during that period. COPD accounts for approximately one-fifth of these visits (U.S. Government Task Force, 1977). Total costs of COPD have been estimated to be 4.55 billion per year in 1972 dollars. Of this figure, an estimated $803,000 were spent for direct costs of hospital treatment, physician service, and prescribed drugs. Indirect costs were estimated at $3.05 billion for disability payments and $645 million for death benefits (U.S. Government Task Force, 1977).

There is no medical cure for COPD. Medical management includes the use of antibiotics, bronchodilators, corticosteroids, and a variety of other medications to prevent influenza and other medical complications. However, mounting evidence suggests that behavioral programs for rehabilitation do produce substantial benefits for COPD patients (Petty and Cherniak, 1981). These programs result in reduced symptoms, improved exercise tolerance, reduction in hospital days, more gainful employment, slowing the progress of disease, and increased survival. Perhaps the most important element in the rehabilitation program is the daily activity and exercise portion. Although COPD patients can benefit from exercise, compliance with exercise programs is usually poor because activity causes discomfort and shortness of breath.

Over the course of the last few years, we have been conducting an experimental trial evaluating the benefits of behavioral programs designed to increase compliance to an exercise regimen for COPD patients. Seventy-five COPD patients were randomly assigned to either experimental or control groups. All patients were given an exercise stress test and an exercise prescription. Experimental groups were given strategies for improving adherence to the regimen while control subjects were simply monitored (Atkins, Kaplan, Timms et al. 1984).

49

Health status information was collected over 18 months and the health decision model was used to translate program benefits into well-year equivalents. By the end of the program there were significant differences between the treated and control patients on the Quality of Well-Being Scale. Figure 5 shows the mean differences between experimental and control subjects over the 18-month period. A total of 7.33 well-years were attributed directly to the program.

Costs of the experimental program were tablulated for expenses directly charged to the program. Expenses incurred by family and friends in helping the patient comply with their program were not considered. Both costs and health effects were discounted to present value using a 5 percent discount rate. A total of $174,100 was spent on the project. Dividing these costs by the benefits produces a well-year at approximately $24,000. Sensitivity analysis demonstrated that using other discount rates (0 percent and 10 percent) for both costs and effects had little impact because the benefits were produced late in the life cycle. In other words, programs designed for the elderly that produce health benefits without delay can

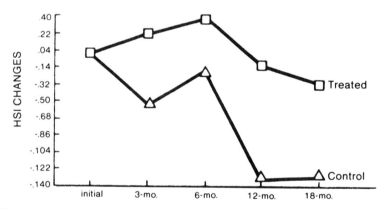

Figure 5. Differences between treated and control groups in COPD study.

be shown to be cost-effective even at high costs because they are not sensitive to discounting assumptions. Comparing the cost-utility figure to other health care programs using the General Health Policy Model, the behavioral adherence program appears reasonably cost-effective as an adjunct therapy for patients suffering from COPD (Toevs, Kaplan, and Atkins, 1984).

One facet of the COPD project is particularly interesting. There is no evidence that the behavioral interventions affected physiological processes. For example, experimental and control groups were not different on tests of pulmonary function or arterial oxygen saturation. The difference seems to be primarily attributable to improved quality of life. The behavioral interventions appeared to affect quality of life directly by making people more active, encouraging them to participate in more activities, and getting them to lead more productive lives. In other words, they appeared to alter the extent to which the disease interfered with daily function. Thus, it was argued that a health care program designed to directly modify quality of life can be shown to be cost-effective when evaluated using the same methods designed for the assessment of medical treatments.

6. Limitations

Despite the many advantages of cost-effectiveness or cost-utility analysis, we must also consider some of its limitations. Several major problems will be listed below. These include

1. In most applications, the analysis is completely dependent upon information provided by epidemiologists. Yet, as demonstrated by the chapters in this volume, there continues to be debate about the determinants of many health outcomes. As policy scientists, we must review the evidence carefully, and summarize current knowledge. When the evidence is inconsistant, it is possible to simulate outcomes under a variety of assumptions. For example, it is possible to consider several different levels of the relationship between diet and heart disease and to estimate the impact these variations have upon the production of well-years.

2. The analysis also requires many assumptions about the efficacy of interventions. To date, there have been very few population based intervention studies. Most behavioral interventions studies are based on small samples of volunteers. Studies such as the Stanford Heart Disease Prevention Program and the North Karelia study are among the very few population based studies that provide estimates of the efficacy of behavioral interventions.

3. As noted in the discussion of the Berwick, Cretin, and Keeler study, assumptions about discount rates can be very important in some cases. Yet, there is no consensus about what the discount rate should be. For example, some analysts prefer to use the inflation rate. Others suggest that discount rate should reflect preferences for different levels of health at different periods in the life cycle. However, analysts are uncertain as to how these preferences should be measured (Kane and Kane, 1982).

51

7. Summary

This paper has reviewed a mathematical health decision model that can be used for policy analysis in health. The model may have great benefit for comparing preventive programs with different specific objectives. Proper use of the model should help policy makers decide between very different policy alternatives.

The value of the model is highly dependent upon quality epidemiologic research. The interface between policy science, epidemiology, and behavioral science must be an important focus for continuing development of this approach.

References

Atkins, C. J., Kaplan, R. M., Timms, R. M., Reinsch, S., and Lofback, K. Behavioral Exercise Programs in the Management of Chronic Obstructive Pulmonary Disease. *Journal of Consulting and Clinical Psychology* 52 (1984) 491-603.

Avogaro, P. Apolipoproteins, the Lipid Hypothesis, and Ischemic Heart Disease. In R. M. Kaplan and M. H. Criqui, (eds.) Behavioral Epidemiology and Disease Prevention. New York: Plenum Press, in press.

Berwick, D. M., Cretin, S., and Keeler, E. B. *Cholesterol, Children, and Heart* Disease: *An Analysis of Alternatives.* New York: Oxford University Press, 1980.

Braunwald, E. Coronary-artery surgery at the crossroads. *New England Journal of Medicine* 297 (1977) 661.

Bush, J. W., Chen, M. M., and Patrick, D. L. Cost-effectiveness using a health status index: Analysis of the New York State PKU screening program. In R. Berg (ed.) *Health Status Indexes.* Chicago: Hospital Research and Education Trust, 1973.

Bush, J. W., Fanshel, S., and Chen, M. M. Analysis of a tuberculin testing program using a health status index. *Social-Economic Planning Sciences* 6 (1972) 49-69.

Bush, J. W., Kaplan, R. M., and Berry, C. C. Sources of variability in a health policy model. *Medical Care,* in press.

Campeau, L. et al. Loss of improvement of angina between 1 and 7 years after aorto coronary bypass surgery. Correlations with changes in vein grafts and in coronary arteries. *Circulation* 60 (suppl. I) (1979) I-1.

Chen, M. M., Bush, J. W., and Patrick, D. L. Social indicators for health planning and policy analysis. *Policy Sciences* 6 (1975) 71-89.

DeBakey, M. E. and Laurie, G. M. Response to commentary of Hultgren et al. on "Aortocoronary-artery bypass assessment after 13 years." *Journal of the American Medical Association* 241 (1979) 2393.

DiMatteo, M. R. and DiNicola, D. D. *Achieving Patient Compliance.* Elmsford, NY: Pergamon Press, Inc.. 1982.

European Coronary Surgery Study Group. Coronary-artery bypass surgery in stable angina pectoris: Survival at two years. *Lancet* (1979) 889.

European Coronary Surgery Study Group. Prospective randomized study of coronary artery bypass surgery in stable angina pectoris. *Lancet* (1980) 491.

Fanshel, S. and Bush, M. W. A health status index and its applications to health services outcomes. *Operations Research* 18 (1970) 1021-1066.

Farquhar, J. W. Community education for cardiovascular health. *Lancet* 1 (1977) 1192-1195.

Haynes, R. B. et al. Improvement of medication compliance in uncontrolled hypertension. *Lancet* 1 (1976) 1265-1268.

Hultgren, H.. W., Takaro, T., Detre, K. M., and Murphy, M. L. Aortocoronary-artery bypass assessment after 13 years. *Journal of the American Medical Asociation.* 240 (1978) 1353.

Jette, A. M. Health status indicators: Their utility in chronic disease evaluation research. *Journal of Chronic Disease* 33 (1980) 567-579.

Kaplan, R. M. Human preference measurement for health decisions and the evaluation of long-term casre. In R. Kane and R. Kane (eds.) *Values and Long-Term Care.* Lexington, Mass.: D. C. Heath (1982).

Kaplan, R. M. and Bush, J. W. Health-related quality of life measurement for evaluation research and policy analysis. *Health Psychology* 1.1 (1982) 61-80.

Kaplan, R. M., Bush, J. W., and Berry, C. C. Health status: Types of validity for an index of well-being. *Health Services Research* 11 (1976) 478-507.

Kaplan, R. M., Bush, J. W., and Berry, C. C. The reliability, stability, and generalizability of a health status index. *American Statistical Association, Proceedings of the Social Statistics Section* (1978) 704-709.

Kaplan, R. M., Bush, J. W., and Berry, C. C. Health status index: Cagtegory rating versus magnitude estimation for measuring levels of well-being. *Medical Care* 5 (1979) 501-523.

Kaplan, R. M. and Ernst, M. S. Do category rating scales produce biased preference weights for a health index? *Medical Care* 21, 2 (1983) 193-207.

Kolata, G. B. Consensus on bypass surgery. *Science* 211 (1981) 42.

Laren, P., Helgeland, A., Hjermann, I., and Holme, I. MRFIT and the Oslo Study. *Journal of the American Medical Asociation.* 249 (1983) 893-894.

Mosteller, F. Innovation and Evaluation. *Science* 211 (1981) 881-886.

MRFIT. Multiple Risk Factors Intervention Trial. *Journal of the American Medical Association.* 248 (1982) 1465-1477.

Murphy, M. L. et al. Treatment of chronic stable angina: A preliminary report of survival data of the randomized Veterans Administration Cooperative Study. *New England Journal of Medicine* 297 (1977) 621.

Office of Technology Assessment, United States Congress. *A Review of Selected Federal Vaccine Immunization Policies: Based on Case Studies of Pneumococcal Vaccine.* Washington, D.C.: U.S. Government Printing Office, 1979.

Patrick, D. L., Bush. J. W., and Chen, M. Toward an operational definition of health. *Journal of Health and Social Behavior* 14 (1973a) 6-23.

Patrick, D. L., Bush, J. W., and Chen, M. Methods for measuring levels of well-being for a health status index. *Health Services Research* 8 (1973b) 228-245.

Pliskin, J. S., Shepard, D. S., and Weinstein, M. C. Utility functions for life years and health status. *Operations Research* 28 (1980) 206.

Reynolds, W. J., Rushing, W. A., and Miles, D. L. The validation of a function status index. *Journal of Health and Social Behavior* 15 (1974) 271.

Riddough, M. A. and Willems, J. S. Federal policies affecting vaccine research and production. *Science* 209 (1980) 563-566.

Salonen, J., Puska, P., and Mustaniemi, H. Changes in morbidity and mortality during comprensive community program to control cardiovascular diseases during 1971-7 in North Karelia. *British Medical Journal* 2 (1979) 1178-1183.

Shapiro, D. and Goldstein, I. B. Biobehavioral perspectives on hypertension. *Journal of Consulting and Clinical Psychology* 50 (1982) 841-858.

Stallones, R. A. Ischemic heart disease and lipids in blood and diet. *Annual Review of Nutrition* 3 (1983) 155-185.

Stewart, A. L., Ware, J. E., Brook, R. H., and Davies-Avery, A. *Conceptualization and Measurement of Health for Adults: Vol. 2 Physical Health in Terms of Functioning.* Santa Monica: Rand Corp., 1978.

Takaro, T. et al. The VA cooperative randomized study of surgery for coronary occlusive disease. II. Subgroup with significant left main lesions. *Circulation* 54 (suppl. III) (1976) III-107.

Toevs, C. D., Kaplan, R. M., and Atkins, C. J. The Costs and Effects of Behavioral Programs in Chronic Obstructive Pulmonary Disease. *Medical Care*, in press.

Warner, K. E. and Hutton, R. C. Cost-benefit and cost-effectiveness analysis in health care. *Medical Care* 10, 11 (1980) 1069-1084.

Weinstein, M. C. and Stason, W. B. *Cost-effectiveness of Coronary Artery Bypass Surgery. Circulation* 66 (suppl. 3) (1982) 56-66.

Weinstein, M. C. and Stason, W. B. Foundations of cost-effectiveness analysis for health and medical practices. *New England Journal of Medicine* 296 (1977) 716.

Weinstein, M. C. and Stason, W. B. Hypertension: *A Policy Perspective.* Cambridge, Mass.: Harvard University Press, 1976.

Weinstein, M. C. and Feinberg, H. *Clinical Decision Analysis.* Philadelphia: W. B. Saunders, 1980.

PART II

EPIDEMIOLOGY

The second section of this book examines epidemiological evidence that individual behavior is related to the development of chronic diseases. Each presentation rigorously evaluates the current state of knowledge with regard to certain associations with behavioral implications, such as diet and heart disease and diet and cancer. Avogaro provides a review of the metabolic evidence that lipoproteins play a causal role in the development of heart disease. He pays specific attention to fractions of HDL and apoproteins. Criqui examines the conflicting evidence on alcohol use and mortality. There appears to be a complex, non-linear relationship between alcohol consumption and mortality, and Criqui carefully evaluates the potential protective and noxious effects of alcohol on cardiovascular disease. The role of high fiber in diet is appraised by Burkitt. Burkitt's findings suggest that fiber plays an important protective role in a variety of disease states. Joossens provides an overview of mortality differences in Belgium. His data reveal that different patterns of coronary heart disease in Belgium may be associated with different dietary patterns and in particular, the use of butter.

General overviews of specific disease states are reviewed in three papers. Kromhout provides a general overview of the role of dietary components in heart disease with specific focus on data collected in the Netherlands. Labarthe reviews the complex issues of the treatment of mild hypertension. The paper by Haines carefully reviews the evidence that different diets may increase or decrease the risk of cancer. Finally, Tchobroutsky discusses the benefits of the control of diabetes. This section also includes one paper in medical care epidemiology. Härtel, Keil, and Cairns present evidence on the relationship between the availability of health care and hypertension control in West Germany.

4

APOLIPOPROTEINS, THE LIPID HYPOTHESIS, AND ISCHEMIC HEART DISEASE

Pietro Avogaro, M.D., Ph.D.

The development of atherosclerosis in man is complex and polifactorial process. Because elevated plasma levels of lipids and abnormal lipoprotein patterns have been identified as relevant risk factors for coronary atherosclerosis, major areas of scientific investigation have included the study of lipid (cholesterol, triglycerides, and phospholipids) and lipoprotein disorders. In recent years, lipoproteins have been classified according to their behavior in density-adjusted ultracentrifugation. Elevated plasma cholesterol and low density lipoproteins (LDL) are positively correlated with prevalence of coronary artery disease (Barr, Russ, and Eden, 1951; Kannel et al., 1961, 1971). Very low density lipoproteins (VLDL) and triglycerides once considered an independent risk factor (Böttiger and Carlson, 1980), are no longer thought to be valuable prognostic or discriminating factor (Heyden et al., 1980). In some groups of patients, the intermediate density lipoproteins (IDL) are significantly higher in atherosclerotic patients (Avogaro et al., 1979). An inverse relationship between coronary heart disease and high density lipoprotein (HDL) cholesterol has great statistical strength (Miller and Miller, 1975; Castelli et al., 1977). The suspicion exists that HDL_2 or the ratio HDL_2/HDL_3 might be a more powerful predictor of protection from heart disease than is total HDL (Anderson et al., 1978). Some investigators, however, have recorded lower levels of both HDL_2 and HDL_3 in new cases of ischemic heart disease (IHD) (Gofman, Young, and Tandy, 1966) and in survivors of myocardial infarction (Avogaro et al., 1979).

Traditionally, monitoring of plasma lipids has included the assay of an isolated lipid component of lipoproteins or the total mass of lipoprotein, whereas little attention has been paid to the protein part of lipoproteins. Recently, due to a substantial improvement in technology, at least ten specific lipid-binding proteins or apolipoproteins (apoLp) have been identified. The various families of apolipoproteins are conventionally identified alphabetically (A, B, C, D, E) (Alaupovic, 1971, 1982) with the individual apolipoproteins in a group being indicated with a Roman numeral (A-I, A-II, C-I, C-II, C-III) or an Arabic numeral (B-100, B-48, B-76, B-24). In a relatively short time, both structural and functional roles of apolipoproteins have been largely identified (Alaupovic, 1982; Schaeffer, Eisenberg, and Levy, 1978). Various methods have been suggested for the study of apolipoproteins. Some of the methods, such as radial diffusion (Outcherlony, 1953), electroimmunoassay (Curry, Alaupovic, and Suenram, 1976), radioimmunoassay (Karlin et al., 1976), are immunochemical. Other methods are nonimmunochemical and are based on the different molecular weight or electric charge of apolipoproteins: among these methods are included molecular sieve chromatography (Kane, Hardman, and Paulus, 1980), sodium duodecyl sulphate polyacrilamide gel electrophoresis (Laemmli, 1970) urea polyacrylamide gel electrophoresis (Kane, 1973), and isoelectric focusing (Gidez, Swaney, and Murnane, 1976). The disposable techniques for the evaluation of apolipoproteins permit fresh evaluation of lipoproteins as a discriminating or pathogenetic factor of atherosclerosis. Berg, Borresen, and Dahlen, 1976 were probably the first to stress a role for apoproteins in human atherosclerosis by recording in a series of survivors of myocardial infarction low levels of proteins (apo)A. This finding was confirmed when Albers and colleagues (1976)[1] showed low apoA-I and apoA-II and Avogaro and colleagues (1978) found low levels of apoA-I in two groups of survivors. Subsequently, Bradby, Valente, and Walton (1978) and Avogaro and colleagues

---

[1]Protein A is the major protein moiety of HDL; it is composed of two peptides, apoA-I and apoA-II. Both peptides have a structural function. ApoA-I is an activator of lecitin: cholesterol acyltransferase and participates in the interconversion of different HDL subfractions.

(1979) recorded increased levels of apoB in two series of atherosclerotic patients.[2] We also identified that apoB is the best single discriminating factor between IHD patients and controls and that the two ratios total Chol/apoB and apoB/apoA-I were the best discriminators between survivors and controls. Initially, there were some difficulties in accepting this new line of thought because survivors of myocardial infarction might be inappropriate examples of human atherosclerosis as they were studied following a heart attack and not before. But the same is true of a large series of angiographically proven coronary patients (Kladetzky, 1980; Riesen, 1982). However, some syndromes characterized by the absence of or by extremely low levels of apoA-I produce no excess of atherosclerotic disease. This is the case for in Tangier disease (Assmann, 1979), despite a report referring a slightly higher prevalence of ischemic heart disease in these patients than in controls (Schaefer et al., 1980), in lecithin-cholesterol-acyltransferase (LCAT) deficiency (Glomset, 1970), in apoA-I variant deficiency (Gustafson et al., 1979) and, more recently, in apoA-I Milano (Franceshini et al., 1980) and fish-eye disease (Carlson, 1982). These are important examples indicating that low levels of apoA-I (and of HDL) and even its absence are not related to a high prevalence of cardiovascular events.

Nevertheless, there are some families with a high prevalence of ischemic heart disease characterized only by extremely low levels of apoA-I (Vergani and Bettale, 1981; Schaefer et al., 1982; Norum et al., 1982). Some of these patients show a clinical picture very close to that considered specific to F. H. Xanthomatosis (Schaefer et al., 1982; Norum et al., 1982). It appears, therefore, that in exceptional cases a severe pattern of coronary disease may be linked to a deficiency of apoA-I. Recently, some IHD patients with an isolated increase of apoB and with normal levels of total-C, total-TG, and LDL-C have been reported (Sniderman et al., 1980; Kwiterovich et al., 1981). In one case of this "hyperapobetalipoproteinemia," binding and internalization of LDL by LDL-receptors was normal (Kwiterovich et al., 1981). A special study has been dedicated to the composition of LDL in patients having "hyperapobetalipoproteinemia." While these subjects do not differ from controls in the composition of "light" LDL (d 1019-1043), they show a lower cholesterol/protein ratio in the "heavy" LDL (d 1043-1052) (Thompson, Teng, and Sniderman, 1982). Patients affected by atherosclerosis of inferior limbs are mainly characterized by a significant increase of VLDL-apoB (Franceschini et al., 1982). Patients with type IV lipoprotein pattern and high levels of LDL-protein B showed a higher prevalence of atherosclerotic heart disease than patients with the same phenotype and normal LDL-protein B (Sniderman et al., 1982). Recently a report

---

[2]ApoB is present in triglyceride-rich lipoproteins (VLDL and IDL) and in LDL; it is the major structural protein of LDL. It exerts an essential action for the uptake of cholesterol-rich lipoproteins and for the biosynthesis of cholesterol.

described the case of a man with a normocholesterolemic tendom xanthomatosis (Vega et al., 1983). Whereas levels of plasma lipids LDL-C and of apoB were normal, an overproduction of apoB was recorded which was considered to be the origin of the patient's xanthomata.

Another apoprotein, apoE, has received much attention. ApoE is present in triglyceride-rich lipoproteins and in HDL. If the concentration of apoE is especially high in the "remnants," it is significantly increased in type III (Havel and Kane, 1973; Utermann, Hees, and Steinmetz, 1977). The major apoE isoproteins seen in plasma are designated apoE-4, apoE-3, and apoE-2 while the apoE alleles are called $[E]^4$, $[E]^3$, and $[E]^2$. The mode of inheritance of the three alleles of apoE provides for six phenotypes: E 4/4, E 3/3, E 2/2, E 4/3, E 3/2 and, E 4/2 (Zannis et al., 1982). It was observed that homozygotics for type III show a special pattern characterized by the absence of E_3 (E 2/2), while heterozygotics show a deficiency of this isoform. In the absence of the isoprotein E_3 (E_4), the "remnants" cannot be bound, internalized, and metabolized. This explains the high levels of "remnants" (and of IDL, cholesterol, and triglycerides) in the plasma of patients with type III. Investigations of the levels of apoE and the prevalence of the various apoE patterns in subjects affected by atherosclerosis have given negative results. The plasma levels of apoE in atherosclerotics do not differ from levels in controls (Avogaro, 1984) and the prevalence of the apoE patterns is similar in patients and in the control population (Utermann, 1983). The only difference concerning apoE between atherosclerotics and controls is due to significantly lower levels of apoE in HDL (Avogaro, 1984).

Do these data mean that in atherosclerotics less apoE is present in HDL_c and that there is a deficiency within a lipoprotein which is essential for the action of LCAT or for the removal of the cholesterol esters? Following the initial studies on the values of apolipoproteins as discriminating risk factors (Albers et al., 1976; Avogaro et al., 1978, 1979; Berg, Borresen, and Dahlen, 1976; Bradby, Valente, and Walton, 1978), little attention has been paid to the study of population groups at risk for IHD. Our group of survivors of myocardial infarction was the first large group analyzed (Avogaro et al., 1982): of 344, only 129 (37.5 percent) had increased values of cholesterol and triglycerides. Of the remaining 215 (62.5 percent) with normal cholesterol and triglycerides, 31.7 percent also had normal values of LDL-C and of HDL-C. But when apolipoproteins A-I and B were analyzed, only two persons appeared to be completely normolipidemic (normal cholesterol triglycerides, LDL-C, HDL-C, apoB, apoA-I and their ratios). Furthermore, not a single case with increased values of cholesterol or triglycerides was recorded with a normal apoproteins pattern.

There is no information concerning the predictive value of apolipoproteins, although a small contribution to this field comes from Ishikawa and colleagues (1978) who recorded low levels of apoA-I in subjects who subsequently died from coronary disease.

Atherosclerosis is not the exclusive factor for the development of ischemic heart disease. Atherosclerotic narrowing of the coronary vessels is present in about 50 percent of men aged more than 50, but IHD is present only in the 5 percent of this population (Maseri, 1982). Moreover, only about 30 percent of survivors of myocardial infarction show a hyperlipoproteinaemic state (Avogaro et al., 1982; Goldstein et al., 1973). Thus, employing the classic lipid parameters, the "lipid hypothesis" is not applicable to 70 percent of patients showing the most serious pattern of coronary disease. But there is no place for a struggle for the lipid versus protein components of lipoproteins, or vice versa.

Because apoproteins seem to relate to IHD on structural, physiological, and clinical grounds, let us think a little more about them. One obstacle to validation of any relation between a given apoprotein and IHD is that large and expensive epidemiological studies have not included apolipoproteins in their protocols. Thus, we are still forced to think and to talk in terms of cholesterol (total, VLDL, IDL, or HDL). Despite some sound arguments, partly outlined in this paper, the lipid hypothesis for atherosclerosis has been hostilely disputed (Mann, 1977; Ahrens, 1982). Actually we shall not be surprised if some of the emerging, exciting fields such a prostaglandins, platelets, and smooth muscle research surpass the old unconvincing leader. Before lipidologists resign, however, the same attention (and the same love) as they dedicated to lipids should be given to apoproteins. Only when this is done will we know whether the lipid hypothesis belongs to the past or is still viable.

References

Ahrens, E. H., Jr. Diet and heart disease shaping public perceptions when proof is lacking. *Arteriosclerosis* 2 (1982) 85-86.

Alaupovic, P. The role of apolipoproteins in lipid transport process. *Research* 12 (1982) 3-21.

Alaupovic, P. Apolipoproteins and lipoproteins. *Arteriosclerosis* 13 (1971) 141-146.

Albers, J. J. et al. Quantitation of apolipoprotein A-I of human plasma high density lipoproteins. *Metabolism* 25 (1976) 633-644.

Anderson, D. W., Nichols, A. V., Pan, S. S., and Lindgren, F. T. High density lipoproteins distribution. Resolution and determination of three major components in a Normal Population sample. *Arteriosclerosis* 29 (1978) 161-179.

Assmann, G. Tangier disease and the possible role of high density lipoproteins in atherosclerosis. *Atherosclerosis Review* 6 (1979) 1-27.

Avogaro, P., Bittolo-Bon, B., Cazzolato, G., Belussi, F., and Pontoglio, E. Apoproteins A and latent atherogenic dyslipoproteinemia. In J. L. De Gennes (ed.) *Latent Dyslipoproteinaemias and atherosclerosis.* New York: Raven Press, 1984. Pp. 147-155.

Avogaro, P., Bittolo-Bon, G., Cazzolato, G., and Quinci, G. B. Are apolipoproteins better discriminators than lipids for atherosclerosis? *Lancet* 1 (1979) 901-903.

Avogaro, P., Cazzolato, G., Bittolo Bon, G., and Belussi, F. Levels and chemical composition of HDL$_2$, HDL$_3$ and other major lipoprotein classes in survivors of myocardial infarction. *Artery* 5 (1979) 495-508.

Avogaro, P. et al. Lipoproteins derangement in human atherosclerosis. In Noseda, G., Fragiacomo, C., Fumagalli, R., Paoletti, R. (eds.) *Lipoproteins and Coronary Atherosclerosis*. Amsterdam: Elsevier Biomedical Press, 1982.

Avogaro, P. et al. Plasma levels of apolipoprotein A-I and apolipoprotein B in human atherosclerosis. *Artery* 4 (1978) 385-394.

Barr, D. P., Russ, E. M., and Eder, H. A. Protein-lipid relationship in human plasma II. In atherosclerosis and related conditions. *Amererican Journal of Medicine* 11 (1951) 480-484.

Berg, K., Borresen, A. L., and Dahlen, G. Serum high-density lipoprotein and atherosclerotic heart disease. *Lancet* 1 (1976) 499-501.

Böttiger, L. E. and Carlson, L. A. Risk factors for ischaemic vascular death for men in the Stockholm Prospective Study. *Arteriosclerosis* 36 (1980) 389-408.

Bradby, G. V. H., Valente, A. J., and Walton, K. W. Serum high-density lipoproteins in peripheral vascular disease. *Lancet* 2 (1978) 1271-1274.

Carlson, L. A. Fish eye disease: A new familial condition with massive corneal opacities and dyslipoproteinaemia. *European Journal of Clinical Investigation* 12 (1982) 41-53.

Castelli, W. P. et al. HDL cholesterol and other lipids in coronary heart disease. The Cooperative Lipoprotein Phenotyping Study. *Circulation* 55 (1977) 767-772.

Curry, M. D., Alaupovic, P., and Suenram, C. A. Determination of apolipoprotein A and its constitutive A-I and A-II polypeptides by separate electroimmunoassays. *Clinical Chemistry* 22 (1976) 315-322.

Franceschini, G. et al. Increased apoprotein B in very low density lipoproteins of patients with peripheral vascular disease. *Arteriosclerosis* 2 (1982) 74-80.

Franceschini, G. et al. A-I Milano apoprotein: Decreased high density lipoprotein cholesterol levels with significant lipoprotein modifications and without clinical atherosclerosis in a Italian family. *Journal of Clinical Investigation* 66 (1980) 892-900.

Gidez, L. I., Swaney, J. B., and Murnane, S. Analysis of rat serum apolipoproteins by isoelectricfocusing. I. Studies on the middle molecular weight subunits of HDL and VLDL. *Journal of Lipid Research* 18 (1976) 69-76.

Glomset, J. A. Physiological role of lecithin cholesterol acyl transferase. *American Journal of Clinical Nutrition* 23 (1970) 1129-1136.

Gofman, J. W., Young, W., and Tandy, R. Ischaemic heart disease, atherosclerosis and longevity. *Circulation* 34 (1966) 679-697.

Goldstein, J. L. et al. Hyperlipidemia in coronary heart disease. 1. Lipid levels in 500 survivors of myocardial infarction. *Journal of Clinical Investigation* 52 (1973) 1533-1543.

Gustafson, A. et al. Identification of lipoprotein families in a variant of hyman plasma apolipoprotein A deficiency. *Scandinavian Journal of Clinical and Laboratory Investigation* 39 (1979) 377-387.

Havel, R. J. and Kane, J. P. Primary dysbetalipoproteinemia: Predominance of a specific apoprotein species in triglyceride rich lipoproteins. *Proceedings of the National Academy of Sciences* 70 (1973) 2015-2019.

Heyden, S., Heiss, G., Hames, C. G., and Burtel, A. G. Fasting triglycerides as predictors of total and CHD mortality in Evans County, Georgia. *Journal of Chronic Disease* 33 (1980) 272-282.

Ishikawa, T. et al. The Tromso study: serum apolipoprotein A-1 concentration in relation to future coronary heart disease. *European Journal of Clinical Investigation* 8 (1978) 179-182.

Kane, J. P. A rapid electrophoretic technique for identification of submit species of apoproteins in serum lipoproteins. *Annals of Biochemistry* 53 (1973) 350-364.

Kane, J. P., Hardman, D. A., and Paulus, H. E. Heterogeneity of apolipoprotein B: Isolation of a new species from human chylomicrons. *Proceedings of the National Academy of Sciences* 77 (1980) 2465-2469.

Kannel, W. B., Castelli, W. P., Gordon, T., and MacNamara, P. M. Serum cholesterol, lipoproteins and the risk of coronary heart disease: The Framingham Study. *Annals of Internal Medicine* 74 (1971) 1-12.

Kannel, W. B. et al. Factors of risk in the development of coronary heart disease. Six year follow-up experience. *Annals of Internal Medicine* 55 (1961) 33-50.

Karlin, J. B., Juhn, D. J., Starr, J. L., and Rubenstein, A. H. Measurement of human high density lipoprotein apolipoprotein A-1 in serum by radioimmunoassay. *Journal of Lipid Research* 17 (1976) 30-37.

Kladetzky, R. G. et al. Lipoprotein and apoprotein values in coronary angiography patients. *Artery* 7 (1980) 191-205.

Kwiterovich, P. et al. Familial hyperapobctalipoproteinemia in a Amish Kindred (Abstract). *Arteriosclerosis* 1 (1981) 87.

Laemmli, U. K. Cleavage of structural proteins during the assembly of the head of bacteriophage T4 *Nature* 227 (1970) 680-685.

Mann, G. V. Diet heart: End of an era. *New England Journal of Medicine* 297 (1977) 644-650. B Maseri, A. Active and quiescent phases in coronary disease. Role of varied and changing susceptibility to dynamic obstruction. *Circulation* 66 (suppl. II, IIB, C, D), 1982.

Miller, C. J. and Miller, N. E. Plasma high-density lipoprotein concentration and development of ischaemic heart disease. *Lancet* 1 (1975) 16-19.

Norum, R. A. et al. Familial deficiency of apolipoproteins A-I and C-III precocious coronary-artery disease. *New England Journal of Medicine* 306 (1982) 1513-1519.

Outcherlony, Ö. Antigen-antibody reactions in gels. VI. Types of reactions in coordinated systems of diffusion. *Acta Path. Microbiology of Scand.* 32 (1953) 231-240.

Riesen, W., Mordasini, R., Salzmann, C., and Gurtner, H. P. Apoproteins in angiographically documented coronary heart disease. In Noseda, C., Fragiacomo, C., Fumagalli, R., and Paoletti, R. (eds.) *Apoproteins in angiographically documented coronary heart disease.* Amsterdam: Elsevier Biomedical Press, 1982.

Schaefer, E. J., Eisenberg, S., and Levy, R. I. Lipoprotein apoprotein metabolism. *Journal of Lipid Research* 19 (1978) 667-687.

Schaefer, E. J., Heaton, H. H., Wetzel, M. G., and Brewer, B. J., Jr. Plasma apolipoprotein A-I absence associated with a marked reduction of high density lipoproteins and premature coronary artery disease. *Arteriosclerosis* 2 (1982) 16-26.

Schaefer, E. J., Zech, L. A., Schwartz, D. E., and Brewer, H. B., Jr. Coronary heart disease prevalence and clinical features in familial high density lipoprotein deficiency (Tangier disease). *Annals of Internal Medicine* 93 (1980) 261-266.

Sniderman, A. D. et al. Association of hyperapobetalipoproteinemia with endogenous hypertriglyceridemia and atherosclerosis. *Annals of Internal Medicine* 97 (1982) 833-939.

Sniderman, A. D. et al. Association of coronary atherosclerosis with hyperapobetalipoproteinemia (increased protein but normal cholesterol levels in plasma low density (B) lipoproteins. *Proceedings of the National Academy of Sciences* 77 (1980) 604-608.

Thompson, G. R., Teng, B., and Sniderman, A. Metabolic conversion of "light" to "heavy" low density lipoproteins (LDL) in subjects with normal and increased plasma LDL-apoB levels (Abstract). *Circulation* 66 (suppl. II) (1982) 158.

Utermann, G. Human apolipoprotein mutants. In Schettler, G., Gotto, A. M., Middlehoff, G., Habenicht, A. J. R., and Jurutka, K. R. (eds.) *Arteriosclerosis VI.* Berlin: Springer-Verlag, 1983.

Utermann, G., Hees, M., and Steinmetz, A. Polymorphysm of apolipoprotein E and occurrence of dysbetalipoproteinaemia in man. *Nature* 269 (1977) 604-607.

Vega, G. L. et al. Normocholesterolemic tendon xanthomatosis with overproduction of apolipoprotein B. *Metabolism* 32 (1983) 118-125.

Vergani, C. and Bettale, G. Familial hypo-alpha-lipoproteinaemia. *Clinical Chim. Acta* 114 (1981) 45-52.

Zannis, V. I. et al. Proposed nomenclature of apoE isoproteins, apoE genotypes and phenotypes. *Journal of Lipid Research* 23 (1982) 911-914.

5

ALCOHOL AND
CARDIOVASCULAR MORTALITY

Michael H. Criqui, M.D., M.P.H.

Although effects of alcohol on the cardiovascular system have been recognized for more than a century (Friedreich, 1861), only in recent years has the complexity of the association of ethanol consumption with cardiovascular disease (CVD) morbidity and mortality begun to be appreciated. Much recent work has detailed the direct effects of ethanol on the myocardium and its conduction system, the striking associations of alcohol with certain physiologic and biological variables influencing atherosclerosis, and the association in epidemiologic studies between ethanol and subsequent morbidity and mortality. In this chapter we intend to review the direct effects of alcohol on the heart; the apparent alcohol effects on certain cardiovascular risk factors such as lipids, blood pressure, thrombogenesis, and stress; alcohol's association with cardiovascular mortality; and the implications of such findings for public health policy.

1. Association of Alcohol and Other Causes of Mortality

Leaving aside cardiovascular disease for the moment, heavy alcohol consumption appears to adversely affect nearly every other cause of mortality; malignant neoplasms, accidents, violence, and even "other" and "unknown" causes, as well as the extensively documented association with cirrhosis of the liver.

Table 1, adapted from the Honolulu Heart Study, shows the association of alcohol with causes of death other than cardiovascular disease and the excess total mortality associated with an average of 40 or more ml of alcohol per day (Kagan et al., 1981). Although this chapter focuses on cardiovascular mortality, we should remember that from a public health viewpoint total mortality is the critical endpoint.

The non-cardiovascular toll exacted by alcohol in most societies in terms of morbidity and mortality, hospitalization, and useful life years lost, not to mention the concurrent psychosocial impact of such human tragedy, is well documented and will not be further detailed here, since it is beyond the scope of this chapter.

Table 1

9 Year Age-Adjusted Mortality Rates per 1000 from All Causes and from Specified Non-Cardiovascular Causes by Level of Ethanol Consumption*

Ethanol ml/day	Cirrhosis	Cancer	Other Non-CVD	All Causes Including CVD
0	1.8	18.4	21.7	77.4
1-6	0.9	20.1	19.9	65.1
7-15	1.0	25.5	13.3	70.0
16-39	1.0	20.2	20.9	57.0
40-59	3.3	42.9	30.3	88.3
60+	14.0	55.9	35.9	122.5

*Adapted from Kagan et al., 1981.

2. Direct Effects of Alcohol on the Heart

During the 19th century several reports suggested alcohol might have a direct toxic effect on the heart, and the exact nature of such effects remains an important research question today. Research has also identified syndromes in which a toxic substance combined with alcohol to produce cardiac damage, such as arsenic-beer drinkers' disease at the turn of the century (Reynolds, 1901) and the more recent cobalt-beer drinkers' disease (Morin and Daniel, 1967). It also seems likely that alcohol can produce or accompany deficiencies of important nutrients, such as thiamine. Thiamine deficiency and its related disorders are not uncommon with alcohol excess. Perhaps the only definitive remedy for the thiamine deficiency-alcohol link is to fortify alcoholic beverages with thiamine, which has been suggested to be cost-beneficial (Centerwall and Criqui, 1978).

Apart from alcohol-related cardiac problems linked to other toxic substances or deficiency states, a considerable body of research suggests that alcohol can directly effect the heart. It is generally accepted that alcohol can be directly toxic to heart muscle (Spann et al., 1968), and it is also now generally agreed that alcoholic cardiomyopathy is a clear-cut syndrome. In one study, 80 percent of patients hospitalized for primary myocardial disease were "heavy drinkers" compared with 28 percent of patients with other diagnoses admitted to the same medical service (Alexander, 1966). Another report documented more frequent regression of myocardial disease in patients quitting than in those continuing to drink (Demakis et al., 1974). Thus, alcohol does seem to contribute directly to myocardial disease.

In addition, ethanol can alter conduction velocity (Gould et al., 1978) and action potential duration (Fisher and Kavaler, 1975). These findings may provide a biologic basis for the known ability of alcohol ingestion to stimulate arrhythmias, even in the apparent absence of underlying cardiac disease. A typical weekend or holiday presentation of acute arrhythmias after heavy drinking has been labeled the "holiday heart" syndrome (Ettinger et al., 1978). One report has documented the provocation of ventricular tachycardia after ethanol consumption in a human subject (Greenspan et al., 1979). Clearly, the potential for life-threatening arrhythmia and sudden death exists in someone abusing alcohol. However, induction of such an arrhythmia with moderate alcohol use presumably occurs much less often.

3. Alcohol and Blood Lipids

Historically, most of the interest in the effect of ethanol on lipids has focused on total cholesterol and triglycerides. Although there have been reports of increased cholesterol with increased use of alcohol (Kozarevic et al., 1980), most studies indicate rather small changes. By contrast, many studies indicate that triglycerides are elevated in drinkers, particularly in heavy drinkers (Lifton and

Scheig, 1978). Most recent epidemiologic information, however, indicates that triglycerides are not an independent risk factor for CHD, but instead are related to cardiovascular disease in univariate analyses because of strong positive correlation of trigylcerides with total cholesterol (Hulley et al., 1980). Thus, the effect of ethanol on triglycerides seems an unlikely pathway for any influence, noxious or beneficial, of ethanol on cardiovascular mortality.

In recent years the issue of the ethanol-cholesterol link has become much more complex because of the slow realization that each of the three lipoprotein classes carrying cholesterol has a different association with cardiovascular mortality. Specifically, low density lipoprotein cholesterol, or LDL cholesterol, appears to be positively related to cardiovascular mortality, and is responsible for the association of total cholesterol with CVD mortality, since about 70 percent of total cholesterol on the average is carried on LDL. Although reported as early as 1951 (Barr, Russ, and Eder, 1951), only recently have we begun to appreciate that the association of high density lipoprotein cholesterol, or HDL cholesterol, to CVD mortality is actually inverse, i.e., higher levels are associated with a lower risk of CVD mortality (Gordon et al., 1977; Miller et al., 1977). Table 2 shows the association of coronary heart disease (CHD) incidence with HDL level in the Framingham Study. Note that higher HDL levels are associated with lower CHD incidence.

Table 2

CHD Incidence per 1000 By Level of HDL Cholesterol,
Framingham Study, Exam 11, Ave. Follow-up of $\simeq$ 4 yrs.*

HDL level, mg/dl	Men	Women
< 25	176.5	0.0
25-34	100.0	164.2
35-44	104.5	54.5
45-54	51.0	49.2
55-64	59.7	39.7
65-74	25.0	13.9
75+	0.0	20.1
All levels	77.1	43.6

*Adapted from Gordon et al., 1977.

Table 3

Mean HDL and LDL By Alcohol Consumption Category in
Subjects Age 50-69*

Alcohol (ml/day)	No. of subjects	Mean level (mg/dl)	
		HDL	LDL
Albany:			
Total	923	49.6	153.5
0	201	46.3	152.9
4-16	372	47.4	155.2
17-42	260	53.3	152.6
43-85	80	54.6	151.4
86+	10	60.6	141.6
Framingham men:			
Total	393	45.7	140.3
0	111	41.4	143.7
4-16	112	44.8	136.4
17-42	111	47.4	144.7
43-85	44	50.1	136.7
86+	15	58.4	123.3
Honolulu:			
Total	1713	45.1	142.8
0	849	42.2	147.0
4-16	320	44.8	148.5
17-42	354	48.3	138.6
43-85	166	52.2	125.8
86+	24	56.7	97.7
San Francisco:			
Total	277	46.6	149.
0	133	44.4	151.7
4-16	90	45.8	156.5
17-42	40	51.7	133.9
43-85	12	57.8	145.6
Framingham women:			
Total	500	58.6	149.8
0	242	54.8	155.1
4-16	169	61.4	145.5
17-42	82	63.5	144.5

*Adapted from Castelli et al., 1977.

The third lipoprotein fraction, very low density lipoprotein cholesterol, or VLDL cholesterol is closely related to the triglycerides level, and though it may increase with increased alcohol, it, like triglyceride, is apparently unrelated to CVD mortality.

LDL cholesterol appears to decrease and HDL increase with increasing amounts of ethanol in population surveys. Table 3 presents data from the Cooperative Lipoprotein Phenotyping Project, which evaluated cross-sectional associations between alcohol and lipoproteins in several populations (Castelli et al., 1977).

Note that HDL is higher with increasing consumption of alcohol and LDL is lower. It should be noted that the alcohol consumption data in the tables in this chapter relating alcohol to risk factors or mortality are self-reported, and this probably reflects some underreporting, particularly in heavy drinkers. Table 4 shows recent data from the Lipid Research Clinics Prevalence Study for the estimated cross-sectional difference in HDL and LDL levels in subjects drinking 43 ml. of alcohol per day compared to none (Criqui et al., 1983). Data are age and sex-specific, and are adjusted by multiple linear regression analysis for the potential confounding effects of age (within 30 year age groups), obesity, blood pressure, physical activity, cigarette smoking, education level, and gonadal hormone use in women.

The higher HDL and lower LDL levels associated with alcohol suggest that alcohol may have a dually protective effect by raising the presumably protective HDL fraction while lowering the atherogenic LDL fraction. This differential effect of alcohol on HDL and LDL, particularly the positive association with HDL, has

Table 4

Multivariable Differences in HDL and LDL
in mg/dl for 43 ml of Ethanol Per Day Compared to None.
The Lipid Research Clinic's Prevalence Study

	HDL	LDL
Men, 20-49	5.5†	-1.1
Men, 50+	8.1†	4.0
Women, 20-49	9.8†	-13.4†
Women, 50+	11.7†	-18.5†

†$p < 0.05$.

Note the significantly higher HDL levels in drinkers. Lower LDL levels with alcohol consumption in this study occurred only in women.

probably stimulated more clinical and epidemiologic interest in the past few years than any other ethanol-related research finding in cardiovascular disease, and has been repeatedly proposed as a possible protective mechanism of moderate alcohol consumption for cardiovascular disease (Kannel, 1977; Yano, Rhoads, and Kagan, 1977).

4. Alcohol and Blood Pressure

An association between alcohol consumption and increased blood pressure has been reported recently by several different authors (Beevers, 1977; Criqui et al., 1981; Klatsky et al., 1977). Klatsky and colleagues reported that subjects drinking two drinks per day, or 30 ml of ethanol, had pressures similar to nondrinkers, but that beginning at three drinks per day (45 ml of ethanol) blood pressure was increasingly higher at higher levels of drinking. This effect produced a doubling of hypertension (SBP $\geq$ 160 or DBP $\geq$ 95) in whites and a 50 percent increase in blacks at six drinks (90 ml of alcohol) per day compared to none. However, the mechanism of action has not been clearly delineated, and the criticism has been raised that heavy alcohol consumption might reflect "stress" in an individual with the "stress" being the true cause of hypertension. In the Lipid Research Clinic Prevalence Study, we confirmed the positive alcohol-blood pressure association and also noted a threshold of about one and one-half to two drinks per day before the relationship was clearly evident (Criqui et al., 1981). We also demonstrated increases in categorically defined hypertension with increased drinking. Further, we had collected data in reported alcohol consumption for two different time periods, the entire week prior to the study and separately for the day prior to the study. This unique data set allowed us to speculate on the mechanism of action.

Table 5 shows multiple regression coefficients for alcohol consumed in the 24 hours prior to the study (Alc 24) and in the previous week minus the previous 24 hours (Alc minus), adjusted for the potential confounding effects of age, obesity, smoking, exercise, education, and gonadal hormone use in women. These data indicate, ml for ml, at least a threefold greater effect of alcohol consumed in the past day compared to the rest of the past week. However, subjects were fasting for 12 hours before the study, so presumably none had drunk alcohol during that period. Thus, much of the blood pressure increase seems likely to be related to alcohol withdrawal. Urinary excretion of epinephrine is greater during withdrawal than during alcohol administration (Ogata et al., 1971) and plasma norepinephrine levels are highest 13 to 24 hours after alcohol cessation (Carlsson and Haggendal, 1967). In addition, plasma arginine vasopressin (Linkola et al., 1978) and plasma renin activity (Linkola, Fyhrquist, and Ylikahri, 1979) are increased during the withdrawal phase. In concordance with these data is the observation that alcoholics undergoing withdrawal may have marked hypertension which disappears without medication following withdrawal (Clark and Friedman, 1983) We thus concluded that the biological plausibility for a withdrawal effect was strong.

73

Table 5

Multiple Regression Coefficients for Alcohol Consumed During
the Previous 24 Hours (Alc 24) and the
Previous Week Except the Previous 24 Hours (Alc Minus)*

	Men, Age 20+ Coefficient	Women, Age 20-49 Coefficient	Women, Age 50+ Coefficient
Systolic BP			
Alc 24	0.027	0.033	0.063
Alc Minus	0.010	0.004	0.017
Diastolic BP			
Alc 24	0.019	0.015	0.025
Alc Minus	0.005	0.004	0.008

*Adapted from Criqui et al., 1981.

Although this is certainly stretching the data a little bit, we also speculated that this evidence indicating the greater relevance of alcohol consumed in the previous 24 hours, along with the biological plausibility of withdrawal, implied a direct physiologic effect, and thus, the possibility that "stress" was increasing alcohol consumption and blood pressure independently seemed less likely.

We should perhaps mention one other potential link between alcohol and blood pressure which has received little if any attention. Studies in animals indicate oral alcohol decreases calcium absorption, and there is reason to believe that some chronic alcoholics may absorb both calcium and vitamin D poorly (Hepner, Roginsky, and Moos, 1976; Krawitt, 1975). Apparently, even young alcoholics show evidence of a decreased bone mass (Saville, 1965). This finding was not present for moderate consumption of wine (McDonald and Margen, 1979). Interestingly, dietary calcium may be inversely related to blood pressure (McCarron, Morris, and Cole, 1982) and a recent randomized clinical trial of calcium supplementation in young adults indicated significantly reduced diastolic pressures in the supplemented group (Belizan et al., 1983). Might calcium deficiency in heavy drinkers stimulate higher blood pressure? Interestingly, vitamin D is also malabsorbed, and vitamin D has been suggested as a risk factor for atherosclerosis (Mann, 1977). Might the alcohol link be of interest here?

5. Alcohol and Thrombosis

There is considerable current interest in agents that might inhibit the clotting process, particularly platelet function. Much of this interest has focused on aspirin, which has been used in clinical trials for the primary prevention of stroke and the secondary prevention of coronary disease (Salzman, 1982). However, little attention has been paid to the possible role of alcohol in this area. It has been documented that ethanol suppresses erythroid progenitor cells, and that this suppression is more marked than what occurs for the granulocytic/macrophage progenitor cells (Meagher, Sieber, and Spivak, 1982). Thrombocytopenia has been repeatedly documented in alcoholics (Lindenbaum and Hargrove, 1968; Post and Desforges, 1968) and has been produced experimentally (Lindenbaum and Lieber, 1969). Haut and Cowan (1974) showed prolongation of the bleeding time and impairment of platelet aggregation during acute alcohol ingestion in human subjects. A recent paper has indicated that ethanol potentiates an aspirin-induced prolongation of bleeding time (Deykin, Janson, and McMahon, 1982). Interestingly, the report on experimental alcohol-induced thrombocytopenia indicated that some subjects had a "rebound" elevation of platelets above control levels after drinking cessation (Lindenbaum and Lieber, 1969), and a report from the Netherlands also documents thrombocytosis after withdrawal (Haselager and Vreeken, 1977). A most interesting paper (Meade et al., 1979) indicates moderate consumption of alcohol was associated with a lower fibrinogen and increased fibrinolytic activity, and fibrinogen was a risk factor for cardiovascular disease in the Northwick Park Heart Study (Meade et al., 1980). Although this area has received inadequate attention, it may be that the above data are relevant to the reported reduction in CHD with moderate alcohol consumption.

6. Alcohol and Stress

Various conditions or states considered to reflect "stress" of one sort or another have been reported to be related to CVD. Of these, probably the best supported by good scientific evidence is the prospective association of Type A behavior with subsequent CVD mortality (Jenkins, 1976). Type A behavior is a specific persistent behavior pattern characterized by enhanced aggressiveness, ambitiousness, competitive drive, and a chronic sense of time urgency. Also intriguing have been the reports indicating that measures of physiological reactivity to stress may predict subsequent CHD (Keys et al., 1971; Sime, Buell, and Eliot, 1979).

Alcohol is used, at least from anecdotal reports, to "calm one's nerves" or "to relax," and few of us would contradict the impression that mild inebriation might provide some refuge from stress. Thus, alcohol could conceivably affect CVD mortality by modulating Type A behavior or deterring the noxious effects of stress. However, probably because it is generally agreed that alcohol creates more problems than it solves and because of the recent focus on ethanol effects on lipid metabolism, this view has engendered little enthusiasm and is given little credence. In addition, the mechanism of action for any such association seems obscure since we know little about the biologic or physiologic pathways for the effect of Type A behavior or other stress related variables on CVD.

7. Alcohol and CVD Mortality

7.1. Case-control studies

Case-control studies have consistently reported an inverse relationship between alcohol and coronary heart disease in men and women (Hennekens et al., 1979; Klatsky, Friedman, and Siegelaub, 1974; Rosenberg et al., 1981; Stason et al., 1976). The study by Hennekens and colleagues (1979) looked at coronary deaths in men and determined alcohol consumption by interviewing wives of both the controls and the deceased cases. Table 6 shows the results from this study for the adjusted relative risks of drinking compared to abstaining.

Note that protection occurred only for light to moderate drinkers of beer, wine, or liquor, defined as less than 60 ml per day. The risk for heavy consumers was similar to nondrinkers.

Table 6

Adjusted Relative Risks for CHD Mortality in Daily Drinkers vs. Non-Drinkers*

Consumption Level and Type	Relative Risk	p-Value
Light to moderate		
beer	0.3	< .001
wine	0.3	< .001
liquor	0.2	< .001
Heavy		
beer	1.0	NS
wine	1.0	NS
liquor	1.1	NS

*Adapted from Hennekers et al, 1979.

7.2. Prospective population studies

Table 7 presents data from the Honolulu Heart Study for the association of alcohol with mortality from coronary heart disease, stroke, and other cardiovascular disease (Kagan et al., 1981).

Note the inverse relationship with coronary heart disease with a reversal at 60 ml per day, the U-shaped relationship with thromboembolic stroke, and the positive relationship to hemorrhagic stroke. These data are quite consistent with what has been observed in other studies. Although rates such as those presented in Table 7 are informative, we should remember that alcohol consumption tends to be associated with cigarette smoking, a behavior highly correlated with cardiovascular disease. Thus, most current studies of alcohol's association with risk factors or mortality (including the Honolulu Heart Study) in addition adjust for cigarette smoking and certain other variables, usually by multivariable analysis.

Table 8 shows data from hospitalizations for a number of cardiovascular conditions at the Kaiser Hospitals in Oakland and San Francisco, California from 1971-1976 (Klatsky, Friedman, and Siegelaub, 1981a).

Note the similarity of the patterns to the Honolulu data in that there is a U-shaped relationship for all cardiovascular disease. The data show increases in hypertension, stroke, venous disease, cardiomyopathy, congestive failure, and

Table 7

9 Year Age-Adjusted Mortality Rates per 1,000 for CHD, Thromboembolic Stroke, and Hemorrhagic Stroke by Level of Ethanol Consumption*

Ethanol, ml/day	CHD	Thromboembolic Stroke	Hemorrhagic Stroke
0	24.6	3.7	3.0
1-6	16.6	5.2	3.2
7-15	20.9	3.0	7.3
16-39	7.8	1.8	6.3
40-59	5.9	3.7	5.5
60+	12.8	5.1	12.7

*Adapted from Kagan et al., 1981.

Table 8

Percent Hospitalizations for Various Cardiovascular Conditions by Alcohol Consumption, Northern California Kaiser-Permanente Facilities, 1971-1976*

Diagnosis	% Incidence by Drinks Per Day			
	0	2 (30 ml)	3-5 (45-75 ml)	6+ (90+ ml)
Hypertension	2.4	2.8	3.4	4.5
Stroke	1.9	2.0	2.3	2.5
All Coronary	8.0	5.9	5.8	5.6
Acute Myo. Infarct.	3.8	3.2	2.6	2.5
Misc. Venous	2.6	2.2	2.3	3.9
Cardiomyopathy	0.1	0.0	0.1	0.4
Cong. Heart Failure	2.1	1.3	1.6	2.3
Arrhythmias	2.6	2.7	3.1	3.6
All Cardiovascular	13.9	12.1	12.3	14.7

*Adapted from Klatsky, Friedman, and Siegelaub, 1981a.

arrhythmias with more drinking, but lower levels of coronary disease. The relationship between alcohol and CVD mortality was also U-shaped in this population (Klatsky, Friedman, and Siegelaub, 1981b). The Yugoslavia Cardiovascular Disease Study also reported an inverse relationship between alcohol and CHD mortality and positive association with stroke (Kozarevic et al., 1980). However, a subsequent report from the same study indicated alcohol was not protective for sudden death (Kozarevic et al., 1982). In a Chicago Study (Dyer et al., 1980), alcohol was positively associated with CHD and CVD only at the level of six or more drinks (90+ ml) per day.

It should be noted that not all studies show an excess of CVD mortality at the highest levels of drinking. Figure 1 is taken from the Whitehall Study of civil servants in London, as reported by Marmot, Rose, Shipley, and Thomas (1981).

Total mortality showed a U-shaped relationship with alcohol, but CVD mortality did not. However, the protective effects of alcohol for CVD mortality plateaued at the modest level of 11 ml of alcohol per day.

Figure 2 shows preliminary data for all cause, CVD, and CHD age-adjusted death rates in men by levels of alcohol consumption in the Lipid Research Clinics (LRC) Follow-Up Study after six years (Criqui et al., 1982).

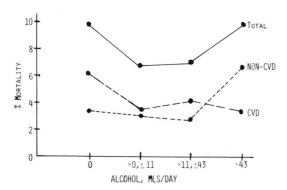

Figure 1. Age-Adjusted Percent Mortality in 10 Years by Alcohol Consumption Level, the Whitehall Study

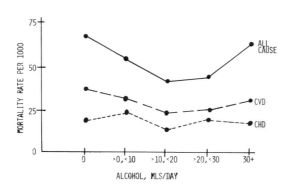

Figure 2. Age-Adjusted Mortality Rates for CHD, CVD, and All Causes by Alcohol Consumption Category, Men. Preliminary Data, The Lipid Research Clinics Follow-Up Study

79

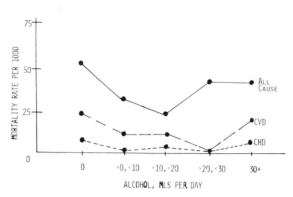

Figure 3. Age-Adjusted Mortality Rates for CHD, CVD, and All Causes by Alcohol Consumption Category, Women. Preliminnary Data, The Lipid Research Clinics Follow-Up Study

The U-shaped curve for CVD mortality is evident, although the result for CHD is less certain. Figure 3 shows the preliminary data for women. A U-shaped curve is evident for both CHD and CVD mortality, as well as all-cause mortality. These data are taken from our ongoing analysis of alcohol and mortality in this study.

In summary, prospective studies published to data tend to support the idea that moderate alcohol consumption, defined as two drinks a day (30 ml) or less, may be protective for CHD, but that higher levels predispose to hypertension, stroke, arrhythmias, cardiomyopathy, and at some point more CHD, although the latter relationship seems to occur at quite high levels of consumption. In concordance with this latter observation, men registered for drinking offenses in Sweden had an excess of coronary disease (Wilhelmsen, Wedel, and Tibblin, 1973), and problem drinkers in an occupational study had excess cardiovascular disease rates (Pell and D'Alonzo, 1973).

8. CHD, Thromboembolic Stroke, and Hemorrhagic Stroke

Table 7 indicates that for hemorrhagic strokes the lowest rates are in nondrinkers and the highest in the heaviest drinkers which is compatible with the idea that hypertension is the dominant risk factor for hemorrhagic stroke. By contrast, the data for thromboembolic stroke show a U-shaped curve with the lowest rates in moderate drinkers. Thus, the alcohol association for thromboembolic stroke is closer to that of coronary disease. Table 9 shows data from the Honolulu Heart Study and seems to indicate that the independent risk factors for thromboembolic stroke are in the middle of a spectrum with coronary heart disease positively associated with cholesterol and negatively with alcohol at one end and hemorrhagic stroke showing the opposite relationships at the other. However, systolic blood pressure is positive for all three endpoints.

Table 9

Systolic Blood Pressure, Age, Cholesterol, and Alcohol
Multivariable Logistic Function Coefficients for
CHD and Stroke Incidence*

Risk Factor	Standardized Coefficients		
	CHD	Thromboembolic Stroke	Hemorrhagic Stroke
Systolic BP	0.434†	0.455†	0.537†
Age	0.292†	0.324†	0.144
Cholesterol	0.202†	-0.032	-0.380†
Alcohol	-0.621†	-0.030	0.263†

†$p < 0.05$.
*Adapted from Kagan et al., 1981.

9. Possible Mechanisms of Action

Given the noxious consequences of alcohol on the cardiovascular system, we must now ask what good alcohol might do to reduce CHD mortality in the moderate consumption range? Among the possibilities outlined earlier, the strong positive association with HDL cholesterol has generated by far the most interest and discussion. This, at least in part, reflects the surprise and excitement among cardiovascular researchers when reports in the 1970s suggested HDL cholesterol might be a protective factor and perhaps stronger in its effect than the traditional risk factors (Gordon et al., 1977; Miller et al., 1977). Simultaneously, reports were being published that alcohol was strongly related to HDL levels and that moderate alcohol consumption was related to lower CHD mortality. This two step causal pathway--alcohol results in higher HDL which prevents atherosclerosis--has nearly achieved the status of dogma, and more than one toast to one's HDL level has been heard at social gatherings of epidemiologists.

Despite the enthusiasm for these findings, there is an obvious problem with this suggested causal pathway. Nearly all studies are in agreement that HDL increases monotonically with alcohol consumption up to even extreme drinking levels, as indicated earlier in Table 3. It has been suggested, in fact, that very high HDLs might alert the clinician to the presence of alcoholism (Kuller and Castelli, 1980). By contrast, the reduction in CHD mortality in most studies is only for moderate drinking. Why doesn't the protection for CHD extend to heavier drinking?

81

It has been suggested that noncardiac deaths in heavy drinkers may be incorrectly assigned a coronary diagnosis (Kuller et al., 1974; Randall, 1980), leading to an artifactual U-shaped curve. In addition, alcohol can precipitate a life threatening arrhythmia, presumably even in the absence of coronary disease.

However, recent laboratory investigations present additional problems for the alcohol = increased HDL = reduced CHD hypothesis. HDL cholesterol can be separated into two major subcomponents, HDL_2 and HDL_3, and only the HDL_2 subcomponent may be protective. Independently Haffner, Appelbaum-Bowden, Hoover, and Haggard, (1982) and Haskell, Krauss, Wood, and Lindgren, (1982) have found alcohol to be more strongly related to the HDL_3 subcomponent than the protective HDL_2, suggesting the inverse alcohol-CHD association may not be mediated by the effect of alcohol on HDL.

In an attempt to address this issue, we looked at the association of alcohol consumption with CHD and CVD mortality in a study with few heavy drinkers (Criqui et al., 1982). We reasoned that if the biologic link between alcohol and CHD/CVD mortality were via HDL, the predictive power of alcohol would be lost in multivariable mortality models which included HDL as a covariate. Figure 4 outlines this schematically.

If the positive alcohol-HDL association explains the inverse HDL-CVD association, the inverse alcohol-CVD association should disappear in a multivariable analysis which includes HDL. If alcohol remains important in such an analysis, its effect must be independent of HDL. Table 10 shows the preliminary results from such an analysis in the LRC Follow-Up Study. The relative risks given for alcohol are derived from Cox proportional hazards models and are for 20 ml of ethanol a day, or one to one and one half drinks, compared to none, with both linear and quadratic terms included for alcohol because of the U-shaped relationship of alcohol with CHD and CVD mortality shown in Figures 2 and 3. The models are adjusted for the possible confounding effects of age,

Figure 4. Schematic of the Alcohol, HDL, and CVD Mortality Relationships.

cigarette smoking, systolic blood pressure, obesity, triglycerides, and non-contraceptive estrogen use in women. Only the models on the right half of the table include the lipoproteins, HDL and LDL.

These data suggest that the effect of alcohol remains even after adjustment for HDL and LDL, and that in men alcohol seems entirely independent of HDL. It appears that the protective effect of moderate (20 ml/day) consumption in women may be partially mediated by HDL and/or LDL. These results, along with the observation that CHD is increased in heavy drinkers despite marked HDL elevations and the finding that alcohol may have a stronger effect on HDL_3 than the protective HDL_2 shed doubt on the significance of an HDL pathway.

Another recent observation in an angiographic study is pertinent to this problem (Hartz et al., 1979). The authors looked at cholesterol, triglycerides, smoking, age, diabetes, and alcohol intake and correlated these factors with myocardial infarction prevalence after adjusting by logistic regression for the degree of angiographic determined occlusion. The association of diabetes and blood lipids with MI was through coronary occlusion, but smoking was positively and alcohol inversely related to MI independent of the degree of occlusion. Since it is thought that high levels of HDL retard atherogenesis (Miller and Miller, 1975), this alcohol effect presumably is independent of an HDL pathway. Similarly, Yano and colleagues (1977) noted that among 226 autopsied cases, alcohol was distinctly inversely related to myocardial infarction, but the inverse alcohol association with the severity of coronary atherosclerosis, assessed independently of myocardial infarction, was weaker and not statistically significant. The authors suggested the major effect of alcohol might not be on atherosclerosis but rather protective against the precipitation of myocardial infarction or coronary occlusion.

Table 10

Estimated Relative Risks (RR) for 20 ml of Alcohol Per Day,
Lipid Research Clinics Follow-Up Study, Preliminary Results

		RR Without HDL and LDL in Model	RR With HDL and LDL in Model
MEN	CHD	0.92	0.93
	CVD	0.88	0.92
WOMEN	CHD	0.47	0.70
	CVD	0.69	0.85

If not HDL or LDL, what might be the pathway for alcohol's effect? We have noted that moderate drinkers may have a less atherogenic diet (Jones et al., 1982), but again such a relationship should result in reduced atherogenesis, not just the acute event.

The overlooked potential effect of alcohol on thrombosis, particularly platelet inhibition, and/or fibrinogen (Meade et al., 1979), may be the best bet for a mechanism of action. Suppression of platelet activity or increased fibrinolytic activity would result in protection from thrombosis precipitating an acute infarction, and would be concordant with the above findings. Thus, it seems that an alcohol effect on thrombosis in many ways fits the observed clinical and epidemiologic data better than an alcohol effect on lipid metabolism and subsequent atherogenesis.

What, then, would be the mechanism for excess CHD mortality in heavy drinking? As discussed, part of the excess may be related to a misattribution of non-coronary deaths to CHD in alcoholics. However, it may also be that alcohol's anti-thrombotic effect is limited to moderate consumption, and some early evidence from a Finnish study suggests this may be so (Salonen, 1983). In this sense alcohol's effect may be similar to aspirin, where a small dose appears to be optimal (Masotti et al., 1979). However, it is also important to remember the temporal sequence in heavy drinkers, which often involves binge drinking followed by withdrawal. Is it possible that the hypertension and thrombocytosis which accompany alcohol withdrawal put the heavy drinker at particularly high risk of CHD and stroke, both of which are increased in heavy drinkers? Interestingly, thrombotic brain infarction has been associated with weekend drinking bouts in Finland (Hillbom and Kaste, 1978).

For the sake of completeness, we might also speculate as to how some of these pathways might be linked. Is it possible that lipids themselves might interact with clotting factors in thrombosis? Increased blood viscosity and plasma fibrinogen, along with a major fibrinolytic inhibitor, α_2 - antiplasmin, have been reported in Type II hyperlipoproteinemia (Lowe et al., 1982) which is characterized by increased LDL, and fibrinogen has been reported to be correlated with cholesterol (Korsan-Bengsten, Wilhelmsen, and Tibblin, 1972). In addition, platelets incubated with "cholesterol-rich" liposomes have been reported to release significantly more arachidonic acid, a stimulator of platelet aggregation, than "cholesterol-poor" platelets (Stuart, Gerrard, and White, 1980).

It also seems possible that "stress" or "emotion" might stimulate clotting through neurophysiologic mechanisms, and alcohol might modulate this effect as well. Admittedly, this suggestion is speculative at best.

10. Ethical and Public Health Implications

Whether the protection offered by moderate alcohol consumption is via lipid, hemodynamic, thrombotic, or stress-related mechanisms, and whatever the pathways for the toxicity of heavier consumption, one finding is consistent, i.e., only modest amounts of alcohol may offer any benefit, if indeed the association is causal.

Would it be wise public health policy to encourage "moderate" alcohol consumption? Consider the problems. It has been reported that the sons of alcoholics separated from their alcoholic parents early in life and raised by foster parents were nearly four times more likely to become alcoholics than were adoptees without alcoholic biologic parents. Conversely being raised by an alcoholic parent does not seem to increase this risk (Goodwin, 1978). Thus, a strong genetic component in alcoholism seems likely, and the ravages of alcoholism, including increased CVD mortality, are well documented and need no repetition here. It thus seems profoundly unwise to recommend one to two drinks a day, because many persons can't stop there, and many more, for sundry reasons, won't even when they can. Some may misuse the epidemiologic data to rationalize excess drinking. It therefore seems that the possible benefit of alcohol is inappropriate information for general public health intervention. Levels of consumption are already high, and according to a recent World Health Organization Report are rising, particularly in developing countries (WHO, 1982). Instead, the information concerning moderate alcohol consumption's possible benefits may be useful only in a limited clinical setting, where a high risk patient or a recent infarction patient considering forfeiting minimal and responsible alcohol consumption might be told that such forfeiture might not be necessary, assuming the absence of such conditions as an alcohol inducible arrhythmia. Beyond such limited advice, we run the risk of placing our patients at peril and of puting the public's health in jeopardy.

References

Alexander, C. S. Idiopathic heart disease I. Analysis of 100 cases with special reference to chronic alcoholism. *American Journal of Medicine* 41 (1966) 213-228.

Barr, D. P., Russ, E. M., and Eder, H. A. Protein-lipid relationships in human plasma. *American Journal of Medicine* 11 (1951) 480-493.

Beevers, D. G. Alcohol and hypertension. *Lancet* 2 (1977) 114-115.

Belizan, J. M. et al. Reduction of blood pressure with calcium supplementation in young adults. *Journal of the American Medical Association* 249 (1983) 1161-1165.

Carlsson, C. and Haggendal, J. Arterial nonadrenaline levels after ethanol withdrawal. *Lancet* 2 (1967) 889.

Castelli, W. P. et al. Alcohol and blood lipids: The Cooperative Lipoprotein Phenotyping Study. *Lancet* 2 (1977) 153-155.

Centerwall, B. S. and Criqui, M. H. Prevention of the Wernicke-Korsakoff syndrome: A cost-benefit analysis. *New England Journal of Medicine* 299 (1978) 285-289.

Clark, L. T. and Friedman, H. S. Alcohol-induced hypertension: Assessment of mechanisms and complications. Presented at the 32nd Annual Scientific Session of the American College of Cardiology, New Orleans, Louisiana, 1983.

Criqui, M. H. et al. Frequency and clustering of other coronary risk factors in dyslipoproteinemias. Presented at the Annual Scientific Sessions of the American College of Cardiology, New Orleans, Louisiana, 1983.

Criqui, M. H. et al. Cigarette smoking, alcohol consumption, and cardiovascular mortality: The Lipid Research Clinics Follow-Up Study. (Abstract) *Circulation* 66 (1982) (Suppl. II), II-235.

Criqui, M. H. et al. Alcohol consumption and blood pressure: The Lipid Research Clinics Prevalence Study. *Hypertension* 3 (1981) 557-565.

Demakis, J. G. et al. The natural course of alcoholic cardiomyopathy. *Annals of Internal Medicine* 80 (1974) 293-297.

Deykin, D., Janson, P., and McMahon, L. Ethanol potentiation of aspirin-induced prolongation of the bleeding time. *New England Journal of Medicine* 306 (1982) 852-854.

Dyer, A. R. et al. Alcohol consumption and 17-year mortality in the Chicago Western Electric Company Study. *Preventive Medicine* 9 (1980) 78-90.

Ettinger, P. O. et al. Arrhythmias and the "holiday heart": Alcohol-associated cardiac rhythm disorders. *American Heart Journal* 95 (1978) 555-562.

Fisher, V. J. and Favaler, F. The action of ethanol upon the action potential and contraction of ventricular muscle. *Recent Advances in Studies on Cardiac Structure and Metabolism* 5 (1975) 415-422.

Friedreich, N. Handbuch der speziellen Pathologie und Therapies. 5th Sect. Krankheiten des Herzens, Ferdinand Enke, Erlangen, 1861.

Goodwin, D. W. Hereditary factors in alcoholism. *Hospital Practice* 13 (1978) 121-130.

Gordon, T. et al. High-density lipoprotein as a protective factor against coronary heart disease: The Framingham Study. *American Journal of Medicine* 62 (1977) 707-714.

Gould, L. et al. Electrophysiologic properties of alcohol in man. *Journal of Electrocardiology* 11 (1978) 219-226.

Greenspan, A. J., Stang, J. M., Lewis, R. P., and Schaal, S. F. Provocation of ventricular tachycardia after consumption of alcohol. *New England Journal of Medicine* 301 (1979) 1049-1050.

Haffner, S., Appelbaum-Bowden, D., Hoover, J., and Haggard, W. Association of high-density lipoprotein cholesterol 2 and 3 with Quetelet, alcohol, and smoking: The Seattle Lipid Research Clinic Population. (Abstract) *CVD Epidemiology Newsletter* 31 (1982) 20.

Hartz, A. J. et al. Risk factors for myocardial infarction independent of coronary artery disease. (Abstract) *Circulation* 60 (1979) (Suppl. II), II-258.

Haselager, E. M. and Vreeken, J. Rebound thrombocytosis after alcohol abuse: A possible factor in the pathogenesis of thromboembolic disease. *Lancet* 1 (1977) 774-775.

Haskell, W., Krauss, R., Wood, P., and Lindgren F. The negative relationship between moderate alcohol intake and coronary heart disease may not be due to elevated serum HDL_2. (Abstract) *CVD Epidemiology Newsletter* 31 (1982) 23.

Haut, M. J. and Cowan, D. H. The effect of ethanol on hemostatic properties of human blood platelets. *American Journal of Medicine* 56 (1974) 22-33.

Hennekens, C. H. et al. Effects of beer, wine and liquor in coronary deaths. *Journal of the American Medical Association* 242 (1979) 1973-1974.

Hepner, G. W., Roginsky, M., and Moos, H. F. Abnormal vitamin D metabolism in patients with cirrhosis. *American Journal of Digestive Diseases* 21 (1976) 527-532.

Hillbom, M. and Kaste, M. Does ethanol intoxication promote brain infarction in young adults? *Lancet* 2 (1978) 1181-1183.

Hulley, S. B., Rosenman, R. H., Bawol, R. D., and Brand, R. J. Epidemiology as a guide to clinical decisions. The association between triglyceride and coronary disease. *New England Journal of Medicine* 302 (1980) 1383-1389.

Jenkins, C. D. Recent evidence supporting psychologic and social risk factors for coronary disease. *New England Journal of Medicine* 294 (1976) 987-994, 1033-1038.

Jones, B., Barrett-Connor, E., Criqui, M. H., and Holdbrook, M. J. Caloric and nutrient intake in drinkers and non-drinkers of alcohol: A community study. *American Journal of Clinical Nutrition* 35 (1982) 135-139.

Kagan, A., Yano, K., Rhoads, G. G., and McGee, D. L. Alcohol and cardiovascular disease: The Hawaiian experience. *Circulation* 64 (1981) (Suppl. III), III-27-III-31.

Kannel, W. B. Coffee, cocktails, and coronary candidates (editorial). *New England Journal of Medicine* 297 (1977) 443-444.

Keys, A. et al. Mortality and coronary heart disease among men studied for 23 years. *Archives of Internal Medicine* 128 (1971) 201-214.

Klatsky, A. L., Friedman, G. D., and Siegelaub, A. B. Alcohol consumption before myocardial infarction. *Annals of Internal Medicine* 81 (1974) 294-301.

Klatsky, A. L., Friedman, G. D., and Siegelaub, A. B. Alcohol use and cardiovascular disease: The Kaiser-Permanente experience. *Circulation* 64 (1981a) (Suppl. III): III-32-III-41).

Klatsky, A. L., Friedman, G. D., and Siegelaub, A. B. Alcohol and mortality: A ten-year Kaiser-Permanente experience. *Annals of Internal Medicine* 95 (1981b) 139-145.

Klatsky, A. L., Friedman, G. D., Siegelaub, A. B., and Gerard, M. J. Alcohol consumption and blood pressure: Kaiser-Permanente multiphasic health examination data. *New England Journal of Medicine* 296 (1977) 1194-1200.

Korsan-Bengsten, K., Wilhelmsen, L., and Tibblin, G. Blood coagulation and fibrinolysis in a random sample of 788 men 54 years old: II. Relations of the variables to "risk factors" for myocardial infarction. *Thrombosis et Diathesis Haemorrhagica* 28 (1972) 99-108.

Kozarevic, D. et al. Drinking habits and coronary heart disease: The Yugoslavia Cardiovascular Disease Study. *American Journal of Epidemiology* 116 (1982) 748-758.

Kozarevic, D. et al. Frequency of alcohol consumption and morbidity and mortality. *Lancet* 1 (1980) 613-616.

Krawitt, E. L. Effect of ethanol ingestion on duodenal calcium transport. *Journal of Laboratory and Clinical Medicine* 85 (1975) 665-671.

Kuller, L. and Castelli, W. High HDL levels can unmask covert alcoholic. *Skin and Allergy News* 11 (1980) 25.

Kuller, L., Perper, J. A., Cooper, M., and Fisher, R. An epidemic of deaths attributed to fatty liver in Baltimore. *Preventive Medicine* 3 (1974) 61-79.

Lifton, L and Scheig, R. Ethanol-induced hypertriglyceridemia prevalence and contributing factors. *American Journal of Clinical Nutrition* 31 (1978) 614-618.

Lindenbaum, J. and Hargrove, R. L. Thrombocytopenia in alcoholics. *Annals of Internal Medicine* 68 (1968) 526-532.

Lindenbaum, J. and Lieber, C. S. Hematologic effects of alcohol in man in the absence of nutritional deficiency. *New England Journal of Medicine* 281 (1969) 333-338.

Linkola, J., and Fyhrquist, F., and Ylikahri, R. Renin, aldosterone, and cortisol during ethanol intoxication and hangover. *Acta Physiologica Scandinavica* 106 (1979) 75-82.

Linkola, J., Ylikahri, R., Fyhrquist, F., and Wallenius, M. Plasma vasopressin in ethanol intoxication and hangover. *Acta Physiologica Scandinavica* 104 (1978) 180-187.

Lowe, G. D. O. Increased blood viscosity and fibrinolytic inhibitor in type II hyperlipoproteinaemia. *Lancet* 1 (1982) 472-475.

Mann, G. V. Diet-heart: End of an era. *New England Journal of Medicine* 297 (1977) 644-650.

Marmot, M. G., Rose, G., Shipley, M. J., and Thomas, B. J. Alcohol and mortality: A U-shaped curve. *Lancet* 1 (1981) 580-583.

Masotti, G. et al. Differential inhibition of prostacyclin production and platelet aggregation by aspirin. *Lancet* 2 (1979) 1213-1216.

McCarron, D. A., Morris, C. D., and Cole, C. Dietary calcium in human hypertension. *Science* 217 (1982) 267-269.

McDonald, J. T. and Margen, S. Wine versus ethanol in human nutrition. III. Calcium, phosphorus and magnesium balance. *American Journal of Clinical Nutrition* 32 (1979) 823-833.

Meade, T. W. et al. Characteristics affecting fibrinolytic activity and plasma fibrinogen concentrations. *British Medical Journal* 1 (1979) 153-156.

Meade, T. W. et al. Haemostatic function and cardiovascular death: Early results of a prospective study. *Lancet* 1 (1980) 1050-1054.

Meagher, R. C., Sieber, F., and Spivak, J. L. Suppression of hematopoietic-progenitor-cell proliferation by ethanol and acetaldehyde. *New England Journal of Medicine* 307 (1982) 845-849.

Miller, G. J. and Miller, N. E. Plasma-high-density-lipoprotein concentration and development of ischaemic heart-disease. *Lancet* 1 (1975) 16-19.

Miller, N. E., Forde, O. H., Thelle, D. S., and Mjos, O. D. The Tromso Heart-Study. High-density lipoprotein and coronary heart-disease: A prospective case-control study. *Lancet* 1 (1977) 965-968.

Morin, Y. and Daniel, P. Quebec beer-drinkers' cardiomyopathy: Etiologic considerations. *Canadian Medical Association Journal* J 97 (1967) 926-928.

Ogata, M., Mendelson, J. H., Mello, N. K., and Majchrowicz, E. Adrenal function and alcoholism. II. Catecholamines. *Psychosomatic Medicine* 33 (1971) 159-180.

Pell S. and D'Alonzo, C. A five-year mortality study of alcoholics. *Journal of Occupational Medicine* 15 (1973) 120-125.

Post, R. M. and Desforges, J. F. Thrombocytopenia and alcoholism. *Annals of Internal Medicine* 68 (1968) 1230-1236.

Randall, B. Sudden death and hepatic fatty metamorphosis: A North Carolina survey. *Journal of American Medical Association* 243 (1980) 1723-1725.

Reynolds, E. S. An account of the epidemic outbreak of arsenical poisoning occurring in beer-drinkers in the North of England and the Midland Countries in 1900. *Lancet* 1 (1901) 166-170.

Rosenberg, L. Alcoholic beverages and myocardial infarction in young women. *American Journal of Public Health* 71 (1981) 82-85.

Salonen, J. T. Personal communication, 1983.

Salzman, E. W. Aspirin to prevent arterial thrombosis. *New England Journal of Medicine* 307 (1982) 113-115.

Saville, P. D. Changes in bone mass with age and alcoholism. *Journal of Bone and Joint Surgery* 47 (1965) 492-499.

Sime, W. E., Buell, J. C., and Eliot, R. S. Psychophysiological (emotional) stress testing: A potential means of detecting the early reinfarction victim (Abstract). *Circulation* 60 (1979) (Suppl. II), 11-56.

Spann, J. F., Mason, D. T., Beiser, G. D., and Gold, H. K. Actions of ethanol on the contractile state of the normal and failing cat papillary muscle (Abstract). *Clinical Research* 16 (1968) 249.

Stason, W. B., Neff, R. K., Miettinen, O. S., and Jick, H. Alcohol consumption and non-fatal myocardial infarction. *American Journal of Epidemiology* 104 (1976) 603-608.

Stuart, M. J., Gerrard, J. M., and White, J. G. Effect of cholesterol on production of thromboxane B_2 by platelets in vitro. *New England Journal of Medicine* 302 (1980) 6-10.

Wilhelmsen, L., Wedel, H., and Tibblin, G. Multivariate analysis of risk factors for coronary heart disease. *Circulation* 48 (1973) 950-958.

World Health Organization. *World Health.* Peter Ozorio (ed.) 5 (1982) 30.

Yano, K., Rhoads, G. G., and Kagan, A. Coffee, alcohol and risk of coronary heart disease among Japanese men living in Hawaii. *New England Journal of Medicine* 297 (1977) 405-409.

6

THE PROS AND CONS OF
ECONOMIC DEVELOPMENT

Denis Burkitt, M.B.

1. Major Causes of Death and Disease

All living creatures including man have had to struggle for survival, and this struggle is predominantly for food to keep alive. Predators are constantly vying with one another for means of subsistence. Man and his ancestors have always been conscious of their macro-predators in the form of wild animals, but more especially of other members of their own species. They have, however, been infinitely more vulnerable to micro-predators in the form of micro-organisms of all kinds and the numerous parasites that prey upon man. Because all but the larger of these have until recently been invisible and consequently unrecognized, man has not been able to mount any conscious effort to combat them. These micro-predators, whether viruses, bacteria, or parasites, depend on man and animals or intermediate hosts for their survival. In man, they can give rise to disease that may or may not be lethal, and in the long run it is better for their survival that death of the host they invade should not be the outcome.

Infective diseases include not only the common and often mild infections in the West but also the major pandemics of diseases like malaria, cholera, smallpox, yellow fever, and trypanosomiasis mainly in tropical regions.

Another major cause of death mainly in areas subject to drought has been starvation.

2. Reduction in Mortality from Infectious Disease in the West

Until about half a century ago, infectious diseases remained the commonest killer world-wide, but by then mortality rates from these causes had been enormously reduced in Western countries. Infectious diseases were, in fact, being replaced by a largely new set of diseases as major causes of morbidity and mortality. Before examining these new diseases and considering the best means of combatting them, it is important to ascertain the means whereby infectious disease was to a large extent conquered as a major cause of death in economically more developed countries and the manner in which endemic epidemic diseases have been reduced in the Third World.

The most striking feature in the pattern of decline in mortality rates from most of the common infectious diseases in Great Britain between the middle of the nineteenth century and the Second World War is that they had almost reached their present low levels before any effective therapy became available, for they preceded the advent of the first drugs to be effective against infectious disease, that is the introduction of sulphonamides in the 1930s and antibiotics in the early 1940s. The experience of Great Britain can be generalized to other economically more developed countries. Consequently, we must look elsewhere than therapy to explain the conquest of these infections.

McKeown (1979) has argued persuasively that the main factors responsible were:

1.   Increased resistance to disease resulting from improved nutrition.
2.   Reduced contact with pathogenic organisms by the provision of clean water, clean milk, and adequate sewage disposal.
3.   Immunization.

The enormously high mortality rates from measles in mal-nourished children in Third World countries is a striking example of the importance of the role of adequate nourishment in combatting infection. The part played by inadequate or contaminated water and lack of satisfactory sewage disposal is emphasized by the widely accepted assessment that mortality rates, particularly in children, could be halved in many developing countries by provision of clean water and sewage disposal. Such measures might do more for the health of many communities than is achieved by all the costly therapeutic medical facilities at present provided. There are, of course, numerous examples of the conquest of disease by such measures, cholera, typhoid fever, and gastro-enteritis in infancy, to name but a few.

By emphasizing prevention, I in no way belittle the enormous importance of providing curative medicine with care and compassion. Curative and preventive measures must go hand-in-hand, but over-sophistication of the former must be cautioned against.

In Western countries, immunization has been largely responsible for the virtual disappearance of diphtheria and poliomyelitis, and has greatly reduced mortality from measles and whooping-cough. In tropical countries, it has largely conquered yellow fever and apparently eliminated smallpox, which must be the greatest triumph of medical science in this century.

In Western countries by the nineteenth century, some degree of immunity against common infectious diseases must have already been acquired in contrast to the almost total lack of such protection in the Amerindinas when they first met European explorers with such devastating results.

The lesson to be learned is that wherever a disease has been reduced in frequency it has been through identification and reduction or elimination of causative factors, except in the case of the infectious diseases that have been controlled by immunization. With the possible exception of some highly contagious infections such as venereal disease and malaria, it is doubtful whether any disease has been significantly reduced in frequency by improving methods of treatment. This lesson will be seen to be highly relevant when considering the common diseases of economically more developed countries today, and when deciding priorities in health care in Third World countries.

3. The Emergency of a New Pattern of Disease

As infectious diseases declined as the major killers, a new pattern of disease emerged in Western countries. It is now generally accepted that there is a formidable list of diseases which always have their maximum frequency in economically more developed countries and are rare, or in some instances unknown, in rural communities in the Third World. These diseases are recognized as being characteristic of modern Western culture and include such important causes of ill-health in Western communities as:

Coronary heart disease: the commonest cause of death

Large bowel cancer: the second commonest cancer death after
lung tumors related to smoking (Lung cancer has until recently
been a predominantly Western disease, but I will limit
consideration to disorders believed to be causally related to diet.)

Breast cancer: the commonest cancer in women

Gall-stones: the commonest indication for abdominal surgery

Diverticular disease: one of the commonest disorders of the
    large intestine

Hemorrhoids: which affect the majority of the population
    at some time

Appendicitis: the commonest indication for abdominal emergency
    surgery

Hiatus hernia: one of the commonest disorders of the stomach

Diabetes: the commonest endocrine disorder

Obesity: the commonest nutritional disorder

Varicose Veins: one of the commonest venous disorders

There is a mass of evidence both epidemiological and experimental incriminating diet as an important cause of these disorders.

There are, in addition, other characteristically Western diseases whose causes remain obscure. Among the best known of these are multiple sclerosis, ulcerative colitis, and Crohn's disease.

3.1 Constipation

Although constipation cannot in itself be considered to be a disease, deranged bowel behavior is a basic cause of many of the diseases listed.

Few subjects relevant to the whole field of health have until recently been surrounded by so much ignorance as has bowel behavior. Regretably it has always been and still is a taboo subject. So great is the ignorance even amongst physicians in the United States, that when inquiring as to the average weight of stool passed per day in several major medical centers, I was given estimates varying from 5g. to 700g.! Many studies of bowel behavior have been conducted in both Western and Third World countries during the past 15 years. Average daily output in young, healthy adults in the West where frequencies of Western diseases are maximum is in the region of 80 to 120 grams and in geriatric patients often under 60 grams. In contrast, figures for Third World communities who experience a minimum of Western Diseases, average daily stool output is 300 to 600 grams.

The time it takes food residue to traverse the alimentary tract can be estimated by measuring the time lapse between swallowing and evacuating radio-opaque pellets. The pellets can be visualized when plastic bags into which stools are voided are x-rayed. This transit-time is around 72 hours in the West, and often over 200 hours in the elderly, whereas it is about 30 hours in the Third World (Burkitt, Walker, and Painter, 1974).

Enough is in fact now known to state that, in different countries and in individuals within a community, the risk of developing any of the Western diseases is more directly related to stool size than to the factors customarily estimated such as serum cholesterol and blood pressure levels.

As will be described below, stool size and consistency and intestinal transit times are profoundly influenced by the amount of fiber, and cereal fiber in particular, in the diet.

Few people are prepared to weigh their stools to estimate the average amount voided daily, but the large volume soft stools associated with high fiber intakes normally float, while small, hard stools resulting from fiber-depleted diets sink. The increased gas production which accompanies adequate fiber intake results in gas entrapped in the stools which accounts for their buoyancy.

4. Epidemiological Features of Western Diseases

Not only are these diseases most frequently observed today in economically most developed communities and least often in rural populations in the Third World, but there is no evidence that any of them was other than relatively rare even in Western countries before the present century. In addition they have until recently been commoner in upper than in lower socio-economic groups. When communities emigrate from low risk to high risk situations the next and subsequent generations are at comparable risk of developing these diseases as are those of other ethnic groups in the host country. Examples are provided by the Japanese who emigrated to Hawaii and California, the Polynesian islanders to New-Zealand, and the Jews from the Yemen, North Africa, and parts of Russia to the new state of Israel. It is of particular interest that black Americans today are at similar risk to white Americans, whereas their distant ancestors cannot have been more prone to develop these diseases than are village Africans today.

These and other observations indicate clearly that although individual and ethnic susceptibilities to particular environmental factors must always play a role in the development of illness, the predominant causes of these diseases must be environmental rather than genetic. Consequently, it must be concluded that these maladies would be largely preventable, could major causative factors in the environment be identified and reduced. The epidemiological and other studies in which these conclusions are drawn will be found in Burkitt and Trowell (1975) and Trowell and Burkitt (1981).

5. The Significance of Relationships

Since all the results of a common cause tend to be related to one another, it can be argued conversely that observed associations between results, in this case diseases, suggests that they share some common causative factor or factors. (Burkitt, 1970). As has been pointed out the frequencies of all of the diseases enumerated in section 3 are related to the environmental factors that are associated with the changes that accompany economic development. It has also been seen that they tend to be associated with one another not only in their geographic but also in their socio-economic distribution. They are also associated with one another in their chronological emergence following impact with Western culture, though they appear in a certain order depending on the period of exposure to environmental factors required before a particular disease becomes clinically manifest. In addition there is an observed tendency for several of these diseases to occur simultaneously in the same patients. All of these observations suggest that they either share some causative factor or, alternatively, that the same factor confers protection against each.

6. Identifying Causes

It is now beyond dispute that Western diseases are caused by some aspect of modern Western life style. It might be argued that their emergence was related to various environmental changes associated with economic development and that any of these might be considered possible causative factors. It is not, however, possible to erect hypotheses making biological sense which incriminate various technical developments such as the introduction of plastic utensils, radios, or bicycles. Since all the diseases listed above can be shown to be directly or indirectly related to the digestive tract and the most important influence on the behavior and content of our intestines is the food we eat, it must seem reasonable to consider changes in diet that throughout the world have preceded the emergence of, or increased frequency of, these diseases before concentrating on other aspects of Western life style.

Plausible hypotheses have now been erected to suggest how dietary changes could be causative of or provide protection against each of these diseases. Not only are these hypotheses consistent with the geographical distribution and historical emergence of these diseases, but in many instances the physiological and other changes postulated to result in alterations of diet have been confirmed both in animal studies and in clinical trials.

7. Dietary Contrasts between Communities at Maximum and at Minimum Risk of Developing Western Diseases

The contrasting patterns of diet consumed in economically developed and in poorer communities together with average daily weight of stools is depicted in Figure 1.

It will be noted that the proportion of energy derived from protein is comparable in both situations, although it is mainly from animal sources in the former and from plant foods in the latter. Moreover, protein intake in Western countries has varied little during the past century, although it was during that period that disease patterns altered dramatically.

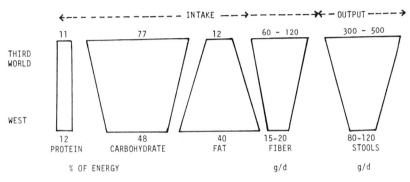

Figure 1: Intake of protein carbohydrate, fat and fiber and output of stool in the West and Third World.

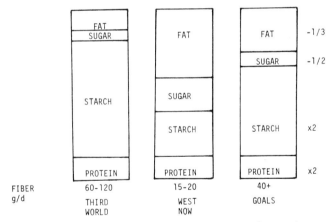

Figure 2: Goals for consumption of fat, sugar, starch and protein compared with the current consumption in Western and Third World Countries.

Economic development has always been associated with a marked reduction in consumption of carbohydrate foods, and since this is accompanied by a huge increase in sugar, the dramatic reduction has been in starch intake. In Western populations this is little more than a quarter of that consumed in developing countries.

Fat intake is reciprocally increased to compensate for the reduction in carbohydrate, and affluent societies consume nearly three times the proportion of fat in their diet as do poorer populations.

The greatest, and we believe the most important, dietary change that has accompanied economic development has been the deliberate removal of fiber from food. This has resulted in most Western diets containing less than 20 grams of fiber per day in contrast to 60 to 120 grams in Third World communities.

8. Nature and Properties of Fiber

Fiber, or preferably "dietary fiber," to distinguish it from the old and almost meaningless term "crude fiber," is the main constituent of plant cell walls. It used to be eroneously referred to as roughage on the misconception that it was abrasive to the bowel lining. The opposite is actually the case, fiber-rich foods producing soft stools and vice versa. It can also be viewed as the component of food that passes through the small intestine largely undigested in contrast to proteins, fats, and sugars that are mostly absorbed from the small gut.

Because fiber provided little nutrition, it was not only neglected but deliberately removed from our food. It is consequently not surprising that it is now widely recognized as the only component of diet of which Western populations are as a whole deficient. It is in fact only during the last 15 years that medical and nutritional scientists have begun to appreciate and understand the profound influence that fiber has on almost every activity that takes place in the intestine.

It would be impossible to deal in any detail with the physiological, bacteriological, and other properties of fiber here, but a brief summary may be appropriate. Fiber holds water in the gut and by this and other means protects against constipation. By so doing it is believed to confer protection against diseases including appendicitis (Burkitt, 1971), diverticular disease of the colon (Painter, 1975), large bowel cancer (Cummings, 1981), the complications of hemorrhoids (Burkitt, 1975), and perhaps varicose veins. By their influence on the metabolism of bile-acids and cholesterol and in other ways, fiber-rich diets are related to low frequencies of coronary-heart-disease (Trowell, 1975) and gall-stones (Heaton, 1978).

Because fiber provides bulk to food without contributing energy, it helps to fill the stomach before excessive calories have been consumed and thus protects against over-eating and resultant obesity (Bolton, Heaton, and Burroughs, 1981).

Fiber also reduces the rate of absorption of energy from the intestine into the body, and by this and other means is believed to be strongly protective against diabetes (Type II) (Jenkins and Worlever, 1982).

Fiber has recently been shown to reduce serum oestrogens by increasing faecal excretion of oestrogens. This may explain the reduced frequency of breast cancer in populations on high fiber diets, including American vegetarians. (Goldin et al., 1982).

9. Dietary Changes Viewed as a Whole

It must be emphasized that increasing fiber consumption is only one aspect of changing diets towards those which are always associated with low frequencies of Western diseases. A high-fiber diet is almost invariably also rich in starch and low in fat and sugar. Adequate fiber has in fact been postulated to be partially protective against all the diseases listed above.

Excessive fat is believed to contribute to the causation of coronary heart disease, large-bowel cancer, breast cancer, obesity, and possibly gall-stones.

Excessive sugar contributes certainly to obesity and dental caries and possibly to diabetes.

Starch appears to be protective against diabetes and starchy foods containing their normal compliment of fiber mitigate against obesity.

Salt, when consumed in excess of the body's needs, is a common cause of high blood-pressure in genetically susceptible individuals.

10. Recommendations for Western Populations

The proportions of various dietary components in situations at minimal and at maximum risk of developing Western diseases are contrasted in figure 2 together with recommendations applicable to the more economically developed Western communities. There can be no doubt that the health of affluent societies could benefit far more from simple dietary changes than from any extension or even continuation of the massive expenditure currently devoted to the cure of disease. Fundamental recommendations include:

1. Approximately doubling consumption of fiber and starch. This would entail perhaps a three-fold increase in consumption of bread or other flour products such as pasta, but made from high-extraction flour and the nearer to whole-meal the better, more fiber rich breakfast cereals, pulses such as peas and beans, and root vegetables like potatoes.
2. Fat should be reduced by about a third. This would entail minimizing fried foods, ensuring that potatoes were neither cooked nor eaten with fat, reducing consumption of dairy products, and reducing meat consumption by choosing fish or fowl in preference to red meat.

3.   Intake of sugar and salt could beneficially be halved. The former would necessitate consuming confectionery and sweetened drinks sparingly, and the latter avoiding highly salted foods like salted fish and ham; and adding no salt at table.

11. The Pros and Cons when East Meets West

11.1 The first impact

There are many historical examples of the devastating disease epidemics that resulted from initial contact between hitherto isolated communities and adventurers from overseas. This happened in a massive scale throughout North and South America when the Amerindians, lacking any natural immunity to common European diseases, were confronted for the first time by infections such as measles, smallpox, tuberculosis, and influenza carried comparatively harmlessly by partially immune invaders from overseas. Whole populations were often decimated in this way (McNeill, 1976).

Similar ill-effects have been recorded from Pacific Island communities following the arrival of the first traders or explorers from the West. Events of this nature must have devastated various European populations in the past as more ancient civilizations have expanded into new territories. The ill effects of impact with more civilized communities in those days was the introduction of non-infectious diseases, against which the invaded population lacked immunity, but a new and potentially as great a danger threatens them today.

11.2 The second impact

The second impact of more developed on less developed populations has been occurring during the last few decades. This has been the introduction of modern Western diseases into developing countries as they are tempted to copy aspects of our life-style, particularly with regard to smoking and eating habits. In all five continents of the world the emergence of these diseases, initially in the urban and more westernized sections of the community, has been documented (Trowell and Burkitt, 1981). These observations are obvious danger signals calling for urgent attention to warn populations and in particular governments in these countries of the drastic dangers to health, with all the associated expenditure involved, which will certainly result if they blindly follow the bad examples set by the West.

12. Can We Share the Good without Exporting the Bad?

Much has already been done to control epidemic and endemic disease in tropical countries. Obvious examples are the successful campaigns against malaria, yellow fever, plague, trypanosomians, and smallpox.

A very high priority must be given to the provision of clean water and the disposal of sewage. Agricultural enterprises must be closely linked with health priorities. And it is of course of the greatest importance that death control cannot be practiced without birth-control if any headway is to be made in improving the per capita nutrition in developing countries.

Energetically carried out immunization programs could be a much more cost-effective way of improving health than over-reliance of modern technology for curing disease. There must of course always be compassion and care shown for the sick with the best affordable medical treatment.

The real danger is that while helping in all such measures, Western countries unwittingly portray their patterns of life-style as inherently superior to those of poorer countries, and thereby tempt the latter to blindly follow their footsteps in matters of diet and habits like smoking. Such a course could prove disasterous and would virtually mean that we in the West were bringing to poorer communities the means of combatting infectious disease with one hand while imposing on them our current Western diseases with the other.

Every effort must be made to share the good lessons we have learned but not to propagate the mistakes we have made by following a life-style that is profoundly detrimental to health.

References

Bolton. R. P., Heaton, K. W., and Burroughs, L. F. The role of dietary fibre in satiety. Glucose and Insulin Studies with fruit and fruit juice. *American Journal of Clinical Nutrition* 34 (1981) 211-217.

Burkitt, D. P. Relationships: A clue to causation. *Lancet* 2 (1970) 1237.

Burkitt, D. P. The aetiology appendicitis. *British Journal of Surgery* 58 (1971) 695-699.

Burkitt, D. P. Appendicitis in Burkitt, D. P. and Trowell, H. C. (eds.) *Refined carbohydrate foods and disease.* London: Academic Press, 1975. Pp. 87-97.

Burkitt, D. P. Varicose veins, deep vein thrombosis and haemorrhoids in Burkitt, D. P. and Trowell, H. C. (eds.) *Refined carbohydrate foods and disease.* London: Academic Press, 1975. Pp. 143-160.

Burkitt, D. P. and Trowell, H. C. (eds.) *Refined Carbohydrate Foods and Disease.* London: Academic Press, 1975.

Burkitt, D. P., Walker, A. R. P., and Painter, N. S. Dietary fibre and disease. *Journal of the American Medical Association.* 227 (1974) 1068-1074.

Cummings, J. H. Colon cancer. Dietary fibre and large bowel cancer. *Proceedings of the Nutrition Society* 40 (1981) 7-14.

Goldin. B. R., Adlercrertz, H., Gorbach, L., et al. Estrogen excretion patterns and plasma levels in vegetarian and omnivorous women. *New England Journal of Medicine* 307 (1982) 1542-1547.

Heaton, K. W. Are gallstone preventable? World Medicine 14 (1978) 21-23.

Jenkins, D. J. A. and Worlever, T. M. S. The Diabetic diet. Dietary carbohydrate and difference in digestibility. *Diabepologia* 23 (1982) 477-484.

McKeown, T. The role of medicine: Dream mirage or nemesis. Princeton: Princeton University Press, 1979.

McNeill, W. H. *Plagues and peoples.* New York: Anchor Press, 1976.

Painter, N. S. *Diverticular disease of the colon.* London: William Heinemann Medical Books Ltd., 1975.

Trowell, H. C. Ischaemic heart disease atheroma and fibrinolysis in Burkitt, D. P. and Trowell, H. C. (eds.) *Refined carbohydrate foods and disease.* London: Academic Press, 1975. Pp. 195-226.

Torwell, H. C. and Burkitt, D. P. *Western diseases, their emergence and prevention.* Cambridge, Mass.: Harvard University Press, 1981.

7

REGIONAL DIFFERENCES IN MORTALITY IN BELGIUM

J. V. Joossens, M.D.
J. Geboers, M.Sc.

1. Historical Background

More than a hundred years ago the northern, Dutch speaking part of Belgium was a poor area, living mostly from agriculture together with some local industry. Many people in the north travelled daily or weekly either to the south, to work in the coal mines, or the north of France. Around 1890, infant mortality was extremely high, up to 28 percent in the two northern provinces West Flanders and Antwerp (Tulippe, 1952).

The tradition of poverty in the north had a profound influence on nutrition. Cheap and, therefore, salted foods dominated: large amounts of bread and potatoes, salted pork, corned beef, salted fish (herring and cod fish), salted vegetables, and cheese. However, all this gradually improved over the years. This process was hastened by the mass introduction of refrigerators around 1950. Cooking was done with pork fat and later on with hardened margarines from fish oils. Peanut oil was the habitual cooking oil in 1950, now corn oil is used primarily.

The southern. French speaking part of Belgium was richer, with a flourishing mining and steel industry. In the south, butter and vegetable oils (peanut and soybean) were preferred for cooking. Gradually, but more quickly after World War II, the economic gap between the two regions of the country disappeared. New industries (e.g., petrochemical, photographic, and steel) flourished in the north, while in the south the coal mining industry totally disappeared and the steel industry began to disintegrate. Some indications of nutritional changes in Belgium as a whole over nearly 130 years are given in Table 1.

More recently, the nutritional gradient has been unconsciously widened by a mass health education program launched in 1968 and supported by the two major universities in northern Belgium. This program tried to reduce the intake of saturated fat and salt in the diet and, at the same time, to increase the intake of polyunsaturated fat. After a few years, this program was also promoted in the south. Behavioral differences (Kornitzer et al., 1979) between the regions may have increased the impact of the health education program in the north.

Table 1

Consumption of certain nutrients in Belgium (g/day)

Year	Bread	Potatoes	Meat	Salt added[*] in grams to bread and potatoes
1853	537	752	29	15.5
1891	709	644	82	16.2
1910	674	728	104	16.8
1921	650	703	117	16.2
1932	476	563	92	12.5
1948	328	398	94	8.7
1957	282	414	154	8.4
1967	250	318	171	6.8
1974	219	289	173	6.1
1979	193	290	203	5.8

Source: National Institute of Statistics, Brussels, quoted up to 1967 by Lederer (1970)

[*] Assuming a constant added amount of 12 g NaCl/kg.

It is with this background that, at ages over 35 years, a difference in mortality rates, which did not exist in 1890, gradually emerged between the two major regions of Belgium. Before 1970, however, nobody was aware of nutritional and mortality differences between the regions. It came as a surprise that serum cholesterol was significantly higher in French speaking than in Dutch speaking people among 42,804 members of the Belgian army (Van Houte and Kesteloot, 1972).

Although genetic differences could not be eliminated, it was suspected from the start that the serum cholesterol differences might be of nutritional origin, since there was no difference in serum cholesterol among the French and Dutch speaking army members living in Western Germany and sharing the same kitchen. Blood pressure was slightly higher in the north, whereas weight, height, and blood group distribution were identical (Kesteloot and Van Houte, 1974). Van Houte and Kesteloot also observed a much higher prevalence of ECG-detected myocardial infarctions in the French speaking men. This was later confirmed by Vastesaeger (1979) and Kornitzer and his colleagues (1979). However, as already mentioned, no data were available on nutrition or on mortality in the two regions of Belgium. The first regional mortality data became available in 1976 (for the year 1972), just before the European Congress of Cardiology in Amsterdam.

Meanwhile data on food intake had been collected since 1971 by groups in Leuven (Joossens et al., 1977a) and in Brussels (Vastesaeger et al., 1974). Butter consumption was much higher in the south, margarine consumption higher in the north. From available data on serum cholesterol in the two regions, and using findings of the Seven Countries Study (Keys, 1970), it was deduced that mortality should be higher in the south. As expected, the 1972 vital statistics showed a nearly 25 percent higher mortality rate from all causes in middle-aged persons in the south.

The publication of this information (Joossens et al., 1977b) had an enormous impact in Belgium and provoked strong reactions by the dairy industry, which considered the whole problem to be a battle in the war between dairy products and margarines. It was indeed true that since 1968 the margarine industry had cleverly based their publicity on health education. From 1977 on, the dairy industry markedly increased their publicity for butter. They were aided by the Common Market through the sales of cheap Christmas butter. The dairy industry insisted on the possible relationship between polyunsaturated fat and cancer, forgetting to mention that saturated fat is even more suspected from this point of view (National Research Council, 1982).

Since then more data have been accumulated on nutrition, psychological factors, smoking and drinking habits, and on mortality in the two major Belgian regions. They will be reviewed in the next paragraphs.

2.   Nutrition, Psychological Factors, Drinking, and Smoking Habits

2.1. The situation around 1974

Data collected around 1974 (Figure 1) consistently indicated a nearly four times higher butter intake and a nearly twice as low margarine consumption when comparing the south to the north of Belgium.

A detailed nutritional survey of the "National Institute of Statistics" (NIS) in Brussels performed during a whole year 1973-74 in 2613 families (Joossens, 1979, 1980) indicated no important regional difference in the intake of vegetables, fruits, potatoes, cheese, meat, and eggs. Neither was a difference noted in total protein, carbohydrates, fiber, potassium, calcium, magnesium, phosphorus, iron, copper, zinc, vitamins B1, B2, B6, C, and β-carotene content of the diet. Consumption of sugar, hardened shortenings, pork and beef fat, vegetable oil, cream, and coffee was somewhat higher in the south. Pre-added salt in processed food (without the amount added in the kitchen) and bread was about 15 percent higher in the north,

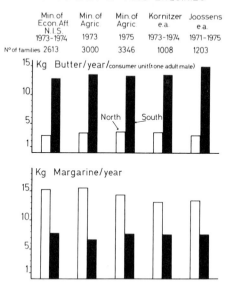

Figure 1.   The reliability and the reproducibility of the household butter and margarine consumption is clearly shown by the results of five totally independent investigators. (Reproduced with permission from *Acta Cardiology* Suppl. **23** (1979).

Table 2

Basic Nutrient Intake in North and South Belgium in 1974 and 1979

	NORTH		SOUTH	
	1974	1979	1974	1979
BREAD	234	205	201	176
MEAT	174	203	165	199
FISH	19	22	14	18
POTATOES	291	276	306	341
VEGETABLES	167	181	172	195
FRUITS	196	197	182	206

Items in g/day/adult person standardized to 2800 kcal.

Source of raw data: National Institute of Statistics

(Tables 2 to 8).

vitamin A and fish consumption 25 percent higher. The major differences in nutritional patterns were butter (8 g/day/adult person in the north versus 36 g in the south) and margarine consumption (43 g in the north versus 22 g in the south). In 1975, the intake of saltfree, dietetic margarine (P/S ratio of 3.4 to 4.4) was, however, three times larger in the north (Kornitzer et al., 1979).

Totals for the nutrients are shown in Tables 2 to 8. The major difference was a higher saturated fat and food cholesterol intake, and a lower polyunsaturated fat consumption in the south (Table 7). Using Keys' formula, a 12.6 mg/dl difference in serum cholesterol between the south and the north was predicted in 1974, a value consistent with the observed one (Joossens et al., 1977b).

The "anatomy" of saturated fat intake showed that from butter alone the daily intake of saturated fat was 4.3 g in the north versus 18.3 g in the south, whereas saturated fat from all other sources was more similar being 49.1 g versus 43.9 g. This implies that the major food difference, i.e., the saturated fat intake, was primarily caused by butter consumption (Table 8). An opposite but smaller difference existed for polyunsaturated fat. In 1974, the P/S ratio was near to 0.54 in the north and 0.35 in the south. In 1960, a P/S ratio of 0.2 was found in the north (Joossens et al., 1966).

Psychological tests showed a significantly lower value on the Bortner scale, a lower extraversion score, and a higher neuroticism score in the South (Kornitzer et al., 1979). This may have contributed to the mortality difference, although it is difficult to quantify.

Table 3

Nutrients Intake in North and South Belgium
in 1974 and 1979

	NORTH		SOUTH	
	1974	1979	1974	1979
FIBER (g)	17	17	16	17
SODIUM (mmol)	140	135	120	117
POTASSIUM (mmol)	80	83	82	90
CALCIUM (mg)	922	922	882	882
MAGNESIUM (mg)	292	340	292	316
COPPER (mg)	2	3	2	2
ZINC (mg)	9	10	9	10
RETINOL (μg)	1267	1137	945	1033
β-CAROTENE (μg)	2366	2435	2336	2687
VITAMIN C (mg)	72	74	73	83

Items are in indicated units/day/adult person standardized to 2800 kcal.

Table 4

Dairy Intake in North and South Belgium
in 1974 and 1979

	NORTH		SOUTH	
	1974	1979	1974	1979
MILK (ml)	180.0	160.0	187.0	174.0
CHEESE (g)	24.0	29.0	24.0	29.0
EGGS (pieces)	2.7	2.2	2.8	2.7
CREAM (ml)	0.8	1.1	1.7	2.4

Items are in indicated units/day/adult person (standardized to 2800 kcal),
except for eggs, for which weekly values are given.

Table 5

Butter, Margarine, Shortenings. and Oils in
North and South Belgium in 1974 and 1979

	NORTH		SOUTH	
	1974	1979	1974	1979
BUTTER	8	13	36	31
DIETETIC MARGARINE	--	8	--	4
OTHER MARGARINES	43	26	21	18
SHORTENINGS AND FATS	3	4	4	4
OILS	9	7	13	11

Items in g/day/adult person (standardized to 2800 kcal), except for oils, which are in ml/day/adult person.

Table 6

Carbohydrate, Protein, Fat, and Alcohol Intake in
North and South Belgium, 1974 and 1979

	NORTH		SOUTH	
	1974	1979	1974	1979
MONO- AND DISACCHARIDES	16.5	16.7	16.3	15.3
STARCH	28.3	27.2	26.5	26.5
PROTEIN	11.0	11.6	10.5	11.2
FAT	41.1	41.5	43.1	43.4
ALCOHOL	3.0	2.9	3.5	3.5

Items are in percent of total calories.

Table 7

Cholesterol and Fat Intake in North and South Belgium, 1974 and 1979

	NORTH		SOUTH	
	1974	1979	1974	1979
POLYUNSAT. FAT	9.3	8.1	6.8	7.1
SATURATED FAT	17.2	17.3	20.0	19.2
P/S-RATIO	0.54	0.47	0.35	0.37
DIET. CHOL.	307	329	375	386

Fats are in percent of total calories, dietary cholesterol in mg/day/adult person standardized to 2800 kcal.

Table 8

Origin of Saturated Fat in North and South Belgium, 1974 and 1979

ORIGIN OF SATURATED FAT	NORTH	SOUTH	NORTH	SOUTH
BUTTER	4.3	18.8	6.9	16.1
OTHER FAT CONTAINING ITEMS	49.0	44.7	46.9	44.1

Saturated fat content of items is in g/day/adult person standardized to 2800 kcal.

Drinking habits were different in the two regions. Wine and aperitifs, especially, were consumed more frequently in the south. The evaluation of smoking habits yielded contradictory results. Some sources of information reported somewhat heavier smoking in the south, others in the north. All in all, the difference must not have been large, since lung cancer in the south was 9 percent higher in middle-aged men and 18 percent lower in women.

2.2. Nutritional trends in Belgium

Since the 1973-74 survey of the NIS, a new survey of 1566 families was performed in 1978-79. Meanwhile, new tables of food constituents have become available (Paul and Southgate. 1978). The data of 1973-74 have therefore been recalculated for blue and white collar workers. and for non-actives, and compared with those of 1978-79 (Tables 2 to 8). All the data are standardized to an energy intake of 2800 kcal/day. The data presented in this paper may therefore differ slightly from previous publications.

Since 1960 butter consumption has always been greater in the south of Belgium (Kornitzer et al., 1979). However, from that year on, butter intake started to decline gradually in both regions with a somewhat faster decrease from 1965 to 1973. After 1974, butter intake increased in the north from 8 g/day/adult person to 13 g in 1979, and decreased in the south from 36 to 31 g in 1979. Margarine intake especially of the harder variety has decreased markedly since 1968. This decline was most pronounced in the north (Kornitzer et al., 1979). Saltfree, dietetic margarine consumption increased faster in the north until 1975. In 1975 it was three times higher in the north than the south, whereas in 1979 a ratio of two was obtained. Corn and soybean oil intake has increased markedly since 1950, especially in the north. Total oil intake is, however, still 50 percent larger in the south.

Summarizing the trends in fat consumption, one can say that the saturated and polyunsaturated fat intake trends converge after 1975. The nutritional gap is therefore closing. As a consequence, the estimated difference in serum cholesterol between the regions, according to the Keys formula, has now reduced from 12.6 to 7.8 mg/dl. This is also consistent with observations on serum cholesterol in the north, where a marked decrease was found from 1967-69 to 1971-75 (Joossens et al., 1977b). Recent unpublished data from our department (1979-81) showed similar or slightly higher serum cholesterol values in the north.

It is rather disappointing to observe that the Lancet paper (Joossens et al., 1977b) did not improve the nutritional situation in the north in terms of fat intake. On the contrary, a slight deterioration was observed. In the south, however, this publication seems to have had a salutary impact. It can be predicted that, if the situation is not reversed in the north, coronary mortality will begin to rise again. There are already indications from total cardiovascular mortality that only little progress has been realized since 1978.

Salt intake was highest in the north of Belgium, being 15.2 g/24h in 1966 (Joossens and Brems-Heyns, 1975). There are some indications that this value was lower in the south. Since 1966, a gradual decrease was observed in the north. Actually (1980-82) values around 9 g/24h are observed and there is practically no difference between the regions at this time (Kesteloot and Geboers, 1983). Fiber intake was identical in both regions in 1974 and in 1979. Total sugar consumption was and remained similar in both regions.

111

3. Mortality

3.1. Regional mortality differences in Belgium

The most important regional mortality differences, at least for middle-aged persons, i.e., a higher mortality in the south than in the north, are found for all causes, total cardiovascular, ischemic heart disease, and stroke mortality as illustrated in Figures 2 to 5. In general, the differences are more pronounced in males, suggesting a higher susceptibility in males in the south. The opposite is true for females. This was also observed for total cancer and for lung cancer (Figures 6 and 7). There is no indication that increasing consumption of polyunsaturated fat since 1968 and the higher amount consumed in the north had any influence on total cancer mortality. Diabetes mortality is also higher in the south (Joossens, 1979). Stomach cancer mortality is, however, much higher in the north (Figure 8) a feature shared by the provinces of the Netherlands nearest to Belgium (Joossens and Brems-Heyns, 1975; Hayes et al., 1982). This may be due to a higher salt intake in these provinces. If so, one would expect a higher stroke mortality in the north (Figure 5). Yet, this is not the case. One can speculate about this discrepancy, but recent epidemiological and pathophysiological investigations have indicated that a low P/S ratio increases blood pressure on a given salt intake and this could then increase stroke mortality in the south (see Puska et al., 1983).

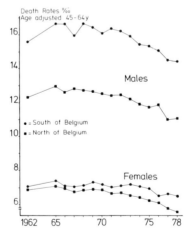

Figure 2.   No significant difference in trend between the two regions for the average of both sexes.

Source of raw data for Figures 2 to 8: National Institute of Statistics, Brussels.

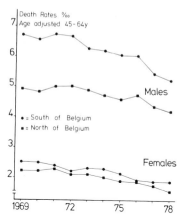

Figure 3. The absolute value of the slope of the decreasing regression line is significantly larger in the south than in the north for the average of both sexes.

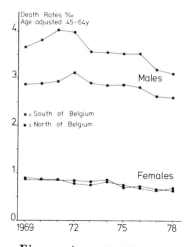

Figure 4. See Figure 3.

This is only true for mortality, not for morbidity. Mortality rates for violent death, though not different at the age of 15 to 24 years, are higher in the south in middle-aged persons. This is consistent with observations that when coronary mortality rates are high (e.g., in the U.S.) the people who die from non-coronary causes also have a high degree of coronary atherosclerosis. This was confirmed in autopsies from war casualties in Korea and Vietnam, as contrasted to results of autopsies of poor Indio populations in Mexico, where coronary atherosclerosis is rare (Tamayo et al., 1961). Similarly, in countries with medium to high levels of

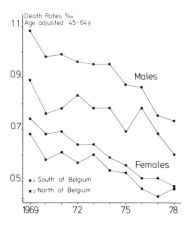

Figure 5.   See Figure 3.

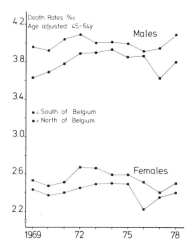

Figure 6.   No clear cut trend in regional differences is observed. Note the inverse relationship of the differences between regions in the two sexes.

coronary mortality, the latter correlates with total non-cancer mortality, both in males and females of middle-aged groups. This technique has also been used (Joossens, 1979, 1980) to indicate that coronary mortality was probably underclassified in the south of Belgium but not in the north, pointing to the non-

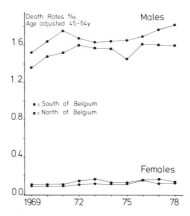

Figure 7.   See Figure 6.

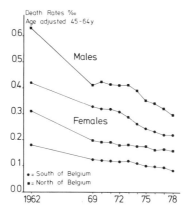

Figure 8.   The values from 1968 to 1978 have been smoothed by a five-year moving mean. The absolute value of the slope of the decreasing regression line (stomach cancer mortality against year, unsmoothed values) is significantly larger in the north than in the south for the average of both sexes. This implies that the regional difference is narrowing.

spurious origin of the difference in coronary mortality. There was no adult mortality difference in 1890 (Tulippe, 1952), but such a difference gradually developed over the decades to near 10 percent in 1947 (Joossens et al., 1977b) and increasing to 25 percent around 1975. There are indications that the differences in total cardiovascular, ischemic heart disease, and stroke mortality are narrowing recently (Figures 3 to 5). This implies that these mortality rates are now decreasing faster in the south than the north. Stomach cancer mortality, however, is declining faster in the north. This is consistent with the observations that the nutritional differences between the regions narrowing. If confirmed, this fact will be a major argument against the purely genetic origin of the north-south mortality differences and will favor nutritional differences as the most important factor.

3.2. Mortality from all causes and saturated fat intake

As shown before, the major regional nutritional difference in the early seventies was that in saturated fat intake. If saturated fat contributes to coronary atherosclerosis and to increased mortality from all causes, it should correlate with the observed all-causes mortality in Belgian provinces. All-causes mortality is preferred to mortality from ischemic heart disease because it is much more reliable. Figure 9 shows all-causes mortality in 1972 plotted against butter intake (butter is 51 percent saturated fat). A similar plot is shown for regions in France in 1967 and for saturated fat from margarine and butter in the two Belgian regions from 1965 to 1975 (Figures 10 and 11). For middle-aged persons, all-causes mortality increases in all cases with about 5 percent per kg of yearly saturated fat intake.

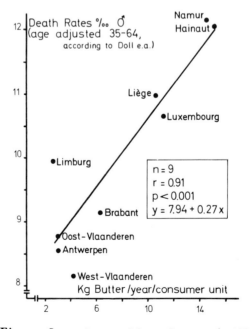

Figure 9.   Source of data: Joossens (1977b).

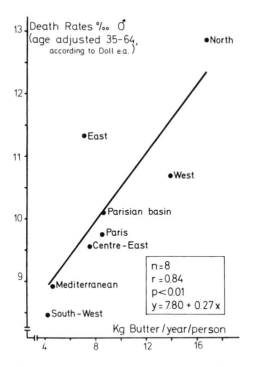

Figure 10. Source of data: MacLennan and Meyer (1977).

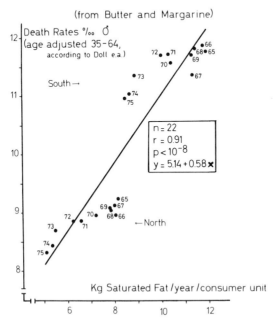

Figure 11. Source of data: National Institute of Statistics and Kornitzer et al., (1979).

3.3. Recent mortality trends in Belgium

Since mortality is decreasing in the two Belgian regions (90 percent of the total population), it must also decrease in Belgium as a whole. This is confirmed, both in middle-aged persons and in elderly. Belgium is the only country of the Common Market where coronary mortality in middle-aged persons has changed significantly (-27 percent) since 1968 (changes are calculated over a ten-year period). There are two other European countries where decreases were observed, namely Finland (-13 percent) and Norway (-7 percent). This can be compared with -31 percent in the U.S. Similar results were obtained for mortality from stroke and stomach cancer, which are highly related. All-causes mortality has changed by -18 percent since 1968 in Belgium for middle-aged persons, in the U.S. by -20 percent. It is important to know that in Belgium stroke mortality for persons aged 75 years and more decreased 36 percent, i.e., more markedly than in 32 other countries. Only Finland and Iceland did better. At this age, medical treatment is much less important. Many doctors even believe that treating hypertension in this age group is harmful. The marked decrease of stroke mortality in elderly in Belgium provides at least a suggestion that life style changes, including a lower salt intake (Joossens and Geboers, 1983), may be important in terms of public health.

Acknowledgement

We are grateful to the National Institute of Statistics, Brussels (Director General E. Rosselle, Mrs. M. Luyck-Draelandts and Mrs. M. Portaels) for the regional mortality data and for the Family Budget Enquest data of 1974 and of 1979.

For 1979 to 1981 mortality data for Belgium were provided by the Ministry of Health (Director General R. Beckers, Mrs. M. Verlinden).

Grants of the N.F.W.O. Brussels and the ASLK Brussels made it possible to build our "Mortality Monitoring System."

WHO provided us with the raw mortality data on tape and through their publications. The use of those data implies no responsibility whatsoever for WHO.

The graphs were made and the text was typed by Mrs. J. Smisdom-Rongy.

To all of them our most sincere thanks.

References

Hayes, R. B., Swaen, G. M. H., Ramioul. L., and Tuyns, A. J. Stomach cancer mortality - geographic comparison in the Netherlands and in Belgium. *European Journal of Cancer Clinical Oncology* 18 (1982) 623-628.

Joossens, J. V. Food pattern and mortality in Belgium. *Acta Cardiology* Suppl. 23 (1979) 133-161.

Joossens, J. V. Epidemiology of coronary heart disease: Lessons from North and South Belgium. *Postgraduate Medical Journal* 56 (1980) 548-556.

Joossens, J. V., and Brems-Heyns, E. Cerebrovascular mortality, gastric cancer mortality and salt (Dutch). *T. Soc. Geneesk.* 53 (1975) 530-542.

Joossens, J. V. and Geboers, J. Community control in different countries - Belgium. (Submitted for publication).

Joossens, J. V., Verdonk, G., and Pannier, R. "Normal" serum cholesterol values in Belgium as related to age and diet: A comparison with other countries. *Acta Cardiology* 21 (1966) 431-445.

Joossens, J. V., Brems-Heyns, E., Raes, A., and Carlier, J. Changing food habits in Belgium. In: K. H. Guenther (ed.), *Community Control of Cardiovascular Diseases*. Potsdam: Society of Cardiology and Angiology of the German Democratic Republic, 1977a.

Joossens, J. V. et al. The pattern of food and mortality in Belgium. *Lancet* 1 (1977) 1069-1072.

Kesteloot, H. and Geboers, J. Personal communication.

Kesteloot, H. and Van Houte, O. An epidemiological survey of arterial blood pressure in a large male population group. *American Journal of Epidemiology* 99 (1974) 14-29.

Keys, A. (ed.) *Coronary Heart Disease in Seven Countries*. AHA Monogr. 29, New York, 1970.

Kornitzer, M., De Backer, G., Dramaix, M.. and Thilly. C. Regional differences in risk factor distributions, food habits and coronary heart disease mortality and morbidity in Belgium, *International Journal of Epidemiology* 8 (1979) 23-31.

Lederer, J. Evolution de la consommation du pain et santé publique. In: J. Lederer (ed.) *Pain et Santé*. Leuven: Nauwelaerts, 1970.

MacLennan, R. and Meyer, F. Food and mortality in France. *Lancet* 2 (1977) 133.

National Research Council. *Diet, Nutrition and Cancer*. Washington, D.C.: National Academy Press, 1982.

Paul, A. A. and Southgate, D. A. T. *The Composition of Foods*. Amsterdam: Elsevier/North-Holland Biomedical Press, 1978.

Puska, P. et al. Controlled, randomized trial of the effect of dietary fat on blood pressure. *Lancet* 1 (1983) 1-5.

Tamayo, R. P.. Brandt. H., and Ontiveros, E. Pathology of atherosclerosis in Mexico. *Archives of Pathology* 71 (1961) 113-117.

Tulippe, O. *Le Vieillissement de la Population Belge*. Brussels: Ed. Art et Technique, 1952.

Van Houte, O. and Kesteloot, H. An epidemiological survey of risk factors for ischaemic heart disease in 42,804 men. I. Serum cholesterol value. *Acta Cardiology* 27 (1972) 527-564.

Vastesaeger, M. Serum cholesterol and spreading fats in Belgium. *Acta Cardiology* Suppl. 23 (1979) 96-124.

Vastesaeger, M. et al. Cholestérolémie, triglycéridémie et prévalence des cardiopathies ischémiques d'expression française et néerlandaise. *Acta Cardiology* 29 (1974) 441-454.

8

DIETARY COMPONENTS AND THE RISK OF CORONARY HEART DISEASE MORTALITY

Daan Kromhout, Ph.D., M.P.H.

Introduction

The so-called diet-heart hypothesis has dominated research on relations between dietary components and coronary heart disease during the last 30 years. This hypothesis states that saturated fats elevate serum cholesterol levels and persons with elevated serum cholesterol levels are at high risk for coronary heart disease (Keys, 1952). The evidence available for the diet-heart hypothesis will be reviewed in this paper. Other areas of current interest are the relations between energy intake, dietary fiber, obesity, and coronary heart disease and relations between dietary minerals, blood pressure, and coronary heart disease. These relations will also be reviewed and recommendations for future research will be made.

1. Diet-Heart Hypothesis

The Dutch internist C. D. de Langen postulated in 1916 that the cholesterol content of the diet influences serum cholesterol and coronary heart disease (CHD) (DeLangen, 1916). He based his hypothesis on the observation that the serum cholesterol level of the Dutchmen was higher than that of the inhabitants of the island of Java, in Indonesia. He also observed that CHD was less frequent in Indonesia than in the Netherlands. Similar observations were made by the Dutch internist Snapper who worked in the 1930s in China (Snapper, 1941).

It took until the late 1950s for geographic differences in diet, serum cholesterol, and CHD to be investigated in a thorough way. Between 1958 and 1964, baseline surveys were carried out in 16 cohorts in 7 countries under the leadership of Dr. Ancel Keys (Keys et al., 1967). A total of 12,763 men aged 40 to 59 were enrolled into this study. The Seven Countries Study showed clearly that *between* the cohorts strong correlations exist between the percentage of energy from saturated fat in the diet, serum cholesterol, and 10-year incidence of coronary heart disease (Keys, 1970; 1980). The Seven Countries Study and the Pooling Project, in which the results of the major cohort studies carried out in the U.S.A. were combined, showed that also *within* cohorts serum cholesterol is a good predictor of coronary heart disease (Keys, 1980; Pooling Project Research Group, 1978).

Epidemiological studies have not shown consistent relations between the fatty acid composition of the diet and serum cholesterol (Garcia-Palmieri et al., 1977; Kahn et al., 1969; Kannel and Gordon, 1970; Kay, Sabry, and Csima, 1980; Nichols et al., 1976; Shekelle et al., 1981; Stulb et al., 1955). These zero or low order correlations may be explained by the large intra-individual variation in both fatty acid intake and serum cholesterol. A study from Israel showed that twenty-two 24-hour recalls scattered over a year are needed in order to obtain accurate information about the saturated fat intake of an individual (Balogh, Kahn, and Medalie, 1971). The intra-individual variation of serum cholesterol in free-living persons amounts to 20 mg/dl (Keys, 1967). Therefore, only zero or low order correlations may be expected when relations between fatty acid composition of the diet and serum cholesterol are investigated in epidemiological studies (Jacobs, Anderson, and Blackburn, 1979; Liu et al., 1978). Metabolic ward studies have shown that under controlled conditions the serum cholesterol level can be increased by saturated fat and decreased by polyunsaturated fat (Hegsted et al., 1965; Keys et al., 1965). The serum cholesterol increasing power of saturated fat is two times stronger than the serum cholesterol decreasing power of polyunsaturated fat.

The serum cholesterol level is also influenced by the cholesterol content of the diet. The serum cholesterol increasing power of dietary cholesterol is less than that of saturated fat (Keys et al., 1965). Recently published results of the Western Electric Study showed that dietary cholesterol was significantly associated with the risk of death from CHD independent of serum cholesterol (Shekelle et al., 1981).

This supports the idea that dietary cholesterol may be related to atherosclerosis through other mechanisms, in addition to influencing serum cholesterol. One of these mechanisms may be an alteration in the structure or composition of lipoproteins.

The next question to be asked is whether changes in the fatty acid composition of the diet are followed by changes in serum cholesterol and coronary heart disease incidence. This question can only be answered in intervention studies. The two most well known intervention studies, The Finnish Mental Hospital Trial and the Los Angeles Veterans Administration Trial, have shown that a diet low in saturated fat and moderately high in polyunsaturated fat, compared to the traditional Finnish and American diet, lowered serum cholesterol and CHD incidence but not total mortality (Dayton et al., 1969; Miettinen et al., 1972). The effectiveness of serum cholesterol in terms of reduced mortality is therefore still debated (Glueck and Connor, 1978; Keys, 1980; Mann, 1977; Stamler, 1980; Werkö, 1979). The intervention studies in progress, for example, the Lipid Research Clinic Trial and the Partial Ileal Bypass Trial, may show that substantial reductions in serum cholesterol, reductions between 30 and 40 percent, are needed in order to reduce total mortality (Long et al., 1983; Rifkind and Levy, 1978). Such reduction can be obtained by drugs such as cholestyramine in the Lipid Research Clinic Trial or by operation as in the Partial Ileal Bypass Trial. By dietary intervention serum cholesterol lowering between 10 and 15 percent may be expected (Rifkind et al., 1983). It is unlikely that antagonists of the diet-heart hypothesis will be convinced by positive results of the Lipid Research Clinic Trial and the Partial Ileal Bypass Trial because in these trials serum cholesterol lowering was not due to changes in diet but to drug treatment and operation. Therefore, the relations between fatty acids composition of the diet, serum cholesterol, and CHD should be judged on the basis of available evidence.

Epidemiological, clinical, and animal experimental studies have shown that the fatty acid composition of the diet is an important determinant of serum cholesterol (Hully et al., 1981). Changes in the fatty acid composition of the diet are accompanied by changes in serum cholesterol and CHD incidence. It can, be concluded, therefore, that the relations between fatty acids in the diet, serum cholesterol, and CHD are probably causal. The results of ongoing intervention studies will provide information about the effects of cholesterol lowering on total mortality.

2. Dietary Determinants of Obesity

It is generally thought that the energy intake of obese persons is higher than that of lean persons. Several cross-sectional studies have shown, however, that obese persons have a lower energy intake than their lean counterparts (Cahn, 1968; Johnson, Burke, and Mayer, 1956; Lincoln, 1972; Maxfield and Konishi, 1966;

123

McCarthy, 1966; Stefanik, Heald, and Mayer, 1959; Wilkinson et al., 1977). The lower energy intake of the obese persons may be the consequence of their obesity. Obese persons may have a low level of energy expenditure and as a result a low energy intake. Relations between energy intake, expenditure, and body fatness should be investigated in prospective studies. Only these studies may show that obesity is preceded by a period of high energy intake compared to energy expenditure.

Relations between diet and body fatness have been investigated in the Zutphen Study (Kromhout, 1983a). This study forms the Dutch contribution to the Seven Countries Study (Keys, 1970, 1980; Keys et al., 1967). In 1960, the prevalence of obesity among the Zutphen men was about 10 percent. This percentage is low compared to a prevalence of about 30 percent found in the U.S. railroad cohort of the Seven Countries Study. It is of interest to study relations between diet and obesity in a relatively lean cohort, because the chance that unreliable data on food intake will be obtained is probably less than in a cohort where obesity is very prevalent.

In the Zutphen Study middle-aged men in the highest quartile of the sum of two skinfold consumed on an average about 400 kcal *less* than men in the lowest quartile. This inverse relationship became even stronger if energy intake was expressed per kg body weight (r = -0.55). In population comparisons of the Seven Countries Study, a strong inverse relationship (r = -0.75) was noted between the average energy intake per kg body weight of a population and its median sum of skinfolds among 14 cohorts at baseline (Keys, 1970). Similarly, longitudinal trends, examined over 10 years follow-up of the Zutphen men, showed that an average 10-year increase of 4 kg of body weight was paralleled by an average *decrease* in energy intake from 43 kcal per kg body weight to 35 (Kromhout, 1983b). Thus, the results from cross-sectional studies within and between populations, and from the Zutphen longitudinal study all showed strong inverse relationships between energy intake per kg body weight and indicators of body fatness. If energy intake per kg body weight is considered an index of physical activity (Keys, 1970; Marr et al., 1970), it can be concluded that obesity increases as physical activity decreases.

The major implication of these results may be that in weight reduction programs and health strategies more emphasis needs to be placed on increased physical activity than on reduced energy intake because obese people are already eating less than their lean counterparts.

It has been suggested that the increased prevalence and incidence of obesity in Western countries since the beginning of this century is due partly to a decreased intake of dietary fiber (Heaton, 1973; VanItallie, 1978). In the Zutphen Study, obese men had a lower intake of dietary fiber and all macronutrients except alcohol (Kromhout, 1983a). Multivariate analyses showed that dietary fiber was not significantly related to the sum of two skinfolds. The lower intake of dietary fiber by the obese men compared to the lean ones may be the consequence of a lower food intake. In such a situation it is impossible to conclude whether the energy intake or the dietary fiber intake is the primary determinant of obesity.

In the Zutphen Study, alcohol was significantly positively related to the sum of two skinfolds and the Quetelet Index after both univariate and multivariate analyses (Kromhout, 1983a). Among the 92 men who drank at least one alcoholic beverage per day the prevalence of obesity based on the sum of skinfolds was 1.5 times as high as that of the 451 non-drinkers. The prevalence of obesity based on the Quetelet Index was three times as high among the men who drank at least one alcoholic drink per day, compared to that of the non-drinkers. In a population based study among men and women aged 30 to 90 years in Southern California, no relation was found between alcohol intake and the Quetelet Index (Jones et al., 1982). The alcohol intake of the American men who used alcoholic beverages was much higher than that of the Zutphen men. That may be an explanation for the observed differences. Much more research is needed to clarify the role of alcohol intake in the etiology of obesity.

3. Energy Intake, Starch, Dietary Fiber, and CHD

Several cohort studies have shown that middle-aged men who developed CHD during a certain period of follow-up consumed during the baseline survey about 200 kcal *less* than survivors (Table 1) (Gordon et al., 1981; Kromhout et al., 1982; Marr and Morris, 1981). The inverse relation between energy intake and CHD became even stronger when the energy intake was expressed per kg body weight. It has been shown by Marr and colleagues (1970) that energy intake per kg body weight is significantly correlated with physical activity. These results may be interpreted as evidence for a protective role of physical activity in the development of CHD. Much more research is needed to prove this hypothesis.

An interesting question to be asked is "What macronutrients are responsible for this difference of 200 kcal." With the exception of the Framingham Study, all studies showed that the carbohydrate and especially the starch intake was significantly higher among survivors compared to men who died from CHD (Table 1) (Gordon et al., 1981; Kromhout, 1982; Marr and Morris, 1981). Foods rich in starch also contain dietary fiber. It may therefore be expected that an inverse relation exists between dietary fiber intake and CHD. Such inverse relations were found in the London and the Zutphen Study (Kromhout et al., 1982; Morris et al., 1977). This relation disappeared in the Zutphen Study after multivariate analyses. In the London Study the inverse relation between dietary fiber and CHD was due to cereal fiber and not to fiber from fruits, vegetables, and pulses. This result was unexpected because laboratory data have shown that the serum cholesterol lowering effect of fiber is due to pectin and not to cereal fiber (Stasse-Wolthuis, 1980). In future studies relations between the different components of dietary fiber and CHD should be investigated.

Table 1

Differences in dietary variables between CHD
cases and non-cases in five prospective studies

Dietary variable		Zutphen	London	Framingham	Honolulu	Puerto Rico
Energy	(kcal)	-223	-213[**]	-253[*]	-170[**]	-172[*]
Energy/kg body weight		- 6[**]		- 4[*]	- 4[**]	- 4[**]
Total protein	(g)	- 8	- 8[**]	- 5	- 2	- 4
Total fat	(g)	- 8	- 11[**]	- 8	- 1	- 3
Total carbohydrate	(g)	- 39[*]	- 24[*]	- 14	- 22[**]	- 28[**]
Alcohol	(g)	+ 7		- 15[**]	- 9[**]	- 3
Starch	(g)	- 26		+ 3	- 14[**]	- 20[**]
Dietary fiber	(g)	- 3.6	- 2.3[**]			

[*] $0.01 < p < 0.05$
[**] $p < 0.01$

The intake of dietary fiber is highly correlated with the intake of certain vitamins, such as vitamins B1, B6, and nicotinic acid, and certain minerals, such as potassium and non-haem iron (Table 2). It may therefore by hypothesized that some of these vitamins and minerals may play a role in the etiology of CHD. This may also be an area of interest for future studies.

Multivariate analyses showed that the inverse relation between energy intake at baseline and CHD disappeared if known risk indicators like serum cholesterol and skinfold thicknesses were included in multivariate models (Gordon et al., 1981; Kromhout and De Lezenne Coulander, 1984). Does that mean that energy intake is not important as a determinant of CHD? This question will be answered by using the results of analyses of the data of the Zutphen Study.

In the Zutphen Study serum cholesterol and skinfold thicknesses are strong predictors of CHD mortality (Keys, 1980). Both risk indicators are significantly inversely related to energy intake per kg body weight (Kromhout, 1983a; Kromhout, 1983c). When a multivariate model, including energy per kg body weight, serum cholesterol and skinfold thicknesses as independent variables, and CHD mortality as dependent variable is used, only serum cholesterol and skinfold thicknesses are important predictors of CHD mortality due to intercorrelations between these independent variables. That result may be interpreted as follows.

Table 2

Correlations between dietary fiber and nutrients present
in vegetable foods in 871 middle-aged men in Zutphen in 1960

	Dietary fiber
Pectin	0.71
"Cellulose"	1.00
Potassium	0.78
Non-haem iron	0.88
Vitamin B1	0.73
Vitamin B6	0.82
Nicotinic acid	0.71
Vitamin C	0.28

All these correlations are statistically significant ($p < 0.001$)

The influence of dietary variables on CHD mortality is mediated through serum cholesterol and skinfold thicknesses. It can be concluded that energy intake per kg body weight is important in the development of CHD not because of its direct influence on CHD but because of its role in regulating serum cholesterol and body fatness.

4. Dietary Minerals, Blood Pressure, and CHD

Since the beginning of this century there has been an interest in the relation between dietary variables, especially sodium and blood pressure (Page, 1973). Between populations, significant relations between sodium intake and blood pressure can be observed (Gleibermann, 1973; Meneely and Batterbee, 1976). Within populations, the intake of sodium is generally unrelated to blood pressure (Schlierf et al., 1980; Simpson et al., 1978), probably due to large intra-individual variation in both sodium intake and blood pressure. Severe restriction of sodium intake to less than 0.2 grams per day is accompanied by a substantial reduction in blood pressure in patients with severe hypertension (Kempner, 1948). Recent studies have shown that moderate restriction of sodium intake to 1.5 to 2.0 grams per day lead to lower blood pressure levels in patients with mild and moderate hypertension (MacGregor et al., 1982; Morgan et al., 1978; Parijs et al., 1973).

127

Besides sodium, potassium plays a role in determining blood pressure levels. Several animal experiments have shown that a high sodium, low potassium intake induces hypertension (Meneely and Batterbee, 1976). In the Evans County Study no difference was found in sodium intake between blacks and whites, but the potassium intake was significantly lower in blacks compared to whites (Grim et al., 1980). The prevalence of hypertension is higher in blacks than in whites. A significant inverse relation between potassium intake and systolic blood pressure was observed in the Zutphen Study in 1970 but not in 1960 and 1965 (Kromhout et al., unpublished results). That may be due to the fact that the average potassium intake of the Zutphen men was high in 1960 but decreased considerably between 1960 and 1970. Recently, two intervention studies have shown that adding 2.5 grams of potassium per day lowered blood pressure significantly (Khaw and Thom, 1982; MacGregor et al., 1982). These results make probable a role for potassium in the regulation of blood pressure.

Langford and Watson (1975) observed an inverse relation between calcium intake and blood pressure in humans, but their results in rats are equivocal. Recently, McCarron, Morris, and Cole (1982) showed in a case-control study that the calcium intake of hypertensives was significantly lower than in normotensives. In the Zutphen Study a consistent inverse relation was observed between the calcium intake and systolic blood pressure in 1960, 1965, and 1970 (Kromhout et al., unpublished results). This relation was somewhat stronger when the calcium/phosphorus ratio instead of the calcium intake was used. The inverse relation between the calcium/phosphorus ratio and systolic blood pressure remained statistically significant after multivariate analyses in two of the three years. A recently reported intervention study showed that in young adults a calcium supplement of 1 gram per day caused a significant reduction of diastolic blood pressure (Belizan et al., 1983). The results of studies on the relation between dietary calcium and blood pressure are promising, but more research is needed in order to establish a definite relationship.

Several epidemiological studies have shown a positive relation between alcohol intake and blood pressure (Stokes, 1982). Such a relation was also found in the Zutphen Study in 1960, 1965, and 1970 (Kromhout et al., unpublished results). Multivariate analyses showed that the regression coefficients between alcohol intake and systolic blood pressure were reduced in all three years and remained statistically significant in 1960 only. That may be due to the strong inverse relation between alcohol intake and the calcium/phosphorus ratio. It may be hypothesized that the relation between alcohol intake and blood pressure is secondary to the relation between calcium/phosphorus ratio and blood pressure.

Several epidemiological studies have shown an inverse relation between water hardness and CHD (Comstock, 1980). The calcium concentration in water is held responsible for this relation. Also, an inverse relation was found between dietary calcium intake and CHD mortality between different regions in England and Wales

(Knox, 1973). Animal experiments showed that in rats on a purified diet rich in saturated fat, the clotting time was prolonged when the dietary calcium intake increased (Renaud et al., 1981b). Comparative studies in humans carried out in France and Scotland showed inverse relations between dietary calcium intake and platelet functions tests (Renaud et al., 1981a, 1981b). It may therefore be hypothesized that dietary calcium influences CHD by its effect on blood platelet aggregation and blood pressure. A thorough investigation of these relations is warranted.

5. Conclusions

Research on diet and coronary heart disease has, until recently, been focused on the effects of fatty acids on serum cholesterol and CHD. In the future, attention should be paid to the effect of energy intake and its relation with physical activity on CHD. Research on relations between the different components of dietary fiber and CHD and between dietary calcium and CHD is also needed. These areas of research seem to be fruitful and may broaden our view on relations between diet and CHD.

References

Balogh, M., Kahn, H. A., and Medalie, J. H. Random repeat 24-hour dietary recalls. *American Journal of Clinical Nutrition* 24 (1971) 304-310.

Belizan, J. M., Villar, J., Pineda, O., Gonzalez, A. E., Sainz, E., Garrera, G., and Sibrian, R. Reduction of blood pressure with calcium supplementation in young adults. *Journal of the American Medical Association* 249 (1983) 1161-1165.

Cahn, A. Growth and caloric intake of heavy and tall children. *Journal of the American Dietetic Association* 53 (1968) 476-480.

Comstock, G. W. The epidemiologic perspective: Water hardness and cardiovascular disease. *Journal of Environmental Pathology and Toxicology* 4 (1980) 9-25.

Dayton, S., Pearce, M. L., Hashimoto, S., Dixon, W. J., and Tomiyasu, U. A controlled clinical trial of a diet high in unsaturated fat in preventing complications of atherosclerosis. *Circulation* 40 (1969) suppl. II.

Ederer, F., Leren, P., Turpeinen, O., and Frantz, I. D., Jr. Cancer among men on cholesterol lowering diets. Experience from five clinical trials. *Lancet* 2 (1971) 203-206.

Garcia-Palmieri, M. R., Tillotson, J., Cordero, E., Costas, R., Jr., Sorlie, P., Gordon, T., Kannel, W. B., and Colon A. A. Nutrient intake and serum lipids in urban and rural Puerto Rican men. *American Journal of Clinical Nutrition* 30 (1977) 2092-2100.

Gleibermann, L. Blood pressure and dietary salt in human populations. *Ecology of Food and Nutrition* 2 (1973) 143-155.

Glueck, C. J. and Connor, W. E. Diet-coronary heart disease relationships reconoitered. *American Journal of Clinical Nutrition* 31 (1978) 727-737.

Gordon, T., Kagan, A., Garcia-Palmieri, M., Kannel, W. B., Zukel, W. J., Tillotson, J., Sorlie, P., and Hjortland, M. Diet and its relation to coronary heart disease and death in three populations. *Circulation* 63 (1981) 500-515.

Grim, C. E., Luft, F. C., Miller, J. Z., Meneely, G. R., Batterbee, H. D., Hames, C. G., and Dahl, L. K. Racial differences in blood pressure in Evans County, Georgia. *Journal of Chronic Diseases* 33 (1980) 87-94.

Heaton, K. W. Food fibre as an obstacle to energy intake. *Lancet* 2 (1973) 1418-1421.

Hegsted, D. M., McGandy, R. B., Myers, M. L., and Stare, F. J. Quantitative effects of dietary fat on serum cholesterol in men. *American Journal of Clinical Nutrition* 17 (1965) 281-295.

Hulley, S. B., Sherwin, R., Nestle, M., and Lee, P. R. Epidemiology as a guide to clinical decisions. II. Diet and coronary heart disease. *Western Journal of Medicine* 135 (1981) 25-33.

Jacobs, D. R., Jr., Anderson, J. T., and Blackburn, H. Diet and serum cholesterol: Do zero correlations negate the relationship? *American Journal of Epidemiology* 110 (1979) 77-87.

Johnson, M. L., Burke, B. S., and Mayer, J. Relative importance of inactivity and overeating in the energy balance of obese high school girls. *American Journal of Clinical Nutrition* 4 (1956) 37-44.

Jones, B. R., Barrett-Connor, E., Criqui, M. H., and Holdbrook, M. J. A community study of calorie and nutrient intake in drinkers and non-drinkers of alcohol. *American Journal of Clinical Nutrition* 35 (1982) 135-139.

Kahn, H. A., Medalie, J. H., Neufeld, H. N., Riss, E., Balogh, M., and Groen, J. J. Serum cholesterol: Its distribution and association with dietary and other variables in a survey of 10,000 men. *Israel Journal of Medical Sciences* 5 (1969) 1117-1127.

Kannel, W. B. and Gordon, T. (Eds.) The Framingham Study: An epidemiological investigation of cardiovascular disease. Section 24: The Framingham Diet Study: Diet and the regulation of serum cholesterol. Washington: U.S. Government Printing Office, 1970.

Kay, R. M., Sabry, A. I., and Csima, A. Multivariate analysis of diet and serum lipids in normal men. *American Journal of Clinical Nutrition* 33 (1980) 2566-2572.

Kempner, W. Treatment of hypertensive vascular disease with rice diet. *American Journal of Medicine* 9 (1948) 441-493.

Keys, A. Coronary heart disease, serum cholesterol and the diet. *Acta Medica Scandinavica* 207 (1980) 153-160.

Keys, A. *Seven Countries: A Multivariate Analysis of Death and Coronary Heart Disease.* Cambridge, Mass.: Harvard University Press, 1980.

Keys, A. Coronary Heart disease in seven countries. *Circulation* 41 (1970) (suppl).

Keys, A. The cholesterol problem. *Voeding* 13 (1952) 539-555.

Keys, A. Blood lipids in men - a brief review. *Journal of the American Dietetic Association* 51 (1967) 508-516.

Keys, A., Anderson, J. T., and Grande, F. Serum cholesterol response to changes in the diet. IV. Particular fatty acids in the diet. *Metabolism* 14 (1965) 776-787.

Keys, A., Aravanis, C., Blackburn, H., Buchem, F. S. P. van, Djordjevic, B. S., Dontas, A. S., Fidanza, F., Karvonen, M. J., Kimura, N., Lekos, D., Monti, M., Puddu, V., and Taylor, H. L. Epidemiological studies related to coronary heart disease: Characteristics of men aged 40-59 in seven countries. *Acta Medica Scandinavica* 460 (1967) (suppl).

Khaw, K. T. and Thom, S. Randomised double blind cross-over trial of potassium on blood pressure in normal subjects. *Lancet* 2 (1982) 1127-1129.

Knox, E. G. Ischaemic-heart-disease mortality and dietary intake of calcium. *Lancet* 1 (1973) 1465-1467.

Kromhout, D. Energy and macronutrient intake in lean and obese middle-aged men (The Zutphen Study). *American Journal of Clinical Nutrition* 37 (1983a) 295-299.

Kromhout, D. Changes in energy and macronutrients in 871 middle-aged men during ten years of follow-up (The Zutphen Study). *American Journal of Clinical Nutrition* 37 (1983b) 287-294.

Kromhout, D. Body weight, diet and serum cholesterol in 871 middle-aged men during ten years of follow-up (The Zutphen Study). *American Journal of Clinical Nutrition* 38 (1983c) 591-598.

Kromhout, D., Bosschieter, E. B., and Lezenne Coulander, C. de. Dietary fibre and 10-year mortality from coronary heart disease, cancer and all causes (The Zutphen Study). *Lancet* 2 (1982) 518-522.

Kromhout, D. and Lezenne Coulander, C. de. Diet, prevalence and 10-year mortality from coronary heart disease in 871 middle-aged men (The Zutphen Study). *American Journal of Epidemiology* May 1984.

Kromhout, D., Bosschieter, E. B., and Lezenne Coulander, C. de. Dietary minerals, alcohol and blood pressure (The Zutphen Study). Submitted for publication.

Langen, C. D. de. Cholestrine-stofwisseling en rassenpathologie. *Geneeskundig Tijdschrift voor Nederlandsch Indië* 56 (1916) 1-36.

Langford, H. G. and Watson, R. L. Electrolytes and hypertension. In *Epidemiology and control of hypertension*. Paul, O. (ed.), Stuttgart: Georg Thieme, 1975. Pp. 119-130.

Lincoln, J. E. Caloric intake, obesity and physical activity. *American Journal of Clinical Nutrition* 25 (1972) 390-394.

Liu, K., Stamler, J., Dyer, A., McKeever, J., and McKeever, P. Statistical methods to assess and minimize the role of intra-individual variability in obscuring the relationship between dietary lipids and serum cholesterol. *Journal of Chronic Diseases* 31 (1978) 399-418.

Long, J. M., Moore, R. B., Matts, J. P., Varco, R. L., and Buchwald, H. The POSCH Group. Program of the surgical control of the hyperlipidemias (Posch). A secondary intervention trial-4-year lipid results. In *Atherosclerosis VI*. Schettler, G., Gotto, A. M., Middlehoff, G., Habenicht, A. J. R. and Jurutka, K. R. (eds.). New York: Springer-Verlag, 1983. Pp. 930-935.

MacGregor, G. A., Markandu, N. D., Best, F., Elder, D., Cam, J., Sagnella, G. A., and Squires, M. Double-blind randomised crossover trial of moderate sodium restriction in essential hypertension. *Lancet* 1 (1982) 351-355.

MacGregor, G. A., Smith, S. J., Banks, R. A., and Sagnella, G. A. Moderate potassium supplementation in essential hypertension. *Lancet* 2 (1982) 567-570.

Mann, G. V. Diet-heart: End of an era. *New England Journal of Medicine* 297 (1977) 644-650.

Marr, J. W., Gregory, J., Meade, T. W., Alderson, M. R., and Morris, J. N. Diet, leisure activity and skinfold measurements of sedentary men. *Proceedings of the Nutrition Society* 29 (1970) 17A-18A.

Marr, J. W. and Morris, J. N. Dietary intake and the risk of coronary heart disease in Japanese men living in Hawaii. *American Journal of Clinical Nutrition* 34 (1981) 1156-1157.

Maxfield, E. and Konishi, F. Patterns of food intake and physical activity in obesity. *Journal of the American Dietetic Association* 49 (1966) 406-408.

McCarron, D. A., Morris, C. D., and Cole, C. Dietary calcium in human hypertension. *Science* 217 (1982) 267-269.

McCarthy, M. C. Dietary and activity patterns of obese women in Trinidad. *Journal of the American Dietetic Association* 48 (1966) 33-37.

Meneely, G. R., and Batterbee, H. D. High sodium-low potassium environment and hypertension. *America Journal of Cardiology* 38 (1976) 768-785.

Miettinen, M., Turpeinen, O., Karvonen, M. J., Elosuo, R., and Paavilainen, E. Effect of cholesterol-lowering diet on mortality from coronary heart disease and other causes. A twelve-year clinical trial in men and women. *Lancet* 2 (1972) 835-838.

Morgan, T., Gillies, A., Morgan, G., Adam, W., Wilson, M., and Carney, S. Hypertension treated by salt restriction. *Lancet* 1 (1978) 227-230.

Morris, J. N., Marr, J. W., and Clayton, D. G. Diet and heart a postcript. *British Medical Journal* 2 (1977) 1307-1314.

Nichols, A. B., Ravenscroft, C., Lamphiear, D. E., and Ostrander, L. D., Jr. Independence of serum lipid levels and dietary habits. The Tecumseh Study. *Journal of the American Medical Association* 236 (1976) 1948-1953.

Page, I. H. "Common" salt and "benign" hypertension. *Modern Medicine* 41 (1973) 54-55

Parijs, J., Joossens, J. V., Linden, L. van der, Verstreken, G., and Amery, A.K.C.P. Moderate sodium restriction and diuretics in the treatment of hypertension. *American Heart Journal* 85 (1973) 22-34.

Pooling Project Research Group. Relationship of blood pressure, serum cholesterol, smoking habit, relative weight and ECG abnormalities to incidence of major coronary events: Final report of the Pooling Project. *Journal of Chronic Diseases* 31 (1978) 201-306.

Renaud, S., Morazain, R., Godsey, F., Dumont, E., Symington, I. S., Gillanders, E. M., and O'Brien, J. R. Platelet functions in relation to diet and serum lipids in British farmers. *British Heart Journal* 46 (1981a) 562-570.

Renaud, S., Dumont, E., Godsey, F., McGregor, L., and Morazain, R. Effects of diet on blood clotting and platelet aggregation. *Nutrition in the 1980's: Constraints on our knowledge.* New York: Alan R. Liss, Inc., 1981b. Pp. 361-381.

Rifkind, B. M., Goor, R., and Schucker, B. Compliance and cholesterol-lowering in clinical trials: Efficacy of diet. *Atherosclerosis VI.* In Schettler, G., Gotto, A. M., Middelhoff, G., Habenicht, A. J. R., and Jurutka, K. R. (eds.). New York: Springer-Verlag, 1983. Pp. 306-310.

Rifkind, B. M. and Levy, R. I. Testing the lipid hypothesis. Clinical trials. *Archives of Surgery* 113 (1978) 80-83.

Schlierf, G., Arab, L., Schellenberg, B., Oster, P., Mordasini, R., Schmidt-Gayk, H., and Vogel, G. Salt and hypertension: data from the "Heidelberg Study." *American Journal of Clinical Nutrition* 33 (1980) 872-875.

Shekelle, R. B., MacMillan Shryock, A., Paul, O., Lepper, M., Stamler, J., Liu, S., and Raynor, W. J. Diet, serum cholesterol, and death from coronary heart disease. The Western Electric Study. *New England Journal of Medicine* 304 (1981) 65-70.

Simpson, F. O., Waal-Manning, H. J., Bolli, P., Phelan, E. L., and Spears, G. F. S. Relationship of blood pressure to sodium excretion in a population survey. *Clinical Science and Molecular Medicine* 55 (1978) 373S-375S.

Snapper, I. *Chinese lessons to western medicine.* New York: Interscience, 1941.

Stamler, J. The established relationship among diet, serum cholesterol and coronary heart disease. *Acta Medica Scandinavica* 207 (1980) 433-446.

Stasse-Wolthuis, M. Influence of dietary fibre on cholesterol metabolism and colonic function in healthy subjects. *World Review of Nutrition and Dietetics* 36 (1980) 100-140.

Stefanik, P. A. Heald, F. P., and Mayer, J. Caloric intake in relation to energy output of obese and non-obese adolescent boys. *American Journal of Clinical Nutrition* 7 (1959) 55-62.

Stokes, G. S. Hypertension and alcohol: Is there a link? *Journal of Chronic Diseases* 35 (1982) 759-762.

Stulb, S. C., McDonough, J. R., Greenberg, B. G., and Hames, C. G. The relationship of nutrient intake and exercise to serum cholesterol levels in white males in Evans County, Georgia. *American Journal of Clinical Nutrition* 16 (1965) 238-242.

VanItallie, T. B. Dietary fiber and obesity. *American Journal of Clinical Nutrition* 31 (1978) S43-S52.

Werkö, L. Diet, lipids and heart attacks. *Acta Medica Scandinavica* 206 (1979) 435-439.

Wilkinson, P. W., Parkin, J. M., Pearlson, G., Strong, H., and Sykes, P. Energy intake and physical activity in obese children. *British Medical Journal* 1 (1977) 756.

9

THE RATIONALE FOR TREATMENT OF "MILD" HYPERTENSION

Darwin R. Labarthe, M.D., Ph.D.

1. Introduction

This review is presented as a reference document for consideration of behavioral interventions in disease prevention, specifically in the treatment and prevention of so-called "mild" hypertension. As background to the review, the matter of defining "mild" hypertension and the evidence of its importance as a public health problem are addressed briefly. Then, the rationale for treatment of "mild" hypertension is reviewed in detail, first as to its scientific basis; second, by reference to the prevailing treatment guidelines; and, third, by discussion of some points of concern raised by recent commentators on the problem of "mild" hypertension and its treatment. In conclusion, by analogy with the evidence which has established the place of pharmacologic intervention in high blood pressure control, research requirements are outlined which must presumably be met if behavioral interventions are to become similarly established.

2. Definition and Importance of "Mild" Hypertension

Because usage varies, the term mild hypertension is not uniformly understood, and attention to its definition becomes necessary here. The concern that this term may be misconstrued as denoting only minor health implications has led to suggestions that it be abandoned. The U.S. Joint National Committee on Detection, Evaluation and Treatment of Hypertension denoted the categories "mild," "moderate," and "severe" by "stratum I, stratum II, and stratum III," respectively (Joint Committee on Detection, Evaluation and Treatment of High Blood Pressure, 1980). Nonetheless, the term mild continues to be used.

Some of the questions to be resolved in adopting a common definition are what levels of blood pressure should be specified, whether to add criteria regarding presence or absence of target organ damage, and whether age-specific criteria should be provided. With respect to the blood pressure levels alone, whether diastolic values, systolic, or both are to be used, and on how many occasions of measurement, are matters subject to debate. These aspects were recently reviewed in detail elsewhere, and their implications for estimation of prevalence, risk, and treatment response have been described. Included in that review were discussions of both the basis for use of the fifth-phase diastolic endpoint (or "disappearance" of the Korotkoff sounds) as the measurement for defining high blood pressure and the need for proper attention to all aspects of the procedure of blood pressure measurement (Labarthe, 1978). For the present purpose, it is sufficient to recognize the problem of inconsistent usage and the need to consult relevant research publications and other documents directly to ensure correct interpretation of reported observations concerning "mild" hypertension.

The importance of "mild" hypertension is readily appreciated, if this range is taken as starting at 90 and ending at 104 mm Hg diastolic pressure (one conventional definition). Then, on single-occasion screening, some 15 percent of adult men and 13 percent of adult women in the United States would be found to be affected. Perhaps only 50 percent of such persons would still be found in this range of pressures upon re-examination within several days; but in the two-stage screening for the Hypertension Detection and Follow-up Program, or HDFP, the 90 to 104 mm Hg range at *second* screening included 57 percent of those initially in the 105-114 mm Hg range, and 31.4 percent of those initially above 114 mm Hg. Thus "mild" in the HDFP represented a range of *initial* screening blood pressures and risks well above those of the casual, single-occasion 90 to 104 mm Hg levels (Labarthe, 1978). Diastolic pressures in this range would characterize some 70 percent of persons with two-stage screening levels of diastolic pressure persistently above 90 mm Hg, a substantial majority of those with confirmed high blood pressure.

The preponderance of premature deaths in persons with high blood pressure is also found in the stratum starting follow-up (after two-stage screening) at pressures in the 90 to 104 mm Hg range, in contrast to the stratum at or above 105 mm Hg (Hypertension Detection and Follow-up Program Cooperative Group, 1979). These and analogous data from diverse studies of both observational and experimental designs support the view that "mild" hypertension is a very significant part of the public health problem of high blood pressure at the present time (Paul, 1980). Its prevalence is generally high but is still higher in some population groups, such as U.S. Blacks, and the risks it conveys are related to morbidity as well as mortality. These aspects of the importance of "mild" hypertension will be illustrated further as we turn to consider its treatment.

3. The Rationale for Treatment of "Mild" Hypertension

The first major stimulus to treatment of persons whose diastolic pressures are found in the 90 to 104 mm Hg range was the second report from the Veterans Administration Cooperative Study Group on Antihypertensive Agents (VA Study), concerning the effects of treatment on morbidity in hypertension, published in 1970 (Veterans Administration Cooperative Study Group on Antihypertensive Agents, 1970). This study, with its major findings reported in 1967, 1970, and 1972, was a nodal point in the history of evaluation of treatment in hypertension (Veterans Administration Cooperative Study Group on Antihypertensive Agents, 1967, 1970, 1972). It was the culmination of decades of clinical research on antihypertensive therapy, which had accelerated in the late 1950s and 1960s with the availability of increasing numbers and types of pharmacologic agents, many of them tested in the Veterans Administration research program. Soon after the 1970 report, and within a span of only 18 months of one another, a series of 6 new trials of therapy in so-called "mild" hypertension were initiated, in the United States, the United Kingdom, Australia, Belgium, and France. In general, this new generation of trials had in common a minimum diastolic blood pressure in the range 88 to 95 mm Hg as a criterion for entry (Labarthe, 1983; Reader, 1980).

The Veterans Administration group had brought the methodologic strategy of the double-blind, random-allocation, placebo-controlled trial into the field of hypertension research and had demonstrated the efficacy of treatment in two strata of the hypertensive population: first, in persons with diastolic pressures in the 115 to 129 mm Hg range (1967 report) and, subsequently, in persons with diastolic pressures in the 90 to 114 mm Hg range (1970 report) (Labarthe, in press). Because of pre-stratification in the original randomization in this study, it was possible to compare active with placebo therapy in the subgroup with qualifying pressures in the 90 to 104 mm Hg range and to show reductions in endpoint event rates of approximately 20 to 50 percent, depending upon age and the presence or absence of cardiovascular-renal abnormalities at entry (Veterans Administration, 1970). This result was consistent with those for the 105 to 114 stratum and for the 115 to 129 stratum as well, although the patients entering at 90 to 114 mm Hg required a longer period of follow-up for detection of a lesser, but important, effect than was found in the 115 to 129 mm Hg stratum.

The finding of treatment benefit in a group entering the VA trial with diastolic pressures as low as 90 mm Hg was coupled with evidence both from prospective epidemiologic studies on risks and from population surveys on the prevalence of such levels to identify elevated blood pressure as a major public health problem. The evidence that a large proportion of persons with such blood pressure levels in screening programs were unaware of their pressure, aware but untreated, or treated but not to desirable levels, added greatly both to the weight of the problem and to the perception that it could be addressed effectively. Thus, even while the need was recognized to confirm the applicability of the VA Study results to the general population--including all hypertensives detectable in entire defined communities, and not only those meeting the specific eligibility requirements for the VA trial--the National Heart and Lung Institute established the National High Blood Pressure Education Program to launch a major campaign of public and professional education about high blood pressure and its management. Concurrently, a demonstration of community applicability of the results of the VA trial was to be attempted through a new trial, the Hypertension Detection and Follow-up Program (HDFP), also organized under the auspices of the National Heart and Lung Institute. This was to be the principal test of the systematic approach to antihypertensive therapy, including not only the drugs themselves but a programmatic effort to remove impediments to long-term adherence to the prescribed regimen (Hypertension Detection and Follow-up Group, 1978b). The additional studies specifically addressing "mild" hypertension and initiated near the same date were noted above and will be described shortly.

Through the 1970s and extending into the early 1980s, two distinct lines of development have occurred which bear directly on a current understanding of "mild" hypertension. On the one hand, the National High Blood Pressure Education Program sought to develop and disseminate recommendations for the public and for health professionals in dealing with high blood pressure. The principal landmarks in that program have been the successive reports released in 1973, 1977, and 1980 containing these recommendations, as updated with current information (Joint Committee, 1980; Moser, 1977; National High Blood Pressure Education Program, 1975). Other agencies and organizations, such as the U.S. Department of Health and Human Services and the World Health Organization (Centers for Disease Control, 1983; U.S. Department of Health and Human Services, 1980; World Health Organization, 1978, 1982, in press), have also issued reports or guidelines for high blood pressure control and for the primary prevention of hypertension. On the other hand, research in progress was aimed at more clearly establishing the scientific basis for specific recommendations especially in the lower range of elevated diastolic pressures. Reports of these studies appeared in the late 1970s and subsequently (Helgeland, 1980; Hypertension Detection,

1979a, 1979b, 1980, 1982, 1982b; The Management Committee of the Australian Therapeutic Trial in Mild Hypertension, 1979, 1980, 1982; Smith, 1977).

As these two lines of development progressed, commentaries appeared in the scientific literature, in review articles, editorials, and correspondence, as well as in other publications. These commentaries are notable in revealing various interpretations of the scientific basis for recommendations concerning "mild" hypertension and some divergence of views as to the appropriateness of the recommendations themselves. Consideration of the issues raised in this further body of material, appearing especially since the late 1970s, will clarify the rationale for intervention and will follow an outline of the recommendations and the reports of the several trials.

4. Results of Trials of Treatment of "Mild" Hypertension

The first of four trials of treatment of mild hypertension to be completed after the VA Study was the United States Public Health Service Hospitals Intervention Trial in Mild Hypertension (USPHS Trial), whose results were reported in 1977 (Smith, 1977). Organized and initiated in the mid-1960s, this was a placebo-controlled, random-allocation trial of drug therapy in 389 persons 21 to 55 years of age at entry, followed for periods of 7 to 10 years. Subjects found at screening to have demonstrable target organ damage or other specified conditions were excluded. Qualifying blood pressures were in the range from 90 to 115 mm Hg diastolic, on repeated pre-randomization visits. Fatal endpoints were few in this study, two in each group, although the combination of all study endpoints (first morbid events) affected 29.0 percent of actively treated participants versus 45.9 percent of those on placebo. Power to detect differences in fatal or other major events was quite low. The authors of the published report suggested that the low event rates in this group might warrant substitution of "hygienic intervention" for drug therapy in persons similar to those studied and that it might be appropriate to defer drug therapy until electrocardiographic abnormalities or a progressive rise in blood pressure were observed.

The second trial to be completed for reporting of all follow-up experience was the Hypertension Detection and Follow-up Program (HDFP), with the main mortality findings published in 1979 (Hypertension Detection, 1979a, 1979b). The HDFP was a trial of systematic management of hypertensives in study clinics (Stepped Care) in comparison with care received from other sources available in the same fourteen community settings (Referred Care). Participants were identified from special household censuses in defined geographic areas (excepting one employed group) so as to represent unselected hypertensives in the population at large. Designed to test a reduction in total mortality of 40 percent, with five years of follow-up for each participant, the HDFP enrolled 10,940 persons aged 30 to 69 years at entry, including black and white men and women, each sex-race group constituting a substantial proportion of the study population. Diastolic blood pressures in the range from 90 to 104 mm Hg at the second screen accounted for 71.5 percent of the whole group, or 7,825 persons.

For this stratum of participants, mortality was reduced by 20.3 percent for Stepped Care versus the Referred Care group. Further treatment comparisons within this stratum indicated relative reductions of five-year mortality by 21.9 percent among participants entering at 90 to 94 mm Hg diastolic pressure, 23.1 percent for those at 95 to 99 mm Hg, and 13.8 percent for those at 100 to 104 mm Hg. Analyses in accordance with the presence or absence of end-organ damage or antihypertensive therapy at entry to the trial later confirmed substantial trends favoring the Stepped Care group as well (Hypertension Detection, 1982b). With respect to treatment status and cardiovascular problems at entry, the greatest relative benefit within the stratum from 90 to 104 mm Hg was for participants free of both end-organ damage and antihypertensive therapy at entry--28.6 percent. The authors of this later, more detailed report on the 90 to 104 mm Hg stratum emphasized that, while benefit accrued to those already demonstrating end-organ damage at entry, their treatment did not achieve the degree of benefit experienced by those in advance of developing these abnormalities. This finding was taken to support emphasis on treatment of those with "mild" hypertension.

The next trial completed, with overall results published in 1980, was the Australian Therapeutic Trial in Mild Hypertension, or Australian National Blood Pressure Study (ANBPS) (The Management Committee, 1980). This, like the VA and USPHS studies, was a placebo-controlled trial. It included 3,427 persons aged 30 to 69 years identified as volunteers in three communities, randomized after exclusion of persons with diastolic pressures outside the range from 95 to 109 mm Hg (or systolic pressure of 200 mm Hg or greater) and those with any evidence of target-organ damage. The study was terminated ahead of schedule due to the judgment that endpoint rates were sufficiently favorable to active treatment to make continuation inappropriate. Analyses were presented in accordance with both "intention to treat" and actual treatment, and in both instances indicated significant reduction of both cardiovascular deaths and all trial endpoints for the group randomized to active treatment. Rates of all endpoints combined were lower with active treatment in each substratum of participants by diastolic pressures at entry (95 to 99, 100 to 104, and 105 to 109 mm Hg). The results were taken to show that even persons with "modest" elevation of blood pressure can benefit with reduced morbidity and mortality as a result of antihypertensive drug therapy. A subsequent analysis addressed the experience of the 1,943 participants in the placebo group and the course of blood pressure changes and other events during their follow-up (The Management Committee, 1982). The relation of entering blood pressure levels to endpoint event rates and to blood pressure levels at three years of follow-up were examined in detail, and the difficulty of predicting these outcomes from any of the variables recorded at entry was emphasized. The authors suggested that, in persons whose initial diastolic pressures over two visits are 95 mm Hg or greater, but whose readings on a third occasion result in an overall average below 95 mm Hg, a limited period of weight reduction and possibly other nonpharmacological intervention should be encouraged.

Finally, the trial conducted in Oslo was reported also in 1980 (Helgeland, 1980). A total of 785 men aged 40 to 49 participated, who were free of target organ damage and had systolic pressures between 150 and 199 mm Hg and diastolic pressures below 110 mm Hg at entry. Subjects were randomly allocated to either active treatment or non-treatment, without placebo. In the study population as a whole, after five years of follow-up, the total numbers of events differed favoring the treated group, but this difference was not statistically significant; power to detect differences was low in this trial. Complications and events considered as "pressure-related" occurred almost exclusively in the control group, while the coronary events occurred somewhat more often (though not significantly so) in the actively treated group. When comparisons were made between treated and control groups with different entry diastolic blood pressure ranges, the treated subjects at entry pressures above 100 mm Hg had a reduced rate of cardiovascular events relative to untreated persons, with a reported difference close to nominal statistical significance, at the 0.05 level. The authors cautioned that conclusions as to mortality reduction should not be drawn from this trial.

These four studies were critically reviewed by the W.H.O./I.S.H. Mild Hypertension Liaison Committee in a report published early in 1982 (W.H.O./I.S.H. Mild Hypertension Liaison Committee, 1982). The main features of each study were presented, including its design, operation, principal results, and the Committee's conclusions and comments. Overall, two of the studies were judged as too small to give firm results on "hard" endpoints (the USPHS and Oslo studies). The remaining studies were both found to give evidence that treatment of persons with "mild" hypertension, even those free of other manifestations, can reduce morbidity (ANBPS) and mortality (HDFP and ANBPS). The last large-scale trial in treatment of "mild" hypertension, still in progress, is the Medical Research Council (MRC) trial in Britain. Expectations were expressed that some of the outstanding questions may be resolved as that placebo-controlled trial, involving some 18,000 participants, is completed. As scheduled, with the last participants entered in early 1982, the study will need to continue until early 1987 to achieve full five-year follow-up of each subject.

5. Recommendations for Treatment of "Mild" Hypertension

With the impetus previously described, the U.S. National High Blood Pressure Education Program undertook, on its establishment in July, 1972, to develop a plan of implementation which would include scientific definitions and standards for the diagnosis and treatment of high blood pressure. A Task Force of non-federal professionals and representatives of federal agencies reported its recommendations in September, 1973 (National High Blood Pressure Education Program, 1975). This was the first of three reports, the others appearing in 1977 and 1980, reflecting the policies to be advanced through the National High Blood Pressure Education Program (Joint Committee on Detection, 1980; Moser, 1977). Other organizations and agencies have of course addressed this question as well; for the present purpose, however, exhaustive review of these is impractical and the three reports cited above must suffice to illustrate the evolution of views on "mild" hypertension and its management.

The "Task Force I" report of 1973 identified the diastolic and systolic blood pressure levels of 95 mm Hg or greater and 160 mm Hg or greater as requiring rescreening before high blood pressure could be confirmed. Notably, at least three visits were suggested, with three blood pressure readings per visit and the second and third of these diastolic readings to be averaged across all three visits. The resulting value for diastolic pressure was to indicate the further measures needed: treatment at or above 105 mm Hg, periodic screening (annually) below 95 mm Hg, and, for persons at pressures from 95 to 104 mm Hg, observation with individualization of the judgment concerning treatment. This latter recommendation was further elaborated to suggest repeated observation combined with management not involving specific drugs, unless otherwise decided by the physician in the individual case. In an Appendix to the report, a fuller statement of this position was provided:

> ...the efficacy (in terms of morbidity and mortality) of antihypertensive drugs is not only unproven in subjects with diastolic pressures below 105 mm Hg, but there is also the certainty that therapy would be expensive and that it would involve risks in terms of unpleasant side effects and potentially adverse effects. The Task Force did feel, however, that there is an urgent need to investigate the benefits of therapy in mild hypertension, even though such investigations will be both difficult and expensive.

Further concerns expressed were the magnitude of the demand on resources by the potential addition of "20 million patients who are currently untreated or inadequately treated," and the question of justifying the shift of resources for this problem at the expense of others in the health care field (National High Blood Pressure Education Program, 1975).

The first successor to Task Force I under the auspices of the National High Blood Pressure Education Program was its Joint National Committee on Detection, Evaluation and Treatment of High Blood Pressure, whose report was published in January, 1977 (Moser, 1977). This revision of the earlier recommendations was based upon interim experience "in detection and referral techniques and in a simplified approach to therapy." For persons with diastolic pressures below 105 mm Hg, it was still judged that there were "no hard data on the benefits of therapy." Recommendations for screening and confirmation were modified to arrange confirmation of sustained high blood pressure within one month of initial screening, with "at least two" (no longer necessarily three) visits, and including all persons whose average diastolic pressures were 90 mm Hg or greater (no longer 95 mm Hg) as confirmed. Initial screening diastolic pressures of 90 to 95 mm Hg (or 140 to 160 mm Hg systolic) would require rescreening under the new recommendations, within two or three months for persons younger than age 50 and within six to nine months for older persons.

In comparison with the Task Force I report, a more explicit statement of considerations in the decision to treat in the 90 to 104 mm Hg range of diastolic pressures (no longer 95 to 104) was given. Presence of other risk factors would lend some weight toward treatment, although use of drugs to control blood pressure was to be a later decision: "For some patients, weight control and a reduced salt intake may lower blood pressure, but if this proves ineffective after three to six months, specific drug therapy may be necessary in addition to diet."

A new Joint National Committee (whose membership only partly coincided with that of its predecessor) was charged in 1978 to review, revise, and augment the 1977 report, under the aegis of the National High Blood Pressure Coordinating Committee. The information used to develop the report of this new Committee, which was published in 1980, included the observations of the HDFP (Joint Committee, 1980). Recommended screening procedures were modified to delete the delayed confirmation schedule for persons over age 50. In addition, nomenclature was revised to suggest use of the designations "stratum I" (DBP 90-104 mm Hg), "stratum II" (DBP 105-114 mm Hg) and "stratum III" (DBP 115+ mm Hg) in preference to "mild," "moderate," and "severe." Commenting specifically on this point, the Committee noted:

> The term "mild" is included in this report because of common usage. It should be emphasized that even for those with so-called mild hypertension, the cardiovascular risk is twice that for individuals with normal pressure. Moreover, in the presence of target organ damage or other independent risk factors, the overall risk is further increased.

With respect to therapy, the Committee statement on "mild" hypertension is especially noteworthy:

> The findings of the Hypertension Detection and Follow-up Program suggest that *long term reduction of blood pressure decreases overall mortality at all levels of hypertension* [emphasis in the original]. This is of special importance with regard to mild hypertension because of its high prevalence in the population. Although reduction in overall mortality has not yet been demonstrated in patients below age 50 with mild hypertension, treatment of these patients reduces the incidence of such hypertensive complications as stroke, congestive heart failure, left

143

ventricular hypertrophy, and progressive rise of blood pressure [citation of the USPHS Trial in the original]. *It is therefore reasonable to reduce blood pressure even in uncomplicated mild hypertension by pharmacologic or nonpharmacologic therapy* [emphasis added].

A further qualification for young persons with uncomplicated mild hypertension suggested special consideration of the disadvantages of long-term therapy in relation to low short-term risks. In general, however, the stated goal of therapy was to maintain the diastolic pressure below 90 mm Hg, and lower as safety and tolerance permit.

In the context of the present conference, it is important to give special attention to this Committee report in relation to non-drug therapy. While noting that the suggestive results of the studies to date required confirmation through larger clinical trials, the Committee offered the following views: (1) Weight reduction and limitation of sodium intake to 2g (5g salt) should be considered as adjuncts to therapy for all persons treated for high blood pressure; (2) in young patients free of hypertensive complications or other risk factors, dietary management is "a reasonable approach" initially with drug therapy to be decided only "after an adequate trial;" and (3) other non-pharmacologic interventions, specifically including biofeedback, psychotherapy, and relaxation, were judged to be still experimental and not to be recommended for sustained control of hypertension (Joint Committee, 1980).

To summarize this series of recommendations as they evolved during the 1970s, the most important changes relevant to Stage I or "mild" hypertension were: first, a reduction in the high blood pressure threshold from 95 to 90 mm Hg diastolic; second, a consideration of other risk factors possibly constituting justification for treatment in this range; and third, the recommendation to initiate treatment with drugs or by non-pharmacologic means in the 90 to 104 mm Hg range, with an explicitly stated goal of blood pressure reduction. Importantly, non-pharmacologic management and the need for additional evidence concerning its role in blood pressure control came to be recognized prominently in the latest Committee recommendations.

3. Commentaries on the Problem of "Mild" Hypertension

As has been shown, the data bearing on the treatment of "mild" hypertension following the landmark report of the VA Study in 1970 were published in 1977, 1979, 1980, and--in the further analyses from the HDFP and the ANBPS--1982. The treatment recommendations reviewed here, as reflected in the Joint National Committee report of 1980, were modified largely on the basis of the HDFP experience, which was exceptionally applicable to the diastolic levels below 95 mm Hg, with 2,941 participants in this substratum, and also with exceptional power for the 95 to 104 mm Hg substratum, with 4,884 participants, all observed for five-year mortality (Hypertension Detection, 1982b). In addition to these publications of specific studies and of authoritative national recommendations, not surprisingly, published commentaries on the trials and on the perceived implications of their results have already appeared in modest numbers. A preliminary search, which is surely not exhaustive, has identified more than fifteen review articles or editorials and numerous letters in the three-year period beginning in January, 1980. These commentaries serve as the basis for the discussion which follows.

Three general issues account for most of the points raised in these several citations: (1) the validity and interpretation of the HDFP and the other trials; (2) the appropriate approach to management of persons with diastolic blood pressures in the stratum I range of 90 to 104 mm Hg; and (3) the role of non-pharmacologic approaches to blood pressure control for such persons, especially as an alternative to use of antihypertensive drugs. Emphasis will be given here, under each subheading, to those specific points which are particularly deserving of consideration, either on their own merit or due to their especially frequent mention.

The HDFP and the Other Trials. The majority of comments referring to any one particular trial have addressed the design, conduct, and results of the HDFP. The most frequently recurring issue concerning the HDFP has been the nature of the intervention under test, Stepped Care management as contrasted with Referred Care. Discussion of this aspect of the trial indicates a substantial misunderstanding of what Stepped Care included and has led to a number of post hoc interpretations of the results, based largely on conjecture. Such misunderstanding has generally resulted in exaggerated attribution of the mortality reduction of Stepped Care to general medical care and, by some implications, not to antihypertensive therapy at all (Abeles and Snider, 1980; Alderman, 1980; Alderman and Madhavan, 1981; Editorial, 1980b; Editorial, 1980c; Freis, 1982; Guttmacher, et al., 1981; Henderson and Tosch, 1980; Kaplan, 1981, 1983; Peart and Miall, 1980). The 1979 reports on the HDFP did acknowledge the possibility of some enhancement of benefit from more frequent medical contact in Stepped Care, which as stated in those reports "cannot be dismissed completely" (Hypertension Detection, 1979a). However, specifically reported information on major cardiovascular risk factors--hypercholesterolemia, cigarette smoking, and overweight--indicated no difference between treatment groups at the conclusion of the trial, and only minimal ancillary advice or treatment of any kind was authorized by the study protocol. Nor has any commentator suggested what aspects of general medical care could plausibly account for the substantial relative reduction in mortality which was observed, especially in comparison with an alternate treatment approach which was very likely to have included more regular clinical contact than that received by the general population.

Importantly, more recent analyses of the HDFP data support strongly the attribution of the benefit of Stepped Care to its paramount component, antihypertensive therapy: Fatal and non-fatal strokes were from 32 percent to 45 percent less frequent among Stepped Care participants from Stratum I to Stratum III, respectively, than in Referred Care; and analyses to test the contributions of blood pressure treatment to mortality differentials over the course of the study confirm that most of the between-group difference is statistically accounted for by the blood pressure treatment (Hardy and Hawkins, 1983; Hypertension Detection, 1982a).

These latter analyses, in particular, diminish the impact of another issue of concern to several commentators, the appearance of a mortality benefit for non-cardiovascular causes, as determined from coded death certificates (Alderman, 1980; Freis, 1982; Guttmacher et al., 1981; Kaplan, 1981, 1983; Peart and Miall, 1980; Traub, 1980). This observation suggests that the analysis by underlying cause fails to take contributing cardiovascular conditions into account, which would be expected, or that survival was prolonged (though to a lesser degree) for noncardiovascular conditions through improved cardiovascular status; it does not seem a sufficient basis for rejecting the evidence of specific, blood-pressure related reduction of cardiovascular events, the predominant result of this study. Nonetheless, this issue does merit further examination.

In addition to the nature of Stepped Care, the heterogeneity of the treatment experience within the Referred Care group has posed difficulty for some commentators in interpretation of the HDFP results (Alderman, 1980; Alderman and Madhavan, 1981; Editorial, 1980c; Grell, 1980; Guttmacher et al., 1981; Madhavan and Alderman, 1981). The necessarily open, unblinded design--with participants and therapists alike fully aware of the treatment group assignment-- left largely to local circumstances the nature of the treatment which might be received through referral to existing sources of care, although this referral process was reinforced in a structured way at each anniversary follow-up examination for the Referred Care group. Referred Care, as a non-standardized category of management, could thus vary freely according to local practitioners' judgments for individual patients. Thus, strata of the HDFP study population defined by any attribute of interest--such as blood pressure at entry, sex, race, or age--might represent quite different types and intensities of management in Referred Care. Therefore, comparisons must be made independently within particular strata between the Stepped Care and Referred Care assignees. This approach alone assures the appropriate *relative* comparisons of event rates in relation to differentials in average blood pressure reduction or other measures of blood pressure control (percentage on treatment or percentage at goal) between treatment classes in the HDFP study population. As some commentators have observed, relative mortality differences would not be expected to be as large in strata with nearly equivalent blood pressure reductions between Stepped and Referred Care groups as when such reductions are widely disparate (Moser, 1981; Relman, 1980). This understanding aids the interpretation of apparent inconsistencies in the magnitudes of the *relative* benefits when particular *stratified* comparisons are made.

Less specific to the HDFP are the questions, what levels of blood pressure were actually treated in these trials, and how, therefore, do their results relate to the general recommendations for detection and management of high blood pressure (Henderson and Tosch, 1980; Levine, 1981; Moser, 1981)? First, it is quite important to keep this consideration in view, because the nominal range of diastolic readings "from 90 to 104 mm Hg" may represent markedly different levels of risk or expectations of treatment effect, as a result of different procedures and conditions for measurement and selection. Second, the inclusion of participants with some treatment history at the time of randomization in the HDFP, and exclusion of such persons from some but not all of the other trials, suggests some greater heterogeneity of "naturally-occurring" blood pressures in the several trials than are indicated by the nominal ranges cited (W.H.O./I.S.H., 1982). This issue, in the case of the HDFP, has been addressed in recent analyses by treatment status at entry and shows maximum relative benefit of stepped Care in persons free of both prior antihypertensive therapy and end-organ damage at entry to the study (Hypertension Detection, 1982b). The general impact of this issue will be discussed further in the following section.

Further question about the benefit of treatment in demographic subgroups (those defined by sex, race, or age, singly or in combination) arises from the general lack of statistically significant differences in such subgroup analyses (Abernethy, 1980; Alderman, 1980; Editorial, 1980c; Guttmacher et al., 1981; Henderson and Tosch, 1980; Kaplan, 1981; Moser, 1981). Accordingly, interpretation of study results in such subgroups and their generalization to the population at large will necessarily remain limited to trends in the data rather than on independent statistical tests in specific strata, unless or until such data become available from future studies. The practical constraints against fulfilling this latter expectation are widely recognized. Difficulties will remain, then, as to judging the benefit of initiating treatment in younger adults with diastolic pressures of 90 mm Hg or greater or perhaps in other groups of particular interest or concern. The available data do not offer uniformly satisfactory answers to questions of this kind.

Other concerns less prominently raised include the failure to demonstrate consistently in every trial a specific reduction in morbidity and mortality from coronary heart disease (Freis, 1982; Moser, 1980); the desirability of initiating other trials, placebo-controlled, in the full range of "mild" hypertension (Moser, 1981; Relman, 1980; Thomson et al., 1981); and the possibility that other trials will not be done, because too large a segment of the medical community will have been convinced of treatment benefit by the imperfect studies already completed or (allowing for the MRC trial) in progress (Henderson and Tosch, 1980).

147

The Management of "Mild" Hypertension. The dominant, though not universal, theme in the collected commentaries on management of "mild" hypertension has been a strong reluctance to treat all persons in the 90 to 104 mm Hg diastolic range, especially with drugs, to achieve sustained blood pressure control at levels below 90 mm Hg. Specific concerns which have been emphasized repeatedly, by the same or other commentators, relate to undesirable features attributed to drugs; the magnitude of the problem as inferred from projected estimates of the prevalence of high blood pressure in this range of diastolic levels; and the large number of persons to be treated relative to the number perceived as likely to benefit. Differing suggestions as to a preferred approach reflect differing points of view on both the relative importance of these concerns and the justification or acceptability of proposed modifications. Such proposals include restriction of treatment eligibility to a lower limit of diastolic pressures of 95 to 100 mm Hg (Editorial, 1980b; Kaplan, 1981; Moser, 1981); extension of pre-treatment observation for various periods, in order to identify those persons whose pressures remain below the suggested criterion level for initiating treatment (Alderman, 1980; Editorial, 1980c; Freis, 1982; Guttmacher et al., 1981; Kaplan, 1981, 1983; Moser, 1981); and elimination of drug therapy altogether from consideration in treatment at these levels (presumably up to 105 mm Hg) (Kaplan, 1981).

Problems with the use of drugs loom large in the view of several commentators, who cite excessive side effects and toxicity, including major metabolic derangements and possible sudden cardiac death as risks of drug treatment, to be weighed against the immediate risks (implied as being very small) of untreated "mild" hypertension (Abeles and Snider, 1980; Guttmacher et al., 1981; Kaplan, 1983). The extent of toxicity which is assumed as the basis for this concern deserves more explicit presentation and discussion, in order that adverse effects of therapy may be assessed with the same care as are its beneficial effects-- and, in addition, the adverse effects of withholding treatment. Alderman, for example, attributes to drug therapy "an increased incidence of sudden death" in the Oslo study; but in that trial only 19 deaths in all occurred; and, of those, 10 and 9 were in the treatment and placebo groups, respectively, with on by 6 and 2 sudden deaths in these groups (Alderman, 1980). Moreover, the authors of that study report observed, "No serious drug-induced disease occurred during the observation period" (Helgeland, 1980). This latter interpretation accords with that of the HDFP investigators, who reported as follows: "In the judgment of the [Toxicity and Endpoint Evaluation] Committee, no deaths were attributable to drug toxicity in the SC [Stepped Care] group. Few hospitalizations occurred because of suspected drug toxicity. Moreover, toxicity problems with the SC regimen rarely prevented achievement of BP control, since alternative medications could be utilized" (Hypertension Detection, 1979a). In this connection, Freis has stated, "It is interesting to speculate on the frequency and severity of the side effects that would result if this advice [to lower diastolic pressure to below 85 mm

Hg] were implemented" (Freis, 1982). Since this was the protocol requirement for the great majority of Stratum I participants in the HDFP, the comments from the HDFP report cited above reduce significantly the need to speculate on this important question.

These brief references are not presented as a final disposition of the matter of drug toxicity and its relevance to antihypertensive therapy. The possibility of adverse effects of interventions of any kind must always be a conscious concern, even when the interventions in question involve drugs which have been in use for many years by millions of persons. The drug trials do not offer sufficient evidence, in themselves, of very long-term or latent effects, owing to their necessary limitations of population size and duration. The unexpected or unanticipated suggestion of an adverse effect must be regarded seriously and evaluated with care. Such, for example, is the very recent suggestion of excessive mortality in Special Intervention participants in the Multiple Risk Factor Intervention Trial (MRFIT), who were hypertensive and had some types of electrocardiographic abnormalities at entry. In this instance, the possibility of an adverse drug effect arose in extensive post hoc subgroup analyses, could not be evaluated statistically as to the likelihood of its being a chance occurrence, and is a matter for further investigation (Lundberg, 1982). Notably, preliminary reports of special analyses of the HDFP experience lend no support to such a risk in approximately similar patients observed in that trial. Investigation of this finding in the MRFIT is understood to be continuing; an alternative explanation may lie in an apparent deficit of deaths from the numbers to be expected in the Usual Care participants in this particular subgroup (Multiple Risk Factor Intervention Trial Research Group, 1982).

The magnitude of the "mild" hypertension problem has been cited repeatedly in discussions of the question of treatment, either to support the withholding or deferral of therapy or to emphasize the potential public health impact on morbidity and mortality if the prevailing treatment recommendations are followed (Editorial, 1980b; Editorial, 1980c; Freis, 1982; Guttmacher et al., 1981; Kaplan, 1981, 1983). Several concerns appear to hinge on this issue of magnitude, such as the societal implications as to costs and allocation of health care resources; the perception of an implied need for an extraordinary effort (presumably on the part of providers) to achieve blood pressure control for a large population at risk; and inconvenience and anxiety (on the part of those with high blood pressure), the latter especially being intensified by the often-cited presumption of a need for lifelong treatment from the time at which it is initiated. (Studies of the course of blood pressure change after interruption of treatment are now in progress.) Although the problems of scale are undeniably serious, this seriousness bears on the underlying condition, its prevalence and its risks, and not only on the question of treatment: The public health problem of "mild" hypertension is indeed a problem of massive scale in most industrialized countries. This being the case, it is especially important to measure the magnitude of the problem with some care. Only in this way can an accurate assessment of the societal burdens attendant upon the condition and any policies for its management be attained.

Unfortunately, it has become common practice to use a measure which is only indirectly related to the one intended and which exaggerates the magnitude of the treatment problem by inflating the value corresponding to this intended measure. This simple oversight results from use of prevalence data for given levels of blood pressure, recorded in single-occasion examinations, when the intention is to reflect the numbers of persons with elevated blood pressure maintained over multiple examinations up to the initiation of treatment. The former measure is the correct one for estimating the prevalence of persons at excessive risk in relation to single-screening blood pressure levels, since this measure is exactly analogous to the methods of the prospective epidemiologic studies on which the available risk estimates are based. In contrast, to estimate the prevalence of persons currently considered to have sustained high blood pressure warranting treatment, that is, at 90 mm Hg diastolic or greater over several examinations, requires that the considerable downward shifts due to intra-individual blood pressure variation and truncated selection for rescreening (or "regression to the mean") be taken into account. Even by taking only first and second occasions of screening into account, on the basis of the two-stage screening experience of the HDFP, a marked decrement in estimated prevalence of "mild" hypertension is readily demonstrated (Hypertension Detection and Follow-up Cooperative Group, 1978a; Labarthe, 1978):

(1) At initial screening of the HDFP target population:

> *20.5%*   of the population were
>   in the 90-104 mm Hg range.

(2) At second-stage screening of persons initially at 95+ mm Hg (note: those 90-94 mm Hg were not rescreened):

> *7.2%*   of the population were now
>   in the 90-104 mm Hg range.

(3) The "mild" hypertensives after two-stage screening included members of the 3 initial screening strata in the following proportions:

Initial Screening		Secondary Screening
95-104 mm Hg	→	67.3% of the HDFP "mild"
105-114 mm Hg	→	26.1% " " " "
115+ mm Hg	→	6.6% " " " "

Specific features of the HDFP screening experience and the particular composition of the participating population preclude unqualified application of these results to other situations or to the U.S. population as a whole: In the HDFP, inclusion of the initial 90 to 94 mm Hg group in rescreening would have increased the final prevalence estimate, but taking the later pre-treatment examinations into account would have decreased it. However, the general conclusion appears sound that multiple screening, as practiced in the evaluation of persons being considered for antihypertensive therapy (and as recommended by Task Force 1 and both Joint National Committees), ensures a major reduction between the population initially selected for rescreening in the 90 to 104 mm Hg range and that eventually considered eligible for treatment and found in this same range. Thus, although the magnitude of the problem from the point of view of treatment eligibility is still large, it is seriously exaggerated by some commonly cited figures. More balanced consideration of all issues should result from their avoidance.

The preceding discussion bears on two other closely related matters. One, the effect of repeated observation upon the identification of persons with sustained "treatable" blood pressure levels, has already been noted. This effect is strongly suggestive that the existing recommendations already provide for substantial "fall-out" of persons whose diastolic pressures do not remain above 90 mm Hg and therefore do not continue under consideration for treatment (Joint Committee, 1980). Further evidence in support of this view is the reports of serial blood pressure readings through screening and early follow-up as for the ANBPS control group, with or without placebo tablets, whose major mean decrement in blood pressure occurred between first screening and the fourth examination at four months' follow-up (The Management Committee, 1982). Whether the number of occasions or the lapse of time was the more relevant factor in the fall in blood pressure is unclear from these data; a matter for straightforward investigation might be to test this question with alternative serial screening strategies in defined populations to aid the development of future recommendations. In any case, the suggestion that long periods (many months, or years) are needed to distinguish between sustained and relatively short-term elevations of blood pressure, raised by some commentators, appears to be incorrect (Alderman, 1980; Guttmacher et al., 1981; Kaplan, 1981, 1983; Moser, 1981).

The further suggestion that initiation of treatment should be deferred until after a definite *increase* in blood pressure is observed, beyond some stipulated threshold at the upper limit of "mild" hypertension, is a more extreme proposal which may by now have been abandoned following publication of the HDFP data on Stratum 1 (Freis, 1982; Hypertension Detection, 1982b). There, the relative benefit of Stepped over Referred Care was shown to be greater for persons without either target-organ damage or current antihypertensive therapy at entry than for those already demonstrating target-organ damage before treatment was initiated.

These latter proposals arise in part from the difficulty addressed on several occasions by Alderman, that of the prognostic "heterogeneity" of the "mild" hypertensive stratum (Alderman, 1980; Alderman and Madhavan, 1981; Madhavan and Alderman, 1981). He has stressed that some persons on no treatment will nonetheless experience a fall in blood pressure, while others may remain hypertensive with or without complications, and still others, even though on treatment, will experience the complications whose prevention was the purpose of the treatment. From these results of follow-up, Alderman has identified distinctions which would be desirable to make predictively, so that the treatment could be offered only to those who would surely benefit and not offered to those who surely would not. However, the limited prognostic value of available baseline data, within the "mild" hypertension group, has been noted, except that risks are increased with the presence of several well-recognized risk factors (Alderman, 1980; Alderman and Madhavan, 1981; Guttmacher et al., 1981; Madhavan and Alderman, 1981; Reader, 1980). (These have been incorporated in the Joint National Committee reports as added indicators both of risk and of the need for treatment in the 90 to 104 mm Hg stratum (Joint Committee, 1980; Moser, 1977.) A point to be recalled is that risks in this stratum relative to those at diastolic pressures *below* 90 mm Hg are clearly distinguishable; this circumstance diminishes the impact of the argument that risks must be differentiated *within* the group of "mild" hypertensives in order to justify treatment. Of course if, in future, evidence is presented which provides a basis for reliable assignment of individual subjects to different management programs, this could be advantageous; whether prolonged withholding of all known approaches to blood pressure control could be defended for any subgroup appears seriously doubtful on the present evidence, however.

An extension of this discussion leads to the broader question of benefit and, as posed by some, the following kind of argument (Alderman, 1980; Alderman and Madhavan, 1981; Freis, 1982; Madhavan and Alderman, 1981): First, the absolute rates of events in untreated "mild" hypertension are very low. ("The majority of untreated patients have survived the trials.") Second, not all who are on treatment are spared these events. ("Some patients, though treated, do not benefit.") Third, benefit accrues exclusively to those patients who would have experienced an event but who, having been on treatment, avoided it. Fourth, treatment is therefore unjustified for the vast majority of those with "mild" hypertension. Fifth, and finally, it is presently unknown--as discussed--how to identify, in advance, the few true potential beneficiaries of treatment; therefore, treatment should not be recommended at all.

This argument is difficult to sustain on many grounds, including the following: (1) Rates of undesirable effects are underestimated by references to "hard" endpoints alone, and broader classifications of unwanted hypertensive complications suggest severalfold higher rates in some trials, such as the USPHS

Trial (Smith, 1977). (2) Few if any medical or public health interventions are universally effective; inadequate or delayed treatment may account for some failures of antihypertensive therapy, while inattention to additional risk factors may add some marginal explanation. However, a criterion of 100 percent effectiveness will not likely be achieved under the best imaginable circumstances, since blood pressure levels are not the sole determinant of any of the events or conditions of concern. (3) Whether reduction of risk constitutes a "benefit" is perhaps an issue judged differently according to the philosophical approach taken. From one point of view, it is in the interest of the individual to reduce his or her chances of adversity, even though absolute protection may not be conferred; it is erroneous to think of risk as characterizing treatment alone, and if treatment risks are less than those of remaining untreated, the individual choice may wisely be for the lesser risk, even in the knowledge that treatment does not provide "zero-risk". Treatment may therefore be seen as fully justified by improvement in the balance of the risks. Thus, the true beneficiaries of treatment represent a much broader spectrum of the "mild" hypertensives than some would imply, and the categorical dismissal of treatment as benefiting only an unknown few becomes most difficult to justify. It should be clear, in any case, that considerations of costs and benefits of alternative treatment policies would be influenced greatly by the assumptions made in the foregoing respects.

The last general issue concerning approaches to treatment has been anticipated in much of the preceding discussion: An undercurrent in many commentaries, which is strongly expressed in some, is the unwelcome prospect of the use of drugs to deal with the massive problem of "mild" hypertension (Editorial, 1980a). Even if the magnitude is much reduced in a proper assessment, as demonstrated above, it remains very large. The problems raised by several commentators in connection with widespread use of antihypertensive drugs have been noted, and many of the other difficulties seem related to these. Suggestions have been raised frequently to place much greater, or even total, reliance on "nondrug" or "non-pharmacologic" approaches to the control of "mild" hypertension. The extent to which the available non-pharmacologic approaches have been shown efficacious and in other ways practical as a major strategy of blood pressure control has been addressed less thoroughly. However, the impression develops, in review of the entire body of material cited, that if the role of drugs were superseded by non-pharmacologic approaches, many of the arguments against treatment would be withdrawn. This leads to the next consideration, that of the alternatives to drug therapy.

Alternatives to drug therapy. The status of "nondrug" therapy, as described in the Joint National Committee Report of 1980 (above), may change as new evidence is obtained (Joint Committee, 1980). Perceptions have differed widely as to the strength of the evidence that sustained effective blood pressure control can be achieved through the non-pharmacologic methods being advocated (weight reduction, dietary sodium restriction, biofeedback, and others). Little attention has apparently been given to such issues as cost, resource allocation, logistical aspects, and patient acceptance for these approaches, relative either to drugs or to one another. The fullest discussion of alternatives to drug therapy is perhaps that of Guttmacher and others, who take the further step to characterize the problem of high blood pressure as essentially one of social causation, therefore most amenable to social interventions to reduce stress and burdens of the social existence of many persons. In their view, the medical approach is inappropriate and in effect diverts the attention of society from the optimum remedy and indeed the root of the blood pressure problem. This viewpoint leads to the only reference in all the literature cited above to the ultimate solution to high blood pressure: primary prevention (Guttmacher et al., 1981).

It is not possible within this review to assess the current status of non-pharmacologic approaches to high blood pressure control, beyond the general comments above. Several additional comments about such approaches do seem appropriate, however. First, the spectrum of these approaches may be thought of quite broadly, to include assistance in adherence to prescribed drug regimens, modification of dietary habits, and psychophysiological interventions such as biofeedback, relaxation techniques, and others (Shapiro et al., 1978). Second, to the extent that any or all of these approaches may enhance sustained, effective blood pressure control at goal levels, even with continued use of drugs, there is an evident climate of receptivity to their use; this is reflected in the conclusions reported by some of the "mild" hypertension trials, in the recommendations at each juncture from the National High Blood Pressure Education Program or Coordinating Committee, and in many of the commentaries reviewed. Third, if such approaches would provide an actual alternative to reliance on drugs while affording maintenance of adequate blood pressure control, their use would predictably be even more enthusiastically welcomed. Fourth, demonstration of efficacy of these non-pharmacologic approaches in the primary prevention of high blood pressure might well provide an important means by which this objective could be pursued on a meaningful scale. Thus at several levels, non-pharmacologic or behavioral strategies may have very significant potential for application in the field of high blood pressure control and prevention.

6.   Research Requirements for Non-pharmacologic or Behavior Approaches

There is some risk of presumption in setting forth requirements for the conduct of research, especially when a discipline or field other than one's own is centrally involved. On the other hand, if the epidemiologist with special interest in studies of blood pressure senses the need for special emphasis on certain aspects of research even within his own discipline, it may be hoped that the good intention of enhancing the quality of his own as well as others' research is understood. In this spirit, several brief comments follow:

(1) The need for clarity in definition and selection of study populations in research on treatment of "mild" hypertension will be evident from the preceding sections of this review. Explicit criteria for blood pressure levels, including measurement techniques, examination sequences, and so forth, are essential, as are indications of any special selective factors such as age, race, sex, target-organ damage, or others.

(2) Characterization of the special type of intervention under test, with details as to its content, administration or delivery, and aspects of standardization, monitoring, and quality control of the intervention itself, is also essential.

(3) Evaluation of the impact of the intervention will presumably be based on blood pressure control as the endpoint or outcome measure; the "hard" endpoint approach should not ordinarily be at issue if blood pressure control is demonstrable. But what constitutes sufficient evidence of blood pressure control must be considered with great care, lest substantial effort be invested in a project perceived as irrelevant to long-term diastolic reductions well below 90 mm Hg, for example. How this outcome is assessed is a further issue in design, including blood pressure measurement techniques, observer blinding, and so forth. Other possible effects of intervention, such as the suggested problems of "labeling" and the "sick role" response are appropriately evaluated in these studies as with drugs.

(4) Interpretation of the results must of necessity address possible confounding factors, especially if drug therapy is in use as part of the experimental setting.

(5) Assuming the successful experimental demonstration of efficacy of alternatives to drugs in reducing blood pressure levels--or, in the preventive mode, in slowing or arresting the expected progression to "treatable" levels over time--the issues of cost, logistical considerations, possible adverse effects, and mass applicability will remain to be addressed.

A research program with these issues clearly in sight may have very great promise of contributing to the ultimate solution of the problem of high blood pressure, especially "mild" hypertension.

References

Abeles, J. H. and Snider, A. H. Drug industry. Mild hypertension therapy. Indust. Follow-up, January 10. 1980. Kidder, Peabody and Co., Inc.

Abernethy, J. D. The pressure to treat (letter). *Lancet* 2 (1980) 364-365.

Alderman, M. H. Mild hypertension: New light on an old clinical controversy. *American Journal of Medicine* 69 (1980) 653-655.

Alderman, M. H. and Madhavan, S. Management of the hypertensive patient: A continuing dilemma. *Hypertension* 3 (1981) 192-197.

Centers for Disease Control. Perspectives in disease prevention and health promotion. Implementing the 1990 prevention objectives: Summary of CDC's seminar. *M.M.W.R.* 32 (1983) 21-24.

Editorial. Lowering blood pressure without drugs. *Lancet* 2 (1980a) 459-461.

Editorial. Millions of mild hypertensives. *British Medical Journal* 281 (1980b) 1024-1025.

Editorial. The pressure to treat. *Lancet* 1 (1980c) 1283-1284.

Freis, E. D. Should mild hypertension be treated? *New England Journal of Medicine* 307 (1982) 306-309.

Grell, G. A. Race and hypertensive complications (letter). *Lancet* 2 (1980) 744-745.

Guttmacher, S., Teitelman, M., Chapin, G., et al. Ethics and preventive medicine: The case of borderline hypertension. *Hastings Center Report* February, 1981. Pp. 12-20.

Hardy, R. J. and Hawkins, C. M. The impact of selected indices and antihypertensive treatment on all-cause mortality. *American Journal of Epidemiology* 117 (1983) 566-574.

Helgeland, A. Treatment of mild hypertension: A five-year controlled drug trial. *American Journal of Medicine* 69 (1980) 725-732.

Henderson, W. G. and Tosch, T. J. Hypertension detection and follow-up (letter). *Journal of the American Medical Association* 244 (1980) 1317-1318A.

Hypertension Detection and Follow-up Cooperative Group. Mild hypertensives in the Hypertension Detection and Follow-up Program. *Annals of the New York Academy of Science* 304 (1978a) 254-266.

Hypertension Detection and Follow-up Cooperative Group. Patient participation in a hypertension control program. *Journal of the American Medical Association* 239 (1978b) 1507-1514.

Hypertension Detection and Follow-up Program Cooperative Group. Five-year findings of the hypertension Detection and Follow-up Program. I. Reduction in mortality of persons with high blood pressure, including mild hypertension. *Journal of the American Medical Association* 242 (1979) 2562-2571.

Hypertension Detection and Follow-up Program Cooperative Group. Five-year findings of the Hypertension Detection and Follow-up Program. II. Mortality by sex-race and age. *Journal of the American Medical Association* (1979) 2572-2577.

Hypertension Detection and Follow-up Program Cooperative Group. Hypertension Detection and Follow-up (letter). *Journal of the American Medical Association* 244 (1980) 1318.

Hypertension Detection and Follow-up Program Cooperative Group. Five-year findings of the Hypertension Detection and Follow-up Program. III. Reduction in stroke incidence among persons with high blood pressure. *Journal of the American Medical Association* 247 (1982) 633-638.

Hypertension Detection and Follow-up Program Cooperative Group. The effect of treatment on mortality in "mild" hypertension. Results of the Hypertension Detection and Follow-up Program. *New England Journal of Medicine* 307 (1982) 976-980.

Joint Committee on Detection, Evaluation and Treatment of High Blood Pressure. The 1980 Report of the Joint National Committee on Detection, Evaluation and Treatment of High Blood Pressure. U.S. Department of Health and Human Services. Public Health Service, National Institutes of Health. NIH Publication No. 81-1088, December 1980.

Kaplan, N. M. Therapy for mild hypertension. Toward a more balanced view. *Journal of the American Medical Association* 249 (1983) 365-367.

Kaplan, N. M. Whom to treat: The dilemma of mild hypertension. *American Heart Journal* 101 (1981) 867-870.

Labarthe, D. R. Evaluation of the treatment of hypertension. *Israeli Journal of Medical Science* 19 (1983) 471-478.

Labarthe, D. R. Problems in definition of mild hypertension. *Annals of the New York Academy of Science* 304 (1978) 3-14.

Levine, S. R. "True" diastolic blood pressure (letter). *New England Journal of Medicine* 304 (1981) 362-363.

Lundberg, G. D. MRFIT and the goals of the Journal. *Journal of the American Medical Association* 248 (1982) 1501.

Madhavan, S. and Alderman, M. H. The potential effect of blood pressure reduction on cardiovascular disease. A cautionary note. *Archives of Internal Medicine* 141 (1981) 1583-1586.

The Management Committee. Initial results of the Australian Therapeutic Trial in mild hypertension. *Clinical Science.* 57 (1979) 449s-452s.

The Management Committee. The Australian Therapeutic Trial in mild hypertension. *Lancet* 1 (1980) 1261-1267.

The Management Committee of the Australian Therapeutic Trial in mild hypertension. Untreated mild hypertension. *Lancet* 1 (1982) 185-191.

Moser, M. Report of the Joint National Committee on Detection, Evaluation and Treatment of High Blood Pressure. A Cooperative Study. *Journal of the American Medical Association* 237 (1977) 255-261.

Moser, M. Antihypertensive drugs and coronary heart disease (letter). *Lancet* 2 (1980) 745.

Multiple Risk Factor Intervention Trial Research Group. Multiple Risk Factor Intervention Trial. Risk factor changes and mortality results. *Journal of the American Medical Association* 248 (1982) 1465-1477.

National High Blood Pressure Education Program. Report to the Hypertension Information and Education Advisory Committee. Task Force I. Data Base. September 1, 1973. U.S. Department of Health, Education and Welfare, Public Health Service, National Institutes of Health. DHEW Publication No. (NIH) 75-593, 1975.

Paul, O. The risks of mild hypertension. *Pharmacology Therapy* 9 (1980) 219-226.

Peart, W. S. and Miall, W. E. M.R.C. Mild Hypertension Trial (letter). *Lancet* 1 (1980) 104-105.

Reader, R. Australian hypertension trial (letter). *Lancet* 2 (1980) 744.

Reader, R. Therapeutic trials in mild hypertension ongoing throughout the world. *Annals of the New York Academy Science* 304 (1978) 309-317.

Relman, A. S. Mild hypertension. No more benign neglect. *New England Journal of Medicine* 302 (1980) 293-294.

Shapiro, A. P., Schwartz, G. E., Redmond, D. P., et al. Non-pharmacologic treatment of hypertension. *Annals of the New York Academy Science* 304 (1978) 222-235.

Smith, W. McF. Treatment of mild hypertension. Results of a ten-year intervention trial. *Circulation Research* 40 Suppl. I (1977) I-98-I-105.

Taylor, R. B. Hypertension detection and follow-up (letter). *Journal of the American Medical Association* 244 (1980) 1317.

Thomson, G. E., Alderman, M. H., Wassertheil-Smoller, S., et al. High blood pressure diagnosis and treatment: Consensus recommendations vs. actual practice. *American Journal of Public Health* 71 (1981) 413-416.

Traub, Y. M. Hypertension detection and follow-up (letter). *Journal of the American Medical Association* 244 (1980) 1317.

U.S. Department of Health and Human Services. Promoting Health/Preventing Disease. Objectives for the Nation. U.S. Department of Health and Human Services, Public Health Service. U.S. Government Printing Office. Washington, 1980.

Veterans Administration Cooperative Study Group on Antihypertensive Agents. Effects of treatment on morbidity in hypertension. I. Results in patients with diastolic blood pressures averaging 115 through 129 mm Hg. *Journal of the American Medical Association* 202 (1967) 116-122.

Veterans Administration Cooperative Study Group on Antihypertensive Agents. Effects of treatment on morbidity in hypertension. II. Results in patients with diastolic blood pressures averaging 90 through 114 mm Hg. *Journal of the American Medical Association* 213 (1970) 1143-1152.

Veterans Administration Cooperative Study Group on Antihypertensive Agents. Effects of treatment on morbidity in hypertension. III. Influence of age, diastolic pressure and prior cardiovasacular disease; further analysis of side effects. *Circulation* 45 (1972) 991-1004.

W.H.O./I.S.H. Mild hypertension Liaison Committee. Trials of the treatment of mild hypertension. An interim analysis. *Lancet* 1 (1982) 149-156.

World Health Organization. Prevention of Coronary Heart Disease. Report of a WHO Expert Committee. Technical Report Series No. 678. World Health Organization, Geneva, 1982.

World Health Organization. Report of the Scientific Group on Primary Prevention of Hypertension. Technical Report Series NO. 686. World Health Organization, Geneva, 1983.

World Health Organization, Geneva, Arterial Hypertension. Report of a WHO Expert Committee. Technical Report Series No. 628, World Health Organization, Geneva, 1978.

10

DIETARY FACTORS AND CANCER

Andrew P. Haines, MB, MRCP, MRCGP

A wide range of epidemiological approaches have been used to study the relationship between diet and cancer. Each of them has some pitfalls in design or interpretation, and it is therefore necessary to include evidence from several different types of study when considering whether an apparent relationship between diet and cancer may be causal.

Interpretation of epidemiological studies of diet and cancer is complicated by the inadequacies of measurement of dietary intake in individuals and by the long latent period between the induction of cancer and its clinical appearance. It is possible, however, that some nutritional factors (e.g., retinoids) may act quite late in the multistage process of carcinogenesis (Sporn and Newton, 1979).

1. Problems of Dietary Measurement in Individuals

One of the major limitations in studies of diet and cancer is the difficulty of measuring dietary intake accurately in large groups of free-living individuals. Many studies have used relatively crude measurements such as recall of food consumption over a recent specified period, usually between 1 and 7 days. This has sometimes been combined with an estimate of the usual frequency of consumption of various food items. Despite the limitations, it is possible to obtain some reproducible data with careful technique. For instance, one study which attempted to determine the reproducibility of dietary interviews showed that for frequency of consumption there was an average of about 90 percent agreement between first and second interviews six months apart for foods eaten regularly but that for food items consumed less often the agreement was less, in some cases as low as 50 percent (Nomura et al., 1976). Several types of dietary record have been used in nutritional and epidemiological studies. Records may be obtained using common household measures as a means of quantifying intake. A menu record may be kept which lists all the food items of interest which have been consumed over a defined period. A pocket-sized, dietary diary may facilitate recording and permit the inclusion of food eaten away from home. Large seasonal variations in diet may necessitate more than one estimate of dietary intake, say in summer and winter.

The commonly used Burke diet history method uses three measurements, the overall eating pattern, the cross check, and the three-day menu (Burke, 1974). The overall eating pattern, consists of a 24-hour recall of actual food consumption and the usual frequency of consumption estimated in common household units (e.g. teaspoons, tablespoons and cups). The cross check is a food frequency questionnaire and the three-day menu is a record of food items consumed over a three-day period.

Weighed intake of food over a period of one week or more gives a more accurate measure of dietary intake, but is time consuming and may result in a transient change of dietary pattern during the measurement period. Because of the high degree of sustained cooperation required, the response rates for weighed intake studies may be relatively low in some groups (Marr, 1971). Analysis of replicate meals may also be used, but is very time consuming and may also result in dietary change. These methods. therefore, have been used infrequently in large-scale studies.

The method of dietary measurement used will depend upon a variety of factors including resources, the sophistication of the population under investigation, and the hypothesis to be tested (for discussion see Bazarre and Myers, 1981). For instance, testing a hypothesis about one particular nutrient which is derived from a narrow range of food items might be considerably easier and require less complex methodology than testing a hypothesis involving the interaction of two nutrients both widely distributed amongst food items. Weighed intakes are generally to be preferred (Marr, 1971). Recent work in a random population sample in a rural community in the United Kingdom suggests that with skilled staff this method may be acceptable to the majority, at least in some communities (Bingham, McNeil, and Cummings, 1981).

Food composition tables, available since 1896, are necessary for the calculation of nutrient intakes from dietary intake studies. However, food composition tables may be inadequate in some areas. For instance, the nutrient content of various foods may depend on a variety of factors, such as light and soil conditions, season, processing and handling, and the like. A weighted estimate based on the differential consumption of the varieties of a food item by the whole population may be calculated. The weighted estimate method may not, however, be ideal for calculating the nutrient intakes of individuals or groups taking an atypical or varying diet. For some nutrients (e.g., some trace elements and dietary fiber) data are still incomplete in many countries.

2. International Studies

Much of the early interest in diet and cancer was stimulated by the observation that cancer incidence varied widely throughout the world and that there were also marked international differences in diet. International or interregional correlations of diet and cancer incidence may be useful in generating hypotheses, but can only provide an indication that a particular dietary factor is worthy of further investigation. The data available from large populations is usually expressed as per capita consumption of various dietary constituents, but frequently does not reflect the variation within the population by age, sex, social class, and the like or wastage during distribution and preparation. There are many differences in life style and environment between countries and regions, some of which may be associated with dietary differences. Variation in cancer incidence which apparently correlates with diet may, therefore, be due to other factors.

3. Case-Control Studies

There are several difficulties associated with case-control studies of diet and cancer. One is that the presence of a tumor may alter dietary patterns and another that recall may be biased in those with cancer depending, perhaps, on their own perceptions about cancer causation. Recall of dietary patterns ten or twenty years before is likely to be inaccurate. Information from spouses has been used in an attempt to provide a check on the quality of dietary data or even as a surrogate for direct assessment. There is some evidence that for frequency of intake of a majority of food items there is good agreement between husband and wife, the exception being for those items eaten away from home (Kolonel and Lee, 1981). Several methods of selecting controls have been used. Hospital patients with non-malignant disease are often used, although their dietary patterns may also be affected unpredictably by their disease. Neighborhood controls are more time consuming to select and interview than hospital controls and because they are likely to be mainly healthy people any dietary differences between them and the index cases may be due to differences in their general state of health rather than specifically related to cancer.

163

4. Migrant Studies

Migrant studies have provided strong evidence that cancer incidence is strongly determined by environmental factors. It was observed that migrants from Japan to the United States showed an increased incidence of cancer of the colon, breast, prostate, and probably endometrium and ovary. There was also a lower stomach cancer incidence compared with Japanese rates. This observation stimulated interest in the role of diet (Dunn, 1975). Although dietary changes toward the pattern in the host country have been well documented in Japanese migrants to the United States (Hankin, Nomura, and Rhoads, 1975), it is evident that migrants undergo many other changes after migration, some of which may influence cancer rates.

5. Special Groups

Recently, a study of enclosed religious orders who eat very little or no meat showed that there was no difference in breast or colerectal cancer mortality between this group and the general population (Kinlen, 1982). However, more than 30 percent had entered the religious orders at age 30 or older. In addition, there was an unexpected excess of oesophageal cancer. Such groups may differ from the population in many ways besides diet, and results from small selected groups may not be generally applicable.

Studies of Seventh Day Adventists in the USA, many of whom are vegetarian, have shown that they tend to have a lower incidence of many tumors than does the general population (Phillips, 1975). But the members of this group also tend not to smoke and to drink little alcohol and possibly differ from omnivores in other aspects of life style.

6. Prospective Studies

A prospective study of cancer incidence in 40,000 men has been undertaken in Norway and the United States (Bjelke, 1974). Because of the very large numbers required for such studies the dietary assessment was based on a questionnaire which is bound to give limited information. Since participants in prospective studies are by and large healthy members of the population, any differences in diet between those who later develop cancer and those who do not are unlikely to be due to the presence of the tumor, providing that those who develop clinical evidence of cancer shortly after entering the study are considered separately from those who develop clinical signs of cancer later.

7. Cancer of Specific Sites

7.1. Gastrointestinal cancer

Nutritional factors seem likely to be particularly important in the causation of gastrointestinal cancers. Cancer of the oesophagus has been linked with alcohol consumption in some Western countries (Tuyns, 1979) and smoking has a marked synergistic effect. A similar relationship has been found in cases of cancer of the oral cavity and larynx (Wynder, Bross, and Feldman, 1957; Wynder et al., 1976). The type of alcohol consumed does not seem to be important except that in parts of France consumption of apple cider and its distillates appears to confer a particularly high risk of oesophageal cancer (Tuyns, 1979). In some areas of extreme risk for oesophageal cancer (parts of Iran and China), alcohol does not appear to play a part in causation. In Iran, no one dietary cause has been found. The consumption of large amounts of hot tea, vitamin and nutritional deficiences (Mahboubi and Aramesh 1980), and a fibrous silica contaminant of food (O'Neill et al., 1980) have all been suggested as possible candidates. In China, possible causal factors include high levels of nitrates and nitrites in food and water leading to nitrosamine formation and contributing influences such as molybdenum deficiency (*Lancet*, 1975). Iron deficiency and possibly deficiency of riboflavin may have played an important part in the causation of Plummer-Vinson syndrome and the associated increased risk of hypopharyngeal cancer which used to be common amongst the women of Northern Sweden (Larsson, Sandstrom, and Westling, 1975).

Stomach cancer incidence is falling in many countries. The association of gastric cancer with pernicious anaemia and partial gastrectomy has led to the suggestion that hypochlorhydria is associated with the intragastric formation of carcinogens, in particular N nitrosamines. N nitrosamines appear to be formed as a result of the overgrowth of bacteria able to reduce nitrate to nitrite due to the rise in pH (Ruddell et al., 1978; Schlag et al., 1980). Nitrite may also be formed from nitrate in saliva by the action of bacteria in the oral cavity (Tannenbaum, Weisman, and Fett, 1976). Green vegetables appear to have a protective affect on the development of stomach cancer as judged from case-control studies (Haenszel et al., 1972; Bjelke, 1973; Graham, Schotz, and Martino, 1972; Stocks, 1957). It is not clear how vegetables exert an apparently protective effect, but it has been shown that vitamin C can inhibit the formation of mutagens from a mixture of nitrite and pickled fish (Weisburger, et al., 1980).

It has been suggested that gastric cancer incidence is related to salt consumption (Joossens and Geboers, 1981). In support of this hypothesis is the observation that there is a close relationship between stroke mortality and gastric cancer mortality both internationally and within individual countries over time. Stroke mortality appeared to decrease before effective antihypertensive treatment was available (Miller and Kuller, 1973). Although there is still controversy about the relationship between hypertension and salt intake within populations, there appears to be a relationship when populations of differing salt intake are compared (Freis, 1976). High salt intake could act by causing atrophic gastritis which in turn favors the synthesis of N nitroso compounds. The increasing use of refrigeration has led to a reduction in the use of salt for the preservation of foods.

International correlations between fat consumption and cancer of the large bowel (CLB) incidence or mortality have generally shown a relationship (Armstrong and Doll, 1975; Knox, 1977; Drasar and Irving, 1973). Case-control studies, however, have not shown such a consistent relationship. Three such studies (Jain et al., 1980; Phillips, 1975; Martinez et al., 1979) have shown that patients with CLB had higher fat consumption that did controls. Another study, in Hawaiian Japanese, showed a greater frequency of consumption of meat, particularly beef, (meat provides a large contribution to dietary fat intake) in cases than controls (Haenszel et al., 1973). However, several others (Higginson, 1966; Modan et al., 1975; Bjelke, 1974; Graham et al., 1978; Haenszel et al., 1980) have shown no significant differences between cases and controls. One small case-control study found that a high fat, low fiber diet was associated with an increased relative risk (2.7) of CLB compared with a high fiber, low fat diet (Dales et al., 1979).

A comparison of the diets in men from a high risk area for CLB (Copenhagen, Denmark) and a low risk area (Kuopio, Finland) showed that the consumption of fat was higher amongst the Finns but that the Finns also consumed more fiber (MacLennan et al., 1977). One epidemiological study using per capita food intake from 20 industrialized nations examined the relationship of dietary cholesterol as well as total fat and fiber to mortality from CLB (Liu et al., 1979). Dietary cholesterol and total fat were directly and fiber inversely correlated with mortality from CLB. However, multivariate analysis showed that dietary cholesterol was the only variable to have a significant relationship when intercorrelations between the three variables were taken into account. There thus appears to be some inconsistency in the results of epidemiological studies of fat and CLB. Some of the studies were not specifically designed to investigate the proposed association and were performed on populations with a relatively homogeneous fat intake. Others did not take into account a possible interaction between fat intake and some components of dietary fiber which could in theory result in the combination of a low fat and high fiber diet providing more protection against the development of CLB than either component alone. There may also be a protective effect of vegetables unrelated to fiber. Some studies have combined colon and rectal cancer, although there is epidemiological evidence that there may be differences in aetiology. In addition, it is possible that the sexes may differ in bowel physiology (Cummings et al., 1979) and that dietary factors implicated in CLB may operate differently in the sexes.

There are several possible biochemical mechanisms by which a high fat diet might increase the risk of CLB. Some bile acids found in faeces may have co-carcinogenic properties (Hill, 1977). A case-control study of faecal bile acid concentrations in patients with newly diagnosed CLB and controls with non-malignant conditions, showed that the CLB cases had significantly higher mean faecal bile acids levels than the controls (Hill et al., 1975). In addition, the cases had a higher rate of carriage of nuclear dehydrogenating clostridia (NDH) in their stools than the controls. These bacteria may be responsible for the formation of unsaturated carcinogens from bile acids. Subsequent case-control studies of faecal bile acids in CLB have given inconsistent results, one has confirmed the earlier findings (Reddy and Wynder, 1977) and others have failed to find a difference (Murray et al., 1980; Moskovitz, et al. 1979). It has also been suggested that dietary cholesterol may be co-carcinogenic (Cruse, Lewin, and Clark, 1979) and non-degradation of faecal cholesterol is found in patients with familial polyposis (Drasar et al., 1975) and certain families who are at high risk of CLB (Lipkin et al., 1981). A prospective study of faecal constituents in CLB is now in progress and may help to clarify the relationship between faecal bile acids, cholesterol, NDH clostridia, and CLB (Haines et al., 1980).

Two studies have found an apparent protective effect of high vegetable consumption (Bjelke. 1974; Hirayama, 1979a) and one a possible protective effect of dietary fiber in colon cancer but not rectal cancer (Modan et al., 1975).

Possible mechanisms for an apparent protective effect of vegetable intake have been suggested. One of the groups of vegetables which is of particular interest is the cruciferae, which includes cabbage, cauliflower, brussel sprouts and broccoli. It has been shown that these vegetables contain indoles which are capable of inducing benzpyrene hydroxylase activity in the small intestine of rats (Wattenburg, 1971). When added to the diet of rats, these indoles reduced the incidence of dimethyl benzathracene induced breast tumors in rats and benzpyrene induced gastric cancers in mice (Wattenburg and Loub, 1978). In humans, several case-control studies have found a lower consumption of cabbage in cases of colon cancer than controls (Graham et al., 1978; Bjelke, 1973; Modan et al., 1975). Two studies, however, found no apparent difference in vegetable consumption between cases of colon cancer and controls (Haenszel et al., 1973; Higginson, 1966).

The suggestion that dietary fiber was a protective factor for cancer of the large bowel was of seminal importance in stimulating research into diet and cancer causation (Burkitt, 1971). Much of the evidence, however, is circumstantial and the interpretation of many studies has been made more difficult by problems in the analysis of dietary fiber. Tables containing values for dietary fiber in food only are available for one or two countries (Cummings, 1981). Dietary fiber is not a homogeneous entity and different components may have different physiological affects. Some components of fiber may be more important than others in protection against cancer. For instance, a study of regional food intake and regional cancer of the large bowel mortality in the United Kingdom showed a strongly negative correlation between the average intakes of the pentose fraction of dietary fiber and mortality rates (Bingham et al., 1979). This fraction has been shown to have a major role in the physiological changes produced by dietary fiber (Cummings et al., 1978). Further investigation of different components of dietary fiber is necessary to clarify their possible role in the protection from CLB.

A recent report from a 10-year follow up of nearly 900 men in the Netherlands on whom food intake data had been recorded by the cross check dietary history method showed that those who died of cancer had significantly lower dietary fiber intakes than those who did not (Kromhout, Bosschieter, and de Lezenne Coulander, 1982). The numbers of cancer deaths were quite small (n = 44), but the relationship appeared to be limited to deaths from lung cancer. There were only 5 deaths from large bowel cancer.

7.2. Breast and endometrial cancer

There have been several studies of the correlation between per capita fat intake in countries or regions and breast cancer incidence or mortality which have demonstrated a significant relationship (Armstrong and Doll, 1975; Drasar and Irving, 1973; Gray, Pike, and Henderson, 1979; Knox, 1977). The correlations with total fat appeared in most instances to be greater than that with specific types of fat. A study in Hawaii found a relationship between individual fat consumption ascertained at interviews and the incidence of breast cancer in different ethnic groups (Kolonel et al., 1981). Three case-control studies have also found higher consumption of fat in breast cancer patients than controls (Lubin et al., 1981; Miller et al., 1978; Phillips, 1975).

There is a strong association between breast cancer and CLB mortality in states within the United States and internationally (Howell, 1976). In addition, patients with breast cancer have a significant excess risk of developing a second primary in the large bowel. It has been known for some time that postmenopausal women who develop breast cancer tend to be taller and heavier than those who do not (de Waard, 1975). A recent study has shown a relationship between oestrogen receptor positive breast cancer and obesity (de Waard, Poortman, and Collette, 1981). In postmenopausal women the production of oestrogens in adipose tissue may represent a major source of oestrogen (McDonald et al., 1978).

There has been a decline in the age of menarche in Western countries in the last century or so and Japanese Americans are taller and have earlier menarche than Japanese women in Japan. The incidence of breast cancer in Japanese women under 60 in California (mainly American born) is about 75 percent of that of white Californians, but amongst those of 60 and over (mainly Japanese born) the Japanese women have only about 25 percent of the incidence of the white Californians (Dunn, 1975). In Japan itself, breast cancer incidence has increased sharply and a prospective study of nearly 150,000 women aged 40 and older showed that those who ate meat daily had a higher standardized mortality ratio for breast cancer compared with non or occasional meat eaters (Hirayama, 1979a). A high correlation was observed between dietary fat intake and adjusted mortality rates for breast cancer in 12 districts in Japan. It is possible that nutritional factors play a role in the etiology of breast cancer, partly because of their effect on the age at menarche, but it has been shown that fat consumption is still correlated internationally with breast cancer incidence even when age at menarche and parity are taken into account (Gray, Pike, and Henderson, 1979).

There is a thirty-fold international variation in the incidence of endometrial cancer. Obesity may increase the risk (Elwood et al., 1977; Wynder, Escher, and Mantel, 1966), and there is also a high correlation internationally between fat consumption and endometrial cancer incidence (Armstrong and Doll, 1975). Seventh Day Adventist women in California, have a lower mortality from breast and endometrial cancers than women who are not Seventh Day Adventists (Phillips, 1975). There is evidence that Seventh Day Adventist vegetarians have lower urinary oestriol and total estrogens, lower plasma prolactin, and higher plasma sex-hormone binding globulin levels than non-vegetarians (Armstrong et al., 1981). These differences could not be explained by differences in bodyweight or adiposity. It is possible that diet may affect breast and endometrial cancer incidence at least in part by influencing hormone metabolism.

7.3. Cancer of the prostate

There is a significant international correlation between fat consumption and mortality from prostatic cancer (Armstrong and Doll, 1975; Howell, 1974). The incidence of cancer of the prostate in four ethnic groups in Hawaii shows an association with the consumption of animal fat and saturated fat within the groups (Kolonel et al., 1981). In Japan, there has been a pronounced increase in fat consumption and mortality from prostate cancer since 1950 (Hirayama, 1979b).

Two case-control studies (Rotkin, 1977; Schuman et al., 1982) have shown an increase in the frequency of consumption of high fat foods in cases. Although diet may have some part to play in the aetiology of prostatic carcinoma, it seems unlikely to explain the much higher incidence in blacks than whites in the United States other factors must also be involved.

7.4. Cancer of the pancreas

Pancreatic cancer has been the object of several studies which have suggested that there is an international association between mortality from pancreatic cancer and intake of fat and protein (Armstrong and Doll, 1975; Stocks, 1970; Wynder, Mabuchi, and Maruchi, 1973). A prospective study in Japan reported an increased relative risk of pancreatic cancer in those who consumed meat on a daily basis (Hirayama, 1977).

Although some studies have suggested that alcohol may be a causal factor (Burch and Ansari, 1968; Ishii et al., 1968), the majority have not confirmed a relationship (Wynder, Mabuchi, and Maruchi, 1973; MacMahon et al., 1981; Haines et al., 1982).

There has also been a report of an association with coffee consumption (MacMahon et al., 1981) with an apparent dose-response gradient for females only. However, this finding was unexpected by the authors and although there is some corroborative evidence from international correlations (Stocks, 1970), the selection of controls has been criticized and the results must be regarded as at best tentative (Feinstein et al., 1981).

8. Vitamin A, Carotene, and Cancer

There has been considerable recent interest in the possible protective effect of vitamin A and pro-vitamin A (Beta carotene) against certain tumors. It has suggested that dietary Beta carotene might be more important in this respect than vitamin A itself (Peto et al., 1981). Preformed vitamin A is converted into retinol in the liver and a variable proportion of Beta carotene is converted to retinol via retinal. It is estimated that a maximum of about one-sixth of Beta carotene is converted to retinol, the proportion decreases as the intake of Beta carotene increases (Goodman, 1979). Increasing dietary vitamin A does not appear to affect plasma retinol levels unless deficiency is present, because the plasma levels are regulated by the liver. The level of Beta carotene in plasma is, however, greatly affected by dietary intake. It is possible that the apparent protective effect of green vegetables may be due to their Beta carotene content. There is evidence from prospective studies that the intake of vitamin A and Beta carotene is lower in those who develop lung cancer, although these results were based on crude estimates of dietary intake (Shekelle et al., 1981; Bjelke, 1975). Retrospective studies suggested a similar relationship for cancer of the bladder (Mettlin and Graham, 1979), larynx (Graham, Mettlin, and Marshall, 1981), esophagus, oral cavity, and cervix (Graham, Mettlin, and Marshall, 1980; Romney, Palan, Duttagupta et al., 1981). In many studies, the majority of the vitamin A intake is actually derived from Beta carotene and it is therefore difficult to test the two hypotheses independently.

Several studies have examined blood levels of vitamin A in patients with cancer and controls without cancer and have found that cancer patients have lower levels (for review see Peto et al., 1981). In many cases, however, the method of selection of controls was not clearly described, and, of course, differences between cancer cases and controls may merely reflect the metabolic effects of cancer. Two prospective studies have shown that the risk of cancer was lower in those whose blood retinol concentrations were at the upper part of the range than in those at the lower end (Kark, Smith, and Hames, 1982; Wald, Idle, and Boreham, 1980).

There have also been a few studies of serum Beta carotene in cancer patients and controls which have generally shown lower levels in patients with cancer, but these have concerned oral (Wahi et al., 1962; Ibrahim, Jafarey, and Zuberi, 1977) or nasopharyngeal (Clifford, 1972) cancer, both of which may markedly affect dietary intake.

Vitamin A, Beta carotene, and other retinoids may be of importance because they may act in the later stages of neoplastic progression. This could, in theory, lead to a reduction in cancer risk within say five years and, therefore, the testing of the value of these substances ramdomized controlled trial may be a practical proposition. Although large doses of vitamin A may cause toxicity, pure Beta carotene appears to be tolerated in very large doses without ill effect (Mathews-Roth et al., 1977). However, there have been some reports of high intake of carrots and other raw vegetables causing amenorrhea (Kemmann, Pasquale, and Skaf, 1983) and neutropenia (Shoenfeld et al., 1982), both of which were associated with hypercarotenemia.

9. Serum Cholesterol and Cancer

There has been much recent debate about the apparent inverse relationship between serum cholesterol and cancer in prospective studies. The relationship has only been found for men and has been found in fifteen of twenty-seven studies in which it has been looked for (Feinleib, 1982). Although some studies have suggested that the relationship was only present for colon cancer (Kagan et al., 1981; Kozarevic, McGee; and Vojvodic, 1981), others have found a relationship with total cancer. Some investigators have suggested that the relationship is due to a metabolic effect of the cancer and is present only in those individuals whose blood samples were taken a few years before diagnosis (Rose and Shipley, 1980). Others have found that the relationship persists even in those whose samples were taken four years or more before the diagnosis of cancer (Williams et al., 1981). It is possible that the relationship between low serum cholesterol and cancer reflects the association between retinol and cholesterol (Kark et al., 1982). There is no firm evidence at the present time that there is a direct causal relationship between cholesterol and cancer.

10. Other Factors

The association between non-nutritive sweeteners and bladder cancer, which caused concern some years ago (Howe, Burch, Miller et al., 1977), has not been consistently confirmed (Kessler and Page-Clark, 1978; Armstrong and Doll, 1975; Hoover et al., 1980; Walker et al., 1982).

Aflatoxin contamination of food in countries such as Thailand and in parts of Africa may cause liver cancer (Van Rensburg et al., 1974), perhaps by reducing resistance to Hepatitis B virus.

Animal studies have suggested that selenium may have an antitumorigenic effect, possibly by protecting against free radical damage. However, the amounts

171

of selenium used far exceeded dietary requirements and the relevance of these experiments to human nutrition is not clear. A limited number of epidemiological studies have suggested an inverse correlation between per capita selenium intake and cancer incidence both between states in the United States (Shamberger, Tytko, and Willis, 1976) and internationally (Schrauzer et al., 1977). There is also a report that blood selenium levels were lower in gastrointestinal cancer patients and those with Hogkins disease compared with controls (Shamberger et al., 1973). Evidence for the role of other trace elements in cancer is somewhat tenuous. For instance, in the case of zinc, there is evidence that blood levels correlate inversely with selenium. Mean zinc concentrations in pooled blood from healthy blood donors at 19 U.S. collection points showed a correlation with cancer mortality rates at various sites (Schrauzer et al., 1977b). Some studies of cancer patients have shown no significant differences in zinc concentration from controls (Strain et al., 1972), whereas others have shown lower concentrations in patients with lung cancer (Davies, Musa, and Dormandy, 1968) and oesophageal cancer (Lin et al., 1977). A small study of zinc levels in plasma taken two to seven years before the clinical development of cancer did not show any differences between cases and controls (Haines et al., 1981). Studies of trace elements and cancer in humans are still at an early stage and are complicated by the necessity of collecting samples without contamination from trace elements in needles, syringes, and storage tubes. In addition, low plasma levels of trace elements are not necessarily an indicator of deficiency.

There has long been interest in the possible role of vitamin C in cancer prevention. Vitamin C appears to impair the formation of nitrosamines from nitrite and secondary amines (Mirvish et al., 1972). Several studies have suggested that patients with cancer have had a lower consumption of vitamin C. Inverse associations between fresh fruit consumption or vitamin C and gastric cancer have been reported (Haenszel and Correa, 1975; Bjelke, 1978; Higginson, 1966; Kolonel et al., 1981). A similar inverse relationship using rather crude measurements of intake has also been described in case-control studies of laryngeal cancer (Graham, Mettlin, and Marshall, 1981) and cervical cancer (Wassertheil-Smoller et al., 1981). Since vitamin C intake is so dependent on fruit and vegetable consumption, it is difficult to know whether differences in consumption between cases and controls merely reflect some other protective effect of fruit and vegetables. Like vitamin C, vitamin E competes for available nitrite and thus blocks the formation of carcinogenic nitroso compounds (Mergens et al., 1979). However, vitamin E is lipid soluble and vitamin C water soluble. There does not appear to be any adequate epidemiological data on the relationship between vitamin E, which is widely distributed in different foods, and cancer.

Despite public concern, there has been little evidence to link food contaminants or additives with cancer in the developed countries. Understandably, there have been few epidemiological investigations because of the difficulty of measuring exposure to these compounds over long periods in human populations. However, the trends of non-smoking-related cancer incidence have not shown any increase following the introduction of additives. Nearly 3,000 substances are added intentionally to food in the United States and another 12.000 chemicals are used in food packaging. It is quite impossible to be sure that low levels of a very large number of compounds could not have some synergistic effect on cancer incidence, but such a mechanism seems unlikely to be important compared with the intake of major nutrients.

11. Conclusions

Despite the problems of establishing a causal relationship between dietary factors and cancer, it seems likely that further understanding of the relationship will be of major public health importance. It has been suggested that dietary factors may account for about 30 percent of tumors in western society (Doll and Peto, 1981).

Given the difficulties of interpretation and investigation in this field, any advances in understanding must come from data derived from several independent types of study. Prospective studies have distinct advantages in testing etiological concepts, but, even some years before the clinical signs of cancer develop, subtle metabolic changes may occur which may mistakenly be thought of as causal. Collaborative studies between biochemists and epidemiologists testing specific hypotheses are likely to be particularly productive provided follow up is undertaken over a long period. Stored biological samples from large populations should, where possible, be linked to cancer registry data to enable incidence as well as mortality from cancer to be studied.

The latency period between the initiation and the clinical development of cancer is not precisely known but may well be 20 years or more. In view of the difficulties of following populations for very long periods it may be more practical to focus on hypotheses involving promoters and inhibitors whose effects may be mediated in the later stages of the development of cancer.

Better dietary measurements are needed so that individuals can be more accurately characterized, and in long term studies allowance may have to be made for changing dietary patterns within the population. These changes, if marked, may make interpretation of results difficult or impossible. Although weighed intake data should probably be used where possible for future population based studies of dietary intake, more work is needed to determine whether this results in changes in diet during the measurement period. Monitoring of biochemical markers of nutrition should be undertaken prior to, during, and after weighed intake studies. Despite their limitations, however, some cruder methods of determining dietary intake do appear to have given useful results in both case-control and prospective studies.

173

The selective use of data from studies and undue confidence in the 'unexpected finding' have led to false alarms in the past and should not be given undue prominence. Interactions between dietary components which may be of physiological importance. e.g., dietary fiber and fat, should be specifically considered in study design. When the intakes of dietary constituents are highly intercorrelated there may be great difficulties in determining which of them (if any) is causally related to cancer. For instance, consumption of green and yellow vegetables may be closely related to Beta carotene and vitamin C intake. It is therefore necessary to consider information from other sources, such as laboratory studies, when attempting to determine the relative importance of closely correlated dietary components. Randomized trials of individual components may also help to elucidate their relevance to cancer prevention, particularly if they act in the later stages of the neoplastic process. There is already a randomized controlled trial of Beta carotene in progress. If dietary manipulations are to be tested, however, an individually randomized study design may not be feasible for a variety of reasons. Since dietary patterns are partly socially determined, large-scale changes might also occur in the control group. The use of communities rather than individuals as the allocation units has been proposed to deal with this problem and to allow the use of a variety of educational techniques within the community (Farquhar, 1978).

In order to have an impact on public health it is particularly important to adopt a strategy that will be seen as positive, urging an increase in the consumption of certain items rather than, yet again, as in the case of smoking, alcohol, and obesity having to urge self-denial and restraint. In this respect the possible protective effect of vegetables and dietary fiber on cancer incidence should be readily acceptable to the public.

References

Armstrong, B., and Doll, R. Environmental factors and cancer incidence and mortality in different countries with special reference to dietary practices. *International Journal of Cancer* (1975) 617-631.

Armstrong, B., and Doll, R. Bladder cancer mortality in diabetics in relation to saccharin consumption and smoking habits. *British Journal Preventive and Social Medicine* 29 (1975) 73-81.

Armstrong, B. K., Brown, J. B., Clarke, H. T., Crook, D. K., Hahnel, R., Masarei, J., and Ratajczak, T. Diet and reproductive hormones: A study of vegetarian and nonvegetarian postmenopausal women. *Journal of the National Cancer Institute* 67 (1981) 761-767.

Bazarre. T. L. and Myers. M. P. The collection of food intake data in cancer epidemiology studies. *Nutrition and Cancer* 1 (1980) 22-45.

Bingham, S., McNeil. N. I., and Cummings, J. H. The diet of individuals: A study of a randomly-chosen cross section of British adults in a Cambridgeshire village. *British Journal of Nutrition* 45 (1981) 23-35.

Bingham, S., Williams, D. R. R., Cole, T. J., and James, W. P. T. Dietary fibre and regional large-bowel cancer mortality in Britain. *British Journal of Cancer* 40 (1979) 456.

Bjelke, E. Epidemiologic studies of cancer of the stomach, colon and rectum; with special emphasis in the role of diet. Vols. I-IV. Thesis University of Minnesota, University Microfilms, Ann Arbor, 1973.

Bjelke, E. Epidemiologic studies of cancer of the stomach, colon, and rectum; with special emphasis on the role of diet. *Scandinavian Journal of Gastroenterology* 9, Supplement 31 (1974).

Bjelke, E. Dietary factors and epidemiology of cancer of the stomach and large bowel. *Aktuel Ernaehrungsmed. Klin. Prax.* Suppl. 2 (1978) 10-17.

Burch, G. E. and Ansari, A. Chronic alcoholism and carcinoma of the pancreas: A correlative hypothesis. *Archives of Internal Medicine* 122 (1968) 273-275.

Burke, B. S. The dietary history as a tool in research. *Journal of the American Dietetic Association* 23 (1947) 1441-1446.

Burkitt, D. P. Epidemiology of cancer of the colon and rectum. *Cancer* 28 (1971) 3.

Clifford, P. Carcinogens in the nose and throat: Nasopharyngeal carcinoma in Kenya. *Proceedings of the Royal Society of Medicine* 65 (1972) 24-28.

Cruse, P., Lewin, M., and Clark, C. G. Dietary cholesterol is co-carcinogenic for human colon cancer. *Lancet* 1 (1979) 752-755.

Cummings, J. H. Dietary fibre and large bowel cancer. *Proceedings of the Nutrition Society* 40 (1981) 7-14.

Cummings, J. H. Southgate, D. A. T., Branch, W. P. T., Houston, H., Jenkins, D. J. A., and James, W. P. T. Colonic response to dietary fibre from carrot, cabbage, apple, bran, and guar gum. *Lancet* 1 (1978) 5.

Cummings, J. H. Stephen, A. M., Wayman, B., and Chapman, G. Influence of age, sex and dose on colonic response to dietary fibre from bread. *Gastroenterology* 76 (1979) 1116.

Dales, L. G., Friedman, G. D., Ury, H. K., Grossman, S., and Williams, S. R. A case-control study of relationships of diet and other traits to colorectal cancer in American Blacks. *American Journal of Epidemiology* 109 (1979) 132-144.

Davies, I. J. T., Musa, M., and Dormandy, T. L. Measurements of plasma zinc II. In health and disease. *Journal of Clinical Pathology* 21 (1968) 363-365.

de Waard, F. Brease cancer incidence and nutritional status with particular reference to body weight and height. *Cancer Research* 35 (1975) 3351-3356.

de Waard, F., Poortman, J., and Collette, H. J. A. Relationships of weight to the promotion of breast cancer after menopause. *Nutrition and Cancer* 2 (1981) 237-240.

Doll, R. and Peto, R. *The causes of cancer.* Oxford: Oxford University Press, 1981.

Drasar, B.S., Bone, M. F., Hill, M. J., and Marks, C. G. Proceedings: Colon cancer and bacterial metobolism in familial polyposis. *Gut* 16 (1975) 824-825.

Drasar, B. S. and Irving, D. Environmental factors and cancer of the colon and breast. *British Journal of Cancer* 27 (1973) 167-172.

Dunn, J. E. Cancer epidemiology in populations of the United States - with emphasis on Hawaii and California - and Japan. *Cancer Research* 35 (1975) 3240-3245.

Elwood, J. M., Cole, P., Rothman, K. J., and Kaplan, S. D. Epidemiology of endometrial cancer. *Journal of the National Cancer Institute* 59 (1977) 1055-1060.

Farquhar, J. The community-based model of life style intervention trials. *American Journal of Epidemiology* 108 (1978) 103.

Feinleib, M. Summary of a workshop on cholesterol and noncardiovascular disease mortality. *Preventive Medicine* 2 (1982) 360-367.

Feinstein, A. R., Horwitz, R. I., Sptizer, W. O., et al. Coffee and pancreatic cancer: The problems of etiologic science and epidemiologic case-control research. *Journal of the American Medical Association* 246 (1981) 957-961.

Freis, E. D. Salt, volume and the prevention of hypertension. *Circulation* 53 (1976) 589-595.

Goodman, D. S. Vitamin A and retinoids: Recent advances. Introduction, background, and general overview. *Federal Proceedings* 38 (1979) 2501-2503.

Graham, S., Dayal, H., Swanson, M., Mittleman, A., and Wilkinson, G. Diet in the epidemiology of cancer of the colon and rectum. *Journal of the National Cancer Institute* 61 (1978) 709-714.

Graham, S., Schotz, W., and Martino, P. Alimentary factors in the epidemiology of gastric cancer. *Cancer* 30 (1972) 927-938.

Graham, S., Mettlin, C., and Marshall, J. In: The Proceedings of the June 1980 Meeting of the Society for Epidemiological Research. Minneapolis.

Graham, S., Mettlin, C., and Marshall, J., et al. Dietary factors in the epidemiology of cancer of the larynx. *American Journal of Epidemiology* 113 (1981) 675-680.

Gray, G. E., Pike, M. C., and Henderson, B. E. Breast-cancer incidence and mortality rates in different countries in relation to known risk factors and dietary practices. *British Journal of Cancer* 39 (1979) 1-7.

Haenszel, W., Berg, J. W., Segi, M., Kurihara, M., and Locke, F. B. Large-bowel cancer in Hawaiian Japanese. *Journal of the National Cancer Institute* 51 (1973) 1765-1779.

Haenszel, W. and Correa, P. Developments in the epidemiology of stomach cancer over the past decade. *Cancer Research* 35 (1975) 3452-3459.

Haenszel, W., Kurihara, M., Segi, M., and Lee, R. K. C. Stomach cancer among Japanese in Hawaii. *Journal of the National Cancer Institute* 49 (1972) 969-988.

Haenszel, W., Locke, F. B., and Segi, M. A case-control study of large bowel cancer in Japan. *Journal of the National Cancer Institute* 64 (1980) 17-22.

Haines, A. P., Meade, T. W., Thompson, S. G., Hill, M., and Williams, R. A prospective study of the relationship of faecal bile acids, neutral steroids and nuclear dehydrogenating clostridia with the development of cancer of the large bowel (CLB). *Cancer Detection and Prevention* Vol. 3, 39 (1980) 323.

Haines, A. P., Thompson, S. G., Basu, T. K., and Hunt, R. Cancer, retinol binding protein, zinc and copper. (letter) *Lancet* 1 (1982) (8262) 52-53.

Haines, A. P., Moss, A. R., Whittemore, A., and Quivey, J. A case-control study of pancreatic cancer. *Journal of Cancer Research and Clinical Oncology* 103 (1982) 93-97.

Hankin, J. H., Nomura, A., and Rhoads, G. G. Dietary patterns among men of Japanese ancestry in Hawaii. *Cancer Research* 35 (1975) 3259-3264.

Higginson, J. Etiological factors in gastro-intestinal cancer in man. *Journal of the National Cancer Institute* 37 (1966) 527-545.

Hill, M. J. In: *Origins of human cancer.* Eds. Hiatt, H., Watson, J., and Winsten, J. New York: Cold Spring Harbor Press, 1977. P. 1640.

Hill, M. J., Drasar, B. S., Williams, R. E. O., Meade, T. W., Cox, A. G., Simpson, J. E. P., and Morson, B. C. Faecal bile acids, clostridia and the etiology of cancer of the large bowel: A case comparison study. *Lancet* 1 (1975) 535.

Hirayama, T. Changing patterns of cancer in Japan with special reference to the decrease in stomach cancer mortality. In: *Origins of human cancer.* Eds. Hiatt, H. H., Watson, J. D., and Winsten, J. A. New York: Cold Spring Harbor Press, 1977. Pp. 55-75.

Hirayama, T. Epidemiological evaluation of the role of naturally occurring carcinogens and modulators of carcinogenesis. In: *Naturally occurring carcinogens-mutagens and modulators of carcinogenesis.* Eds. Miller, E. C., et al. Baltimore: University Park Press, 1979a. Pp. 359-380.

Hirayama, T. Diet and cancer. *Nutrition and Cancer* 1 (1979b) 67-81.

Hirayama, T. A large-scale cohort study on the relationship between diet and selected cancers of the digestive organs. In: *Gastrointestinal Cancers, Endogenous Factors: Banbury Report* 7. Cold Spring Harbor, N.Y.: Cold Spring Harbor Laboratory, 1981.

Hoover, R., Strasser, P. H., Child, M., et al. Artificial sweeteners and human bladder cancer. *Lancet* 1 (1980) 837-840.

Howe, G. R., Burch, J. D., Miller, A. B., Morrison, B., Gordon, P., Weldon, L., Chambers, L. W., Fodor, G., and Winsor, G. M. Artificial sweeteners and human bladder cancer. *Lancet* 2 (1977) 578-581.

Howell, M. A. Factor analysis of international cancer mortality data and *per capita* food consumption. *British Journal of Cancer* 29 (1974) 328-336.

Howell, M. A. The association between colorectal cancer and breast cancer. *Journal of Chronic Disease* 29 (1976) 243-261.

Ibrahim, J., Jafarey, N. A., and Zuberi, S. J. Plasma vitamin 'A' and carotene levels in squamous cell carcinoma of oral cavity and oro-pharynx. *Clinical Oncology* 3 (1977) 58-63.

Ishii, K., Nakamura, H., Ozaki, N., et al. Epidemiological problems of pancreas cancer. Jpn. *Journal of Clinical Medicine* 26 (1968) 1839-1842.

Jain, M., Cook, G. M., Davis, G., Grace, M. G., Howe, G. R., and Miller, A. B. A case-control study of diet and colo-rectal cancer. *International Journal of Cancer* 26 (1980) 757-768.

Joossens, J. V. and Geboers, J. Nutrition and gastric cancer. *Nutrition and Cancer* 2 (1981) 250-261.

Kagan, A., McGee, D. L., Yano, K., et al. Serum cholesterol and mortality in a Japanese-American population. *American Journal of Epidemiology* 114 (1981) 11-20.

Kark, J. D., Smith, A. H., Switzer, B. R., Hames, C. G. Serum vitamin A (retinol) and cancer incidence in Evans county, Georgia. *Journal of the National Cancer Institute* 66 (1981) 7-16.

Kark, J. D., Smith, A. H., and Hames, C. G. Serum retinol and the inverse relationship between serum cholesterol and cancer. *British Medical Journal* 284 (1982) 152-154

Kemmenn, E., Pasquale, S. A., and Skaf, R. Amenorrhea associated with carotenemia. *Journal of the American Medical Association* 249 (1983) 926-929.

Kessler, I. I. and Page-Clark, J. Saccharin, cyclamate and human bladder cancer. No evidence of an association. *Journal of the American Medical Association* 240 (1978) 349-355.

Kinlen, L. J. Meat and fat consumption and cancer mortality: A study of strict religious orders in Britain. *Lancet* 1 (1982) 946-949.

Knox, E. G. Foods and diseases. *British Journal of Preventive Social Medicine* 31 (1977) 71-80.

Kolonel, L. N. and Lee, J. Husband-wife correspondence in smoking, drinking, and dietary habits. *American Journal of Clinical Nutrition* 34 (1981) 99-104.

Kolonel, L. N., Hankin, J. H., Lee, J., Chu, S. Y., Nomura, A. M. Y., and Hinds, M. W. Nutrient intakes in relation to cancer incidence in Hawaii. *British Journal of Cancer* 44 (1981) 332-339.

Kolonel, L. N., Nomura, A., Hirohata, T., Hankin, J. H., and Hinds, M. W. Association of diet and place of birth with stomach cancer incidence in Hawaii Japanese and Caucasians. *American Journal of Clinical Nutrition* 34 (1981) 2478-2485.

Kozarevic, D. J., McGee, D. L., and Vojvodic, N. Serum cholesterol and mortality. The Yugoslavia cardiovascular disease study. *American Journal of Epidemiology* 114 (1981) 21-28.

Kromhout, D., Bosschieter, E. B., and de Lezenne Coulander, C. Dietary fibre and 10-year mortality from coronary heart disease, cancer, and all causes. The Zutphen study. *Lancet* 2 (1982) 518-522.

Lancet Editorial. Cancer of the Oesophagus in China. *Lancet* 1 (1975) 1413.

Larsson, L. G., Sandstrom, A., and Westling, P. Relatioship of Plummer-Vinson disease to cancer of the upper alimentary tract in Sweden. *Cancer Research* 35 (1975) 3308-3316.

Lin, H. J., Chan. W. C., Fong, Y. Y., and Newberne, P. M. Zinc levels in serum, hair and tumors from patients with esophageal cancer. *Nutrition Reports International* 15 (1977) 635-643.

Lipkin, M., Reddy, B. S., Weisburger, J., and Schechter, L. Nondegradation of fecal cholesterol in subjects at high risk for cancer of the large intestine. *Journal of Clinical Investigation* 67 (1981) 304-307.

Liu, K., Stamler. J., Moss, D., Garside, D., Persky, V., and Soltero, I. Dietary cholesterol, fat, and fibre, and colon-cancer mortality. *Lancet* 2 (1979) 782-785.

Lubin, J. H., Burns, P., E. Blot, W. J., Ziegler, R. G.. Lees, A. W., and Fraumeni, J. F. Dietary factors and breast cancer risk. *International Journal of Cancer* 28 (1981) 685-689.

MacLennan, R., Jensen, O. M., Mosbech, J., Buschard, K., Dejgard, J., Bardram, H.. Tvedegard, E., Vuori, H., Kokko, S., Karjalainen, S., Laurell, G., Ryden, A. C., Schwan, A., Williams, R., Drasar, B. S., Hill, M. J., James, W. P. T., Cummings, J. H., and Southgate, D. A. T. Dietary fibre, transit-time, faecal bacteria, steroids, and colon cancer in two Scandinavian populations. Report from the International Agency for Research on Cancer Intestinal Microecology Group. *Lancet* 2 (1977) 207-211.

Mahboubi, E. O. and Aramesh, B. Epidemiology of esophageal cancer in Iran, with special reference to nutritional and cultural aspects. *Preventive Medicine* 9 (1980) 613-621.

Marr, J. W. Individual dietary surveys purposes and methods. *World Review of Nutrition and Dietetics* 13 (1971) 105-164.

Martinez, I., Torres, F., Frias, Z., Colon, J. R., and Fernandez, N. Factors associated with adenocarcinomas of the large bowel in Puerto Rico. In: *Advances in medical oncology, research and education.* Ed. Birch, J. M. Vol. 3. Epidemiology. Elmsford, N.Y.: Pergamon Press, 1979.

Mathews-Roth, M. M., Pathak, M. A., Fitzpatrick, T. B., Harber, I. C., and Kass, E. H. Beta carotene therapy for erythropoietic protoporphyria and other photosensitivity diseases. *Archives of Dermatology* 113 (1977) 1229-1232.

McDonald, P. C., Edman, C. D., Hamsell, D. L., Porter, J. C., and Siiteri, P. K. Effect of obesity on conversion of plasma androstenedione to estrone in postmenopausal women with and without endometrial cancer. *American Journal of Obstetrics and Gynecology* 130 (1978) 448-455.

MacMahon, B., Yen, S. Trichopoulos, D., Warren, K., and Nardi, G. Coffee and cancer of the pancreas. *New England Journal of Medicine* 304 (1981) 630.

Mergens, W. J., Vane, F. M., Tannenbaum, S. R., Green, L., and Skipper, P. L. *In vitro* nitrosation of methapyrilene. *Journal Pharmaceutical Sciences* 68 (1979) 827-832.

Mettlin, C. and Graham, S. Dietary risk factors in human bladder cancer. *American Journal of Epidemiology* 110 (1979) 255-263.

Miller, A. B., Kelly, A., Choi, N.W., Matthews, V., Morgan, R. W., Munan, L., Burch, J. D., Feather, J., Howe, G. R., and Jain, M. A study of diet and breast cancer. *American Journal of Epidemiology* 107 (1978) 499-509.

Miller, G. D. and Kuller, L. H. Trends in mortality from stroke in Baltimore, Maryland: 1940-1941 through 1968-1969. *American Journal of Epidemiology* 98 (1973) 233-242.

Mirvish, S. S., Wallcove, L., Eagen, M., and Shubik, P. Ascorbate-nitrite reaction. Possible means of blocking the formation of carcinogenic N-nitroso compounds. *Science* 177 (1972) 65-68.

Modan, B., Barell, V., Lubin, F., Modan, M., Greenberg, R. A., and Graham, S. Low fiber intake as an etiologic factor in cancer of the colon. *Journal of National Cancer Institute* 55 (1975) 15-18.

Moskovitz, M., White, C., Barrett, R. N., Stevens, S., Russell, E., Vargo, D., and Floch, M. H. Diet, fecal bile acids, and neutral sterols in carcinoma of the colon. *Digestive Disease and Science* 24(10) (1979) 746-751.

Murray, W. R., Blackwood, A., Trotter, J. M., Calman, K. C., and Mckay, C. Faecal bile acids and clostridia in the aetiology of colorectal cancer. *British Journal of Cancer* 41 (1980) 923-928.

Nomura, A., Hankin, J. H., and Rhoads, G. G. The reproducibility of dietary intake data in a prospective study of gastrointestinal cancer. *American Journal of Clinical Nutrition* 29 (1976) 1432-1436.

O'Neill, C. H., Hodges, G. M., Riddle, P. N., Jordan, P. W., Newman, R. H., Flood, R. J., and Toulson, E. C. A fine fibrous silica contaminant of flour in the high oesophageal cancer area of North East Iran. *International Journal of Cancer* 26 (1980) 617-628.

Peto, R., Doll, R., Buckley, J. D., and Sporn, M. B. Can dietary carotene materially reduce human cancer rates? *Nature* 290 (1981) 201-208.

Phillips, R. L. Role of life-style and dietary habits in risk of cancer among Seventh-Day Adventists. *Cancer Research* 35 (1975) 3513-3522.

Reddy, B. S. and Wynder, E. L. Metabolic epidemiology of colon-cancer. Fecal bile acids and neutral sterols in colon cancer patients and patient with adenomatous polyps. *Cancer* 39 (1977) 2533-2539.

Romney, S. L., Palan, P. R., Duttagupta, C., Wassertheil-Smoller, S., Wylic, J., Miller, G., Slagle, N. S., and Lucide, D. Retinoids and the prevention of cervical dysplasias. *American Journal of Obstetrics and Gynecology* 141 (1981) 890-894.

Rose, G. and Shipley, M. J. Plasma lipids and mortality: A source of error. *Lancet* 1 (1980) 523-526.

Rotkin, I. D. Studies in the epidemiology of prostatic cancer: Expanded sampling. *Cancer Treatment Reports* 61 (1977) 173-180.

Ruddell, E. S. J., Bone, E. S., Hill, M. J., and Walters, C. L. Pathogenesis of gastric cancer in pernicious anaemia. *Lancet* 1 (1978) 521-523.

Schlag, P., Ulrich, J., Merkle, P., Böckler, R., Peter, M., and Herfarth, C. Are nitrite and N-nitroso compounds in gastric juice risk factors for carcinoma in the operated stomach? *Lancet* 1 (1980) 727-729.

Schrauzer, G. N., White, D. A., and Schneider, C. J. Cancer mortality correlation studies-III. Statistical associations with dietary selenium intakes. *Bioinorganic Chemistry* 7 (1977a) 23-34.

Schrauzer, G. N., White, D. A., and Schneider, C. J. Cancer mortality correlation studies-IV. Associations with dietary intakes and blood levels of certain trace elements, notable Se-antagonists. *Bioinorganic Chemistry* 7 (1977b) 35-56.

Schuman, L. M., Mandel, J. S., Radke, A., Seal, U., and Halberg, F. Some selected features of the epidemiology of prostatic cancer: Minneapolis-St. Paul, Minnesota case-control study, 1976-1979. In: *Trends in cancer incidence: Causes and practical implications.* Ed. Magnus, K. Washington: Hemisphere Publishing Corp., 1982. Pp. 345-354.

Shekelle, R. B., Liu, S., Raynor, W., et al. Dietary vitamin A and risk of cancer in the western electric study. *Lancet* 2 (1981) 1185-1189.

Shamberger, R. J., Tytko, S. A., and Willis, C. E. Antioxidants and cancer. Part VI. Selenium and age-adjusted human cancer mortality. *Archives of Environmental Health* 31 (1976) 231-235.

Shamberger. R. J.. Rukovena, E., Longfield, A. K., Tytko, S. A., Deodhar, S., and Willis. C. E. Antioxdants and cancer I. Selenium in the blood of normals and cancer patients. *Journal of the National Cancer Institute* 50 (1973) 863-870.

Shoenfeld, Y., Shaklai, M., Ben-Barach, N., Hirschorn. M., and Pinkhas, J. Letter: Neutropenia induced by hypercarotenaemia. *Lancet* 1 (1982) 1245.

Sporn, M. B. and Newton, D. L. Chemoprevention of cancer with retinoids. *Federation Proceedings* 38 (1979) 2528-2534.

Stocks, P. Cancer icidence in North Wales and Liverpool region in relation to habits and environment. Suppl. to Part II of the British Empire Cancer Campaign 35th Annual Report, 1957.

Stocks, P. Cancer mortality in relation to national consumption of cigarettes, solid fuel, tea and coffee. *British Journal of Cancer* 24 (1970) 215-225.

Strain, W. H., Mansour, E. G., Flynn, A., Pories, W. J., Tomaro, A. J., and Hill, O. A. Letter: Plasma-zinc concentration in patients with bronchogenic cancer. *Lancet* 1 (1972) 1021-1022.

Tannenbaum, S. R., Weisman, M., and Fett, D. The effect of nitrate on nitrite formation in human saliva. *Food Cosmetics Toxicology* 14 (1976) 549-552.

Tuyns, A. J. Epidemiology of alcohol and cancer. *Cancer Research* 39 (1979) 2840-2843.

Van Rensburg, S. J., Van Der Watt, J. J., Purchase, I. F. H., Pereira Coutinho, L., and Markham, R. Primary liver cancer rate and aflatoxin intake in a high cancer area. *South African Medical Journal* 48 (1974) 2508 a-d.

Wahi. P. N.. Bodkhe, R. R., Arora. S. et al. Serum vitamin A studies in leukoplakia and carcinoma of the oral cavity. *Indian Journal of Pathology and Bacteriology* 5 (1962) 10-16

Wald, N., Idle, M., Boreham, J. et al. Low serum-vitamin A and subsequent risk of cancer-preliminary results of a prospective study. *Lancet* 2 (1980) 813-815.

Walker, A. M., Dreyer, N. A., Friedlander, E. et al. An independent analysis of the national cancer institute study on non-nutritive sweeteners and bladder cancer. *American Journal of Public Health* 72 (1982) 376-381.

Wassertheil-Smoller, S., Romney, S. L., Wylie-Rosett, J., Slagle, S., Miller, G., Lucido, D., Duttagupta, C., and Palan, P. R. Dietary vitamin C and uterine cervial dysplasia. *American Journal of Epidemiology* 114 (1981) 714-724.

Wattenburg, L. W. Studies of polycyclic hydrocarbon hydroxylases of the intestine possibly related to cancer. Effect of diet on benzopyrene hydroxylase activity. *Cancer* 28 (1971) 99-102.

Wattenburg, L. W. and Loub, W. D. Inhibition of plycyclic hydrocarbo-induced neoplasia by naturally occurring indoles. *Cancer Research* 38 (1978) 1410-1413.

Weisburger, J. H., Marquardt, H., Mower, H., Hirota, H.. Mori, H., and Williams, G. Inhibition of carcinogenesis: Vitamin C and the prevention of gastric cancer. *Preventive Medicine* 9 (1980) 352-361.

Williams, R. R., Sorlie, P. D., Feinleib, M., et al. Cancer incidence by levels of cholesterol. *Journal of the American Medical Association* 245 (1981) 247-252.

Wynder, E. L., Bross, I. J., and Feldman, R. M. A. A study of the etiological factors in cancer of the mouth. *Cancer* 10 (1957) 1300-1323.

Wynder, E. L., Covey, L. S., Mabuchi, K., and Muslunski, M. Environmental factors in cancer of the larynx: A second look. *Cancer* 38 (1976) 1591-1601.

Wynder, E. L., Escher, G. C., and Mantel, N. An epidemiological investigation of cancer of the endometrium. *Cancer* 19 (1966) 489-520.

Wynder, E. L., Mabuchi, K., Maruchi, N., et al. Epidemiology of cancer of the pancreas. *Journal of the National Cancer Institute* 50 (1973) 645-667.

11

THE CONTROL OF DIABETES

Georges Tchobroutsky, M.D.
Fabienne Elgrably, M.D.

Diabetes is a chronic disease characterized by hyperglycemia. There are two main types: the first is a relatively rare form (about 20 percent of the patients) in which the patients must inject insulin subcutaneously at least once a day. The second one is almost always observed in overweight adults whose treatment usually consists of losing weight through caloric restriction.

Controlling diabetes primarily means controlling blood sugar levels. Achieving such a normalization of glycemia is a very difficult or even an impossible and frustrating task. In the insulin-treated diabetic patient--because of the nonphysiological technic of treatment--blood glucose levels vary between hyper and hypoglycemic levels. Chronic hyperglycemia is probably very dangerous in the long-term time range and acute hypoglycemias may severely handicap the daily life of the insulin treated patient and represent a possible cause--even if very rare--of death. The overweight or obese diabetic adult has to strive to lose weight--every day for years and years. This is for the majority of patients an unrealistic goal, as it is for almost all obese subjects, diabetic or not.

I will mainly confine my presentation to the insulin treated patient since Dr. R. Kaplan's chapter (this volume) deals with behavioral programs for the management of the other type of diabetes. The following items will be discussed:

1 - What does controlling diabetes mean?
2 - Why must diabetes be controlled?
3 - Which degree of control can be or should be achieved?
4 - Who controls diabetes?
5 - How is diabetes controlled?
6 - Risks and handicaps in controlling diabetes.
7 - How diabetics behave with diabetes, its treatment and control.
8 - How the control of diabetes can be improved by acting on the tools, the patients and nondiabetic people (family, doctors and the community).

1. What does Controlling Diabetes Mean?

Achieving control of blood sugar levels is the main goal in controlling diabetes (Table 1). In a short-term time range. i.e., in day-to-day living, high blood sugar levels may cause symptoms such as thirst, polyuria and fatigue. In the more severe forms of the disease the lack of insulin may be responsible for severe metabolic symptoms (muscle wastage, exhaustion, ketoacidosis) leading to death if insulin is not given every day in the appropriate amount. All these symptoms may be relieved with a "minimal" therapy including at least one daily insulin injection with relative attention to diet rules including regularly scheduled meals. Such a minimal treatment generally allows the patient to manage his disease with the least possible daily risk but blood sugar levels are far from being normalized.

Hypoglycemic insulin reactions may occur mainly in the case of unscheduled muscular exercise or lack of carbohydrate intake.

Achieving a better control of daily blood sugar level variations is probably advisable for the long-term time range (and is mandatory for the pregnant diabetic woman). It appears extremely probable that chronic hyperglycemia is responsible for the microvascular complications of diabetes (Tchobroutsky, 1981). These include retinal and glomerular changes that may lead to loss of vision or death due to chronic renal failure.

There is no definite, scientific proof of such a relationship between the control of blood sugar and the development of diabetic microangiopathy (Siperstein, Foster, Knowles et al., 1977), but a great number of indirect arguments make such a relationship so probable that almost all physicians at least in Europe do agree that control is worthwhile. Some of this evidence is summarized in Table 2.

Table 1

Ideal (and Practical) Goals for Control in Diabetic Patients

Keep (near to) Normal:

1) Blood Glucose (How close to normal?)
2) Peripheral Plasma Insulin (Idem?)
3) Plasma Cholesterol and Triglycerides (Idem?)
4) Hemorrheological Parameters (via 1 and 2?)
5) Blood Pressure
6) Body Weight
7) Avoid Cigarette Smoking.

Avoid the risks and handicaps due to the treatment itself.

Hyperglycemia is also detrimental to peripheral nerves and might play a role in accelerating atherosclerosis.

Since 75 percent of the adult diabetics die from atherosclerosis--mainly due to cardiac infarction--control of the diabetic also implies controlling atherosclerosis risk factors, i.e., at least hypertension, smoking and hypercholesterolemia.

Such risk factor control may add its own frustrations and consequences to the therapeutic control of diabetes itself.

2. Why Should Diabetes be Controlled?

In a day-to-day management, i.e., in the short or middle term time range, controlling the disease allows the patient, who is free of symptoms, to work, play sports and generally speaking, to live an--almost--normal life. Insulin treated patients cannot practice certain sports or do certain work involving death hazards for the diabetic or for others due to possible hypoglycemic (insulin) reactions leading to loss of consciousness or odd behaviors.

This level of control is achieved by nonsophisticated treatments that do not claim to achieve normal or near-to-normal blood glucose levels. The patient and his doctor are usually satisfied except when hypoglycemic reactions occur or if a life-threatening episode of ketosis develops when under stress.

For the long-term range it is probably wise to strive to achieve a blood glucose control as near normal as possible every day, throughout the diabetic's life (Table 3). Such a therapeutic attitude, in conjunction with the control of atherosclerotic risk factors is--at the present time--the best insurance for the

185

Table 2

Hyperglycemia is (at least in part) Responsible for the Development of the Microvascular Complications of Diabetes

1) Monozygotic twins and triplets, discordant for diabetes, are also discordant for the clinical complications of the disease.

2) Secondary diabetes in man is accompanied by specific complications.

3) Functional and anatomical anomalies do not exist before the actual onset of diabetes and throughout its course.

4) The frequency and severity of the lesions are correlated in epidemiological studies with the degree of hyperglycemia.

5) A normal kidney transplanted to a diabetic human develops diabetic lesions.

6) The control of diabetes in man limits the development of the microangiopathy.

7) An injured kidney coming from a diabetic animal returns to normal after transplantation to a nondiabetic animal. (Such data were also reported in man.)

8) Experimental diabetes is complicated by secondary lesions.

9) Early diabetic microangiopathic changes are reversible in the animal treated by insulin or by transplantation of all or part of the pancreas.

10) Some biochemical, functional and enzymatic changes are reversible by insulin.

Adapted from Tchobroutsky (1981)

diabetic who wants to reach old age with no particular renal, ocular, neurological, or atherosclerotic handicaps. Such a long-term attitude is not self evident and obliges the patient and his doctor to believe in its efficiency. This is also not easy to achieve.

Under certain conditions characterized by a limited period of time such as during pregnancy, or before pregnancy, achieving blood glucose levels as normal as possible is a mandatory goal (Stowers, 1981). Normoglycemia or near normoglycemia is the best way to improve birth conditions, diminishing the risk of perinatal death and fetal malformations.

Table 3

Why must Control be Tight?

To prevent or limit vascular and neurological lesions by limiting:

- Plasma and Tissue Sugar Accumulation and Protein Glycosylations (Accelerated Aging) (?)
- Counter-Insulin Hormone Secretions (GH ?)
- Atherosclerotic Changes
- Increased Capillary Permeability/Microvascular Circulatory Changes
- Hemorrheological Abnormalities

To decrease fetal and neonatal mortality and morbidity.

To improve daily life without increasing the frequency of hypoglycemic reactions.

For the very motivated diabetic subjects, such as the pregnant woman, such a goal can be achieved. But pregnancy lasts only nine months. For some physicians this kind of control might be the desirable one for decades.

3. Which Degree of Control is to be Achieved?

This question implies several others:

- Which degree of control may actually be achieved under routine conditions of treatment with available tools?
- Which degree of control must be achieved for short-term goals?
- Which degree of control is desirable for long-term goals?

Control under daily conditions of treatment with the tools now available, including pumps for continuous insulin administration, cannot reproduce "normality." Insulin is not given in the portal vein and the amount of the hormone delivered is not adapted minute to minute to blood sugar changes as in the normal nondiabetic person. The lack of homeostatic insulin delivery is responsible for the actual situation of the insulin treated diabetic patient whose blood sugar levels vary between hypo and hyperglycemia. A small, portable artificial pancreas for continuous use will probably not be available within the next two decades or may never be available.

The most sophisticated pumps for continuous delivery are not artificial pancreas since blood glucose measurements are not performed by the device, which is imprecisely preprogrammed.

187

Table 4 summarizes the short-term treatment objective: for short-term goals only a fair or poor degree of control is necessary in order for the patient to lead a quite normal daily life.

For long-term goals (or clinical situations in which near-to-normal control is mandatory such as in pregnancy) the main but unresolved question is: what level of hyperglycemia is dangerous? We do not know whether there is a continuum in the deleterious effect of high blood sugar levels upon blood vessels and nerves or if there exists some critical level. At least we would like to know which levels are critical in the development of severe complications. Since we do not know how to recognize which diabetics have high risks for severe microangiopathy nor which levels are dangerous, we must strive to achieve blood glucose control as near

Table 4

"Short"-Term Goals in the Control of Diabetes

Clinical Situations	Short Life Expectancy < 15 yrs.? Social, Cultural, Economic, Psychiatric Handicaps
Aims	No Symptoms No or Very Slight Insulin Reactions Near to Normal Daily Life
Degree of Control to be Achieved	Fair: -No Osmotic Diuresis -No Hypoglycemia -No Ketosis
Difficulties	Moderate
Means	Minimal Education (Selection of objectives according to the goals) Minimal Observance of the Diet One Daily Insulin Injection Limited but Positive Cooperation from the Patient

Table 5

"Long"-Term Goals in the Control of Diabetes

Clinical Situations	Long Life Expectancy > 15 yrs.? Pregnancy (before and during) Severe/Evolutive Microangiopathy
Aims	Delay or Prevention of late Complications Decrease or Perinatal Mortality and malformations

Degree of Control
to be Achieved
	Very Good - Near to Normal Daily Blood
	    Glucose Levels
	  - Normal Blood Lipids
	  - Normal Blood Pressure
	  - Ideal Body Weight
	  - No Smoking

Difficulties	Very Great

Means	Intensive Education and Continuing
	  Support by the Medical Team
	Physical Exercise
	Early Insulin Treatment
	Multiple Daily Injections or Pumps
	Careful Adaptation of Doses
	Strict Observance of the Diet
	Very Close Supervision of the Disease by the
	  Patient Himself

normal as possible in all diabetics with long life expectancy and no social or cultural handicaps (Table 5). Such an attitude—over the decades—is a difficult and frustrating one—since the main actor of this control is the patient himself.

4. Who Controls Diabetes?

Controlling hypertension is the responsibility of the physician but the patient must control all the rest: he has to refrain from smoking, change diet habits in order to normalize plasma lipids and body weight and control blood sugar levels. But he also has to adapt insulin dosages, exercise and meals (and snacks), day after day, in order to keep blood sugar levels as close to normal as possible and avoid hypoglycemic (insulin) reactions.

Doctors, nurses, dieticians and social workers primarily strive to teach the patient and to check that he manages his own disease and treatment well. The diabetic patient on insulin must become an expert in insulin therapy and diet and must be able to perform chemical manipulations daily, including urinary analysis and self blood glucose measurements.

The daily management of his therapy rests on his own shoulders. Since mild hyperglycemia keeps him free from symptoms and hypoglycemia may sometimes severely handicap him, it is easy to understand which choice is the first spontaneous one, not only for him but for the average physician not trained in diabetic care.

5. How is Diabetes Controlled?

The patient, and the medical staff, must use the appropriate tools in order to achieve the desirable goals. This implies an anticipatory therapeutic behavior based on the analysis of the past or present control achieved. Urine analysis gives information on the last few hours but these results depend upon several variables including the time elapsed between the last micturition and the present one, the neurovesical function and the renal threshold for glucose.

Self blood glucose monitoring provides very good information by giving an immediate figure to the patient (Symposium on Home Glucose Monitoring, 1980); this implies blood sampling but new devices permit easy finger punctures. Methods for obtaining tight control are listed in Table 6.

Table 6

Means (Tools) for Tight Control in Insulin Treated Diabetics

---

1) Insulin-Therapy
   Intensive Conventional, i.e., Multiple Daily Injections
   Open-Loop Delivery (Pumps)
2) Careful Attention to Diet
3) Exercise
4) Patient's Education
5) Home Monitoring
6) Frequent Consultations with Diabetologists
7) Drugs? (Lowering Cholesterol, Platelet Aggregation...)

However even with all this information the patient has to decide what his treatment will be for the next few hours. The well trained patient does not change his treatment every day but must change according to his planned activities (sports, dining out, abnormal professional activity and so on). For many reasons including various conditions of insulin resorption from its S.C. deposit, an unexpected physical activity or emotional disturbance, the diabetic patient is continuously--or feels he is continuously--at the mercy of a hypoglycemic insulin reaction.

Some diabetic patients are more unstable than others and infants or children are obviously more difficult to treat, particularly if the parents must take care of their child (that is usually mandatory below 7 years and becomes less true later). After the age of 12 13 the diabetic child is usually able to take care of himself.

Managing diabetes and properly using the therapeutic tools is not easy. Even well informed patients who think they do not make "mistakes" experience a lot of mild hypoglycemic reactions (Goldgewicht, Slama, Papoz, and Tchobroutsky, 1983).

6. Risks and Handicaps in Controlling Diabetes in Insulin Treated Patients

These are relatively poorly known (Table 7). With several colleagues in Paris we have tried to analyze some parameters· in this field. It must be emphasized that the control of blood glucose achieved in conventionally treated people--even with the use of multiple (2 or 3) daily insulin injections and home blood sugar analysis--was far from normal.

Table 7

Potential Risks of Tight Control

More and more severe hypoglycemic reactions including death hazards.
Social and professional limitations and handicaps.
Obsessional behavior and anxiety.
Gain of body weight.
Increased financial burden.
Increased atherosclerosis.

The first study we did (Basdevant, Costagliola, Lanoe et al., 1982) was aimed at comparing two groups of insulin-treated outpatients (one followed up at the Hôtel-Dieu Hospital and the second mainly supervised by general practitioners).

The two populations were comparable in age, age at diagnosis, sex, level of education, overall activity and socio-professional and economic status. Outpatients followed up in the diabetic unit had better blood glucose control, with about the same number of hypoglycemic reactions as patients followed up in general practice. This better control was associated with more social activity and fewer visits to the physician, despite the fact that the hospital outpatients spent more money on their diet and had more daily insulin injections.

We concluded from that study that attempts to improve control in insulin-treated patients are associated with a more active life and with no increase in the frequency of hypoglycemic reactions. Tables 8 and 9 give the main results of that study.

Another study was aimed at discovering the causes of hypoglycemic reactions and how often severe ones occurred in insulin-dependent diabetics (Goldgewicht et al., 1983). One hundred and seventy-one out-patients answered a questionnaire which also inquired about their feelings, opinions and fears with respect to insulin reactions. Hypoglycemic reactions were common: a mild episode occurred at least once a month in 58 percent of patients, and at least one severe reaction (defined by the need of assistance) during the past year was described by 26 percent. Both were positively related to the duration of the disease. However, the occurrence of mild and severe attacks was not related. In addition, patients prone to mild hypoglycemia seem to be somewhat different from patients prone to severe attacks in their attitudes towards the disease. For example, mild reactions are more frequent in patients devoted to perfect control, whereas severe episodes were observed more frequently in those who did not think that controlling diabetes is a difficult task. The social consequences of any reaction, even mild, were important in 30 percent of the patients. Twenty-one percent of the patients said that the short-term risks of a hypoglycemic reaction, even if correctly treated, were high and 6 percent said that the fear of having a reaction was unbearable. In addition to rational explanations, emotional factors

For patients treated with S.C. injections intensively for months (pregnant women) or for those wearing a pump for weeks or months, the control achieved is much better but still not normal (normality is perhaps not necessary) and the risk of hypoglycemia remains. In pump holders several deaths have been recorded. Some are certainly due to hypoglycemic comas and secondary to improper prescription of the tool or a lack of "diabetic education."

We have recently done a study aimed at comparing the medical and psycho-social implications of pump-therapy in out-patients who were not highly selected and in patients undergoing intensified conventional therapy. Among 15 patients

Table 8

Treatment Control and Hypoglycemic Reactions in the Two Groups of Patients

	General Practitioners (n = 64)[a]	University Hosp. (Hotel-Dieu) (n = 83)[a]	p
Fasting Blood Glucose (mmol/l)	9.6 ± 3.1	8.1 ± 3.6	< .05
Post-Prandial Blood Glucose (mmol/1)	12.0 ± 4.6	8.9 ± 3.8	< .001
Glycosuria (g/24h)	16.0 ± 15.0	11.0 ± 12.0	NS
N° of Daily Insulin Injections:			
1	42	9	
2	50	61	< .001
3	8	30	
Hypoglycemic Reactions During The Last Year:			
None	8	8	
Some During the Year	16	22	
Slight Some Each Month	24	28	NS
Some Each Week	52	42	
None	73	69	
Severe With Hospitalization	11	13	NS
With Glucagon Injections	16	18	

Results expressed as mean ± SD
[a]Expressed as percentage of total patients
Adapted from Basdevant et al., 1983.

treated for 8 months with this continuous subcutaneous insulin infusion (CSII) with a pump, 5 thought they were considered to be "handicapped" people by their colleagues and/or by their relatives. Almost all continued sports and/or physical activities and all appreciated the greater freedom for eating provided by the pump. Ten said that they had fewer and milder hypoglycemic reactions than before using pumps.

Table 9

Social and Professional Life Data in the Two Groups of Patients

	General Practitioners (n = 64)[a]	University Hosp. (Hotel-Dieu) (n = 83)[a]	p
Index of Social Life[b]	28.56 ± 5.02	25.70 ± 5.41	< .01
Dining out with Friends:			
Never	31	31	
Sometimes/Year	50	31	< .05
Sometimes/Month	8	21	
Sometimes/Week	11	17	
Dining at Home with Friends:			
Never	40	29	
Sometimes/Year	42	32	< .05
Sometimes/Month	10	26	
Sometimes/Week	8	13	
Do you Practice Gardening or other Leisure Activities?			
Never	50	37	
Sometimes/Year	19	6	< .05
Sometimes/Month	8	16	
Sometimes/Week	23	41	
Do you think you could have Difficulty or have you had Difficulty in getting a new Job?			
Percentage of "Yes"	44	69	< .01
N° of Medical Consultations/Yr	7.9 ± 5.5	3.7 ± 3.0	< .001
Percentage of People Hospitalized at least once	86	94	NS
Does the special diet entail some additional expenses (Patient's View)			
Percentage of "Yes"	48	67	< .05
Do you know about Glucagon?			
Percentage of "Yes"	81	99	< .001

[a]Expressed as percentage of total patients
[b]Results expressed as mean ± SD. This index ranging from nine (very active) to 36 (very inactive) was compiled based on nine questions about social and leisure activities.
[c]Results expressed as mean ± SD
Adapted from Basdevant et al., 1982.

Table 10

Percentage of Diabetic Patients who Experienced Hypoglycemic Reactions

	Hypoglycemic Reactions		Severe (Within Last 5 Years)
	Mild (Per Month)	Severe (Within Last Year)	
No Answer	16	7	10
No Reaction	26	67	48
One or More	58	26	42
1	11	17	10
2-5	28	7	
6-10	9	1	5
> 10	10	1	3

All figures are percentages of the 172 subjects who returned the questionnaire. Adapted from Goldgewicht et al., 1983.

Our conclusions from that study made on this small group of patients were that CSII might lead to social handicaps (because of the relatively large size of the external device) but did not influence sleep, sexual life or physical activities while diminishing the number and severity of hypoglycemic reactions and allowing more freedom for eating habits (Grimm, Haardt, Levy, and Slama, 1983).

As discussed previously a serious medical controversy is centered around the desirable degree of control. This includes some discussion on the risk of late nocturnal hypoglycemia (Unger, 1982) and of course on the need to know what levels of blood glucose are dangerous or critical.

Since it is very probable that the nocivity of hyperglycemia lies in its action on circulating and tissue proteins, the deleterious action of hyperglycemia will be related not only to the (high) levels reached but also to the duration of hyperglycemic periods: the accumulation of sugar or sugar derivates in tissue (glyco) proteins with slow turnover induces changes in the physiological and anatomical properties of some organs (micro vessels in kidney and retina, lens, etc.... (Burn, 1981)). This "accumulative" point of view might offer more flexibility for the management of diabetic patients since it will be possible to take into account several parameters in deciding what degree of control is needed, during the night for instance in order to limit some risks in the short-term time range.

Table 11

Relationship between the Frequency of Hypoglycemic
Reactions and other Parameters (172 Patients)

	Mild Reactions Per Month[a] Correlation-Coefficient		Severe Reactions Per Year Correlation-Coefficient	
	(r)	(p)	(r)	(p)
Duration of Diabetes	0.22	< .01	0.16	< .05
Duration of Insulin	0.29	< .001	0.25	< .01
Age	-0.02	NS	-0.13	NS
Body Mass Index	-0.13	NS	-0.16	< .05
Number of Daily Urine Analyses	0.15	NS	0.18	< .05
Number of Home Blood Glucose Analyses Per Month	0.24	< .01	0.07	NS
Number of Insulin Injections				
1 Daily	2[b]	< .01	0.4[b]	NSD
2 or 3 Daily	4		0.4	

[a]The correlation coefficient (r) between the frequencies of mild and severe reactions is 0.15 (NS).

[b]average number of reactions.

The p value results from a Student's t-test between the two groups.

NS: not significant.

Adapted from Goldgewicht et al., 1983

7. Diabetics' Feelings and Behavior

Some information is now at hand on patient feelings and behavior. Patient education is now one of the most important technics for achieving better control and/or making the patient use the therapeutic tools properly (Etzwiller, 1978). However, our own studies showed us that doctors' and patients' feelings are not always identical.

For instance an agreement is usually possible on the role and necessity of home blood glucose measurements. The patients understand well when their own daily comfort is concerned. We observed that nearly 80% of the patients who had been taught to use this technic in the ward continued to use it at home with a mean use of 10 to 12 stix per week (Sallee, Elgrably, and Slama, 1983). But this is not the case when doctors try to recommend the use of multiple daily insulin injections. Another example is that of the definition of what is a "severe" hypoglycemic reaction. For the physician this is the case when a coma occurs or if glucagon injection or hospitalization is necessary or generally speaking if the need of external assistance is mandatory. For some diabetic patients on insulin (about 30%) a severe hypoglycemic reaction is a reaction that may lead to disturbing social repercussions (Goldgewicht et al., 1983). Another point of disagreement is the question of refined sugars in the diet. In one enquiry we discovered that almost all patients were eating foods containing refined sugars, but not every day, and in small amounts (Dehlinger, Laffitte, Truffe et al., 1984).

When we questioned diabetics on their attitudes about diabetes, its management and the control they achieved, we were surprised to discover very often that some had a pessimistic point of view (Goldgewicht et al, 1983). Tables 12 and 13 summarize these findings.

8. How can Control be Improved?

Several lines of actions are possible. We must improve the quality of the tools we use and/or make a good choice among them according to each specific patient. For instance some prefer pumps, others multiple injections, others will perhaps prefer some kind of implantable devices.

Patient education is one of the most important technics for improving control. A great deal of literature is now available in this field and reviews are regularly published (Elgrably, Denys, Traynard, Tchobroutsky, 1981; Hamburg and Inoff, 1982; Mazzuca, 1982; Surwitt, Scovern, Feinglos, 1982).

Table 12

Feelings of the Patients about the Quality of their Self Care
(172 Insulin Treated Patients)

- Do their Best to			
Improve Control	87%		
Stick to Diet	72%		
- Urine Analysis		- Feelings about the Quality of	
		Control Achieved	
Never	4%	Very Bad	17%
Irregularly	12%	Bad	15%
Regularly 2/Day	27%	Good	28%
3/Day	57%	Very Good	40%

Adapted from Goldgewicht et al, 1983.

Table 13

Feelings Opinions and Fears about Hypoglycemic
Reactions (172 Patients)

Causes (Several Answers)	
- Difficulty of "Playing the Game"	90%
- Carelessness	11%
- Emotions	14%
- Life is Difficult	6%
- "Mysterious"	11%
Short-Term Risks (Treated)	
-None or Very Low	73%
- High	21%
- Unbearable	6%
Long-Term Risks	
- Hypoglycemia	19%
- Hyperglycemia	55%
- Both	19%
- ?	7%

Adapted from Goldgewicht (1983)

Individual and group approaches are used. The action upon the nonspecialized medical community is of paramount importance but is very difficult and probably not successful at the present time.

The main difficulty probably lies in the discrepancy between the well being of the diabetic patient--which is achieved by means of a minimal treatment--and the necessity of tight daily control to prevent, at least in part, or minimize the long-term complications of the disease. The lack of "scientific" proof allowing one to be scientifically sure that hyperglycemia is harmful has contributed to the well known "control controversy" (Ingelfinger, 1977). It explains why specialists and laymen in North America have planned a therapeutic trial aimed at demonstrating such a link between chronic hyperglycemia and microvascular lesions (American Diabetes Association, 1982).

Last but not least the need for information for the nonmedical nondiabetic community is obvious. The diabetic patient very often feels he is a rejected member of the "normal" community.

198

9. Conclusions and Summary

Controlling diabetes usually means lowering blood sugar levels. To which degree this lowering is necessary in order to prevent the long-term diabetic complications is not known. Some authors are even discussing the relationship between these two parameters despite the great amount of indirect evidence showing that the duration and the severity of chronic hyperglycemia is responsible for the microvascular and neurological changes observed in diabetic patients. However a very large consensus suggests that the ideal goal is to maintain blood glucose levels as near to normal as possible in the patients with long life expectancy as long as no handicaps make this goal unrealistic. In some situations, such as a pregnancy (one in the near future or one already existing), this goal is mandatory.

But the available tools for treatment do now allow normalization of blood sugar levels in the insulin treated patient everyday over the years. The risks and handicaps of the treatment itself may be serious and must be taken into account.

Since the patient is in fact his own doctor in his daily life it is of paramount importance to teach and educate diabetic subjects in order to make them feel free and responsible, under the and the support guidance of the medical team. The feelings and behavior of the patients, their consequences upon the quality of treatment and the degree of control achieved have been discussed.

Controlling diabetes is a difficult, frustrating daily task but is justified for the vast majority of diabetic patients. This is not only true for the individual patient but also for the whole community since uncontrolled diabetes in one of the most expensive diseases in developed countries.

References

American Diabetes Association, Organization Section. Proposed protocol for the clinical trial to assess the relationship between metabolic control and the early vascular complications of insulin-dependent diabetes. *Diabetes* 31 (1982) 1132-1133.

Basdevant, A., Costagliola, D., Lanoe, J. L., Goldgewicht, C., Triomphe, A., Metz, F., Denys, H., Eschwege, E., Fardeau, M., Tchobroutsky, G. The risk of diabetic control: A comparison of hospital versus general practice supervision. *Diabetologia* 22 (1982) 309-314.

Bunn, H. F. Nonenzymatic glycosylation of protein: Relevance to diabetes. *American Journal of Medicine* 70 (1981) 325.

Dehlinger, C., Laffitte, A., Truffe, P., Slama, G., Costagliola, D., Celani, E. Goût sucré et consommation de sucres simples chez 88 diabétiques de type I et de type II. *Diabete et Metabolisme* (1984)

Elgrably. F., Denys, H., Traynard, P. Y., Tchobroutsky, G. Evolution des idées sur l'éducation desa diabétiques. Le point en 1981. *Medecine Hygiene* 40 (1982) 46-51.

Etzwiller, D. D. Education of the patient with diabetes. *Medical Clinics of North America* 62 (1978) 857-866.

Goldgewicht, C., Slama, G., Papoz, L., Tchobroutsky, G. Hypoglycemic reactions in 172 type-1 (insulin-dependent) diabetic patients. *Diabetologia* 24 (1983) 95-99.

Grimm, J. J.. Haardt, M. J., Levy, A., Slama, G. Psychosocial implications of CSII in non-highly selected patients. In Artificial insulin delivery systems pancreas and islet transplantation. Study group of the EASD. Second Workshop IGls (Austria), 1983, p. 22

Hamburg. B. A.. Inoff, G. E. Relationships between behavioral factors and diabetic control in children and adolescents: A camp study. *Psychosomatic Medicine* 44 (1982) 321-339.

Ingelfinger, F. J. Debates on diabetes. *New England Journal of Medicine* 296 (1977) 1228-1229.

Sallee, F. X., Elgrably, F., Slama, G. Analyse du suivi à domicile de l'autosurveillance glycémique chez 113 diabétiques insulino-traités. *Diabete et Metabolisme* 10 (1984) 44-47.

Siperstein, M. D., Foster, D. W., Knowles, H. C., Jr. et al. Control of blood glucose and diabetic vascular disease. *New England Journal of Medicine* 296 (1977) 1060-1062.

Stowers, J. M. Assessment and management of diabetic pregnancy. In Diabetes Mellitus. M. Brownlee (ed.), Current and Future Therapies, Vol. 5, Garland STPM Press (New York and London), 1981. Pp. 151-176.

Surwitt, R. S., Scovern, A. W., Feinglos, M. N. The role of behaviour in diabetes care. *Diabetes Care* 5 (1982) 337-342.

Symposium on home blood glucose monitoring. *Diabetes Care* 3 (1980) 57-127.

Tchobroutsky, G. Metabolic control and diabetic complications. In Diabetes Mellitus. M. Brownlee (ed.), Current and Future Therapies, Vol. 5, Garland STPM Press (New York and London), 1981. Pp. 1-39.

Unger, R. H. Nocturnal hypoglycemia in aggressively controlled diabetes. *New England Journal of Medicine* 306 (1982) 1294.

12

MEDICAL CARE UTILIZATION AND SELF-REPORTED HEALTH OF HYPERTENSIVES
- Results of the Munich Blood Pressure Study -

U. Härtel, M.A.
U. Keil, M.D., Ph.D.
V. Cairns, Ph.D.

1. Introduction

In the Federal Republic of Germany (FRG), data on the utilization of medical care based on representative samples of defined populations are scarce. Medical care use has been investigated mainly by analyzing data from routine medical records, such as information for sickness funds, disability insurance, and general practices. The use of such data to study disease etiology is clearly limited (Keil, 1977). Furthermore, these data only provide information on individuals who have already been using medical care, so that a representative picture of the medical care utilization pattern of the general population cannot be obtained.

The Munich Blood Pressure Study (MBS) was based on a random sample of the adult population of Munich. The main goal was to study the prevalence, awareness, treatment, and control of hypertension. Based on these data, a community-wide hypertension education and control program has been implemented (Keil et al., 1982; Keil et al., 1983). A secondary goal is to study the range and determinants of medical care use in the population. The association between the frequency of physician visits and the degree of awareness, treatment, and control of hypertension in the Munich population was of particular interest and will be emphasized in this presentation. There are many variables which might explain part of the observed medical care utilization patterns. However, the analyses here will be restricted to perceived "need factors" (illness variables) although data on "predisposing" and "enabling" variables (Aday and Andersen, 1975) were also collected in the MBS.

With regard to the association between physician visits and hypertension control, it is commonly assumed that hypertensives who visit a physician frequently have a much greater chance of having their high blood pressure detected and treated than hypertensives who rarely see a physician. For example, the finding that more hypertensive women than men are aware of their high blood pressure and are under treatment is generally explained by the fact that women visit physicians more often than men (Wagner, Warner, and Slome, 1980). Hypertension is one of the very few chronic conditions for which medical treatment has been shown to be effective in reducing morbidity and mortality (Hypertension Detection and Follow-up Program Cooperative Group, 1979, 1982). Therefore, a basic supposition of this presentation is that it is useful and necessary to detect hypertension and to bring it under control.

2. The Munich Blood Pressure Study

2.1 Study design

The Munich Blood Pressure Study is a panel study with data collected from the same subjects in 1981 and 1982. Of the 1.3 million inhabitants of Munich, 524,328 met the criteria of the study population: between 30 and 69 years of age, residents of Munich, and German citizens. From these, a random sample of 3,400 was drawn. Of the 3,400 sampled citizens, 202 were lost in the time interval between sampling and the interview. Losses were for various reasons, such as migration, death, and institutionalization. Of the remaining 3,198 people, 2,216 participated in the study, which corresponds to a response rate of 69.3 percent-- 71.8 percent of the eligible men and 67.2 percent of the eligible women.

2.2 Data collection instruments

The first round of data collection of the Munich Blood Pressure Study lasted from December 1980 to May 1981. These are the data upon which most of the findings in this paper are based. The subjects went to one of seven examination centers for a standardized interview, three blood pressure measurements, and height and weight recordings. The questionnaire related to utilization of medical care, history of high blood pressure, attitudes and knowledge concerning high blood pressure and other cardiovascular disease risk factors, family medical history, medication history (including antihypertensive drugs), smoking and drinking habits, and socioeconomic status.

2.3 Blood pressure measurement and categories of high blood pressure

The Hawksley Random Zero Sphygmomanometer was used for the Blood Pressure (BP) measurements. BP recording took place after the completion of the interview, which lasted on average for 30 minutes, so that each person had been in a sitting position for about 30 minutes before the first BP recording was made under the strictly standardized conditions recommended by the American Heart Association. There were intervals of three minutes between each measurement. Each time the first, fourth, and fifth phases of the Korotkoff sounds and the pulse rate were recorded. Three cuff sizes (13 x 23cm, 13 x 28cm, 14 x 35cm) were used according to the circumference of the right upper arm of the participant. The BP data presented in this text are based on the first and fifth phase of the Korotkoff sounds. All results concerning blood pressure values are based on the mean of the second and third BP measurements. The classification of normotensive, borderline, and hypertensive BP levels based on WHO criteria is displayed in Table 1.

Table 1

Classification of normotensive, borderline and hypertensive blood pressure levels (WHO criteria). Munich Blood Pressure Study I (MBS 1980/81)

Normotension	SBP	< 140 mm Hg	and
	DBP	< 90 mm Hg	
Borderline	SBP	140-159 mm Hg	and/or
Hypertension	DBP	90- 94 mm Hg	
Hypertension	SBP	≥ 160 mm Hg	and/or
	DBP	≥ 95 mm Hg	

2.4 Definition of awareness, treatment, and control of hypertension

Awareness, treatment, and control of hypertension are defined as follows:

a.   "Controlled hypertensives" are people who are aware of their high BP, take antihypertensive medication, and have normotensive or borderline BP values when measured after the interviews.

b.   "Aware, treated, uncontrolled hypertensives" are aware of their high BP and take antihypertensive medication, but have hypertensive BP values.

c.   "Aware, untreated hypertensives" are aware of their high BP, take no antihypertensive medication, and have hypertensive BP values.

d.   "Unaware hypertensives" are unaware of their high BP, do not take antihypertensive medication, and have hypertensive BP values. These are the newly detected hypertensives.

Estimates of the level of awareness, treatment, and control of hypertension in the population of Munich are based on three measurements: *first*, a detailed history of the drug consumption during the week preceding the examination; *second*, the question "Have you ever been told you have elevated or high blood pressure?" and, *third*, blood pressure.

2.5 Measurement of medical care utilization

In this paper, medical care utilization refers to "physician use." Two variables are used: "whether a physician was visited in the last four weeks" and the "frequency of physician visits in the year preceding the examination." No categorization according to the different medical specialists was made. However, visits to the dentist were not considered as physician visits. Each hospitalization was counted as one physician visit. All information concerning medical care utilization is based on the subjects' answers to selected questions in the interviews.

2.6 Measurement of health

Objective and subjective measures of the individual's state of health were obtained. The objective measure was the recording of blood pressure. Two questions were asked during the interview to determine the individuals self-reported state of health: 1. "How do you rate your present state of health?" Possible answers: "excellent, good, fair, poor." 2. "Have you ever been diagnosed or treated by a physician for any of the following diseases?" To answer this question the participants were given the following list of chronic conditions: diabetes mellitus, hyperlipidemia, congestive heart failure, myocardial infarction, angina pectoris, other heart diseases, stroke, chronic renal diseases, gastric ulcer, other chronic diseases.

3. Results

3.1 Prevalence rates

Table 2 shows the prevalence of hypertension for men and women by ten-year age groups. Hypertensive BP values were found in 14.0 percent of the participants; 17.7 percent of men and 10.7 percent of women had hypertensive BP values. A particularly large difference in the prevalence rates of hypertension between the sexes is found in age group 30 to 39; in this age group 10.3 percent of men but only 2.7 percent of women have hypertensive BP levels. About every third Munich citizen (32.5 percent in the age group 30 to 69 had elevated BP values, that is, borderline or hypertensive BP values ($\geqslant$ 140/90 mm Hg).

3.2 Awareness, treatment, and control

Figure 1 shows clearly that the awareness, treatment, and control of hypertension is much more likely for female than male hypertensives. Among hypertensives 41.9 percent of females and only 21.9 percent of males are controlled. Figure 1 also shows that only 15.7 percent of female hypertensives are unaware of their hypertension, whereas 38.0 percent of male hypertensives are unaware of their hypertension. When analyzing these data by ten-year age-sex groups, the best "control" status of hypertension is for women age 60 to 69; the worst control status

Table 2

Prevalence of hypertension (WHO criteria). Munich Blood Pressure Study I (MBS 1980/81)

	Prevalence (%)	95% - Confidence Interval (%)
Total	14.0	12.6 - 15.5
Men	17.7	15.4 - 20.1
30-39 Y.	10.3	6.9 - 13.7
40-49 Y.	19.3	15.0 - 23.6
50-59 Y.	22.8	17.4 - 28.1
60-69 Y.	20.9	15.0 - 26.8
Women	10.7	9.0 - 12.5
30-39 Y.	2.7	0.8 - 4.5
40-49 Y.	8.5	5.4 - 11.5
50-59 Y.	14.9	11.0 - 18.9
60-69 Y.	18.2	13.3 - 23.1

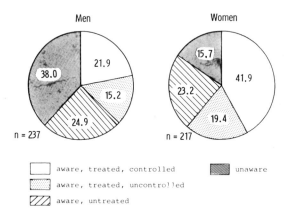

Figure 1: Awareness, treatment and control of hypertension in male and female hypertensives, with control defined as SBP < 160 mm Hg and DBP < 95 mm Hg. Age group 30-69. Munich Blood Pressure Study I (MBS 1980/81).

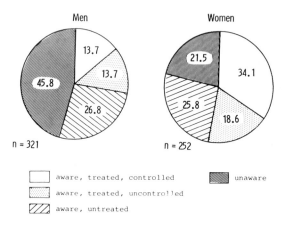

Figure 2: Awareness, treatment and control of hypertension in male and female hypertensives, with control defined as DBP < 90 mm Hg. Age group 30-69. Munich Blood Pressure Study I (MBS 1980/81).

of hypertension is for men age 30 to 39. Using the "control" criteria DBP below 90 mm Hg as recommended by the U.S. National High Blood Pressure Education Program (Joint National Committee Report, 1984) it can be seen that in Munich only 34.1 percent of the hypertensive females and 13.7 percent of hypertensive males have controlled BP values (Figure 2).

3.3 Actual hypertension

Since there are controlled hypertensives, it is appropriate to calculate prevalence rates of "actual" hypertension. Actual hypertensives consist of those subjects with measured hypertensive BP values plus "controlled" hypertensives according to the definition given above. Table 3 provides prevalence rates of "actual" hypertension by age and sex. The prevalence rate of actual hypertension for the age group 30 to 69 is 22.7 percent for men and 18.5 percent for women. The corresponding figures from the Hypertension Detection and Follow-up Program (HDFP) in the USA are 19 percent for white men and 18 percent for white women. However, in the HDFP study the definition of hypertension is based solely on the DBP values with a cut-point of ninety-five millimeters of mercury (mm Hg) and, therefore, the percentage of actual hypertension is slightly inflated in our data, when compared with the HDFP study data.

Table 3

Prevalence of "actual" hypertension[+] by age and sex.
Munich Blood Pressure Study I (MBS 1980/81)

	N	Prevalence (%)	95% - Confidence Interval (%)
Total	2216	20.5	18.8 - 22.2
Men	1042	22.7	20.2 - 25.2
30-39 Y.	302	11.3	7.7 - 14.9
40-49 Y.	326	21.8	17.3 - 26.3
50-59 Y.	232	31.9	25.9 - 37.9
60-69 Y.	182	31.9	25.1 - 38.7
Women	1174	18.5	16.3 - 20.7
30-39 Y.	298	3.4	1.4 - 5.4
40-49 Y.	319	12.9	9.2 - 16.6
50-59 Y.	315	25.1	20.3 - 29.9
60-69 Y.	242	36.0	29.9 - 42.1

[+] "actual" hypertension: controlled hypertensives (SBP < 160 mm Hg, DBP < 95 mm Hg) plus subjects with measured hypertensive BP values (WHO criteria).

Table 4

Physician visits within the last 4 weeks by awareness,
treatment, and control status of hypertensives and
normotensives by sex. Munich Blood Pressure Study I
(MBS 1980/81)

	Total	Physician Visit Within the Last 4 Weeks	
	N	N	%
Men			
Controlled Hypertensives	52	39	75.0
Aware, Treated, Uncontr. Hypert.	36	22	61.1
Aware, Untreated Hypertensives	59	22	37.3
Unaware Hypertensives	90	28	31.1
Normotensives	805	261	32.4
Women			
Controlled Hypertensives	91	73	80.2
Aware, Treated, Uncontr. Hypert.	42	36	85.7
Aware, Untreated Hypertensives	50	23	67.6
Unaware Hypertensives	34	14	41.2
Normotensives	957	441	46.1

3.4 Medical care utilization of hypertensives and normotensives

The pattern of physician visits by hypertensives and normotensives is shown in Table 4. Here normotensives include borderline hypertensives. Seventy-five percent of men with controlled hypertension reported making a physician visit during the preceding four weeks. 61.1 percent of the treated but uncontrolled hypertensive men reported seeing a physician during the preceding four weeks. On the other hand, only 31.1 percent of unaware hypertensive men had seen a physician during the preceding four weeks. The figures for normotensive men are very similar to those for unaware hypertensive men. Twice as many treated hypertensive men as unaware hypertensive and normotensive men had seen a physician during the previous four weeks. Among women, the groups of hypertensives and normotensives show patterns of physician visits very similar to those of men. 80.2 percent of controlled hypertensive women and 41.2 percent of unaware female hypertensives had seen a physician during the previous four weeks. This percentage of physician visits for unaware female hypertensives is very similar to that of normotensive women.

Table 5 provides data on how often the different groups have seen a physician within the last year. The data are generally the same as those in Table 4. Treated hypertensives and "unaware hypertensives"--both men and women--are the extreme groups with regard to the categories "more than ten physician visits" or "no physician visit." Thus, the chance of detecting and treating elevated BP is relatively high for individuals frequently using physician services. However, there is a high percentage of treated hypertensives who visit their physician more than ten times a year but fail to show controlled BP values ($33/99 = 33.3$ percent).

Rather than discuss the various reasons for failures in therapeutic success, we will expand on the question of why the "controlled hypertensives" visit their physician more frequently than all the other groups studied. Assuming that a successful antihypertensive treatment usually does not require more than ten physician visits per year and that high blood pressure alone normally causes only minor complaints, one can suspect that the treated hypertensives visited their

Table 5

Frequency of physician visits within the last year by awareness, treatment, and control status of hypertensives and normotensives by sex. Munich Blood Pressure Study I (MBS 1980/81)

	Total	Physician Visits Within the Last Year			
		More Than 10		None	
	N	N	%	N	%
Men					
Controlled Hypertensives	52	27	51.9	---	0.0
Aware, Treated, Uncontr. Hypert.	36	12	33.3	1	2.8
Aware, Untreated Hypertensives	59	12	20.3	11	18.6
Unaware Hypertensives	90	4	4.4	32	35.6
Normotensives	805	101	12.5	176	21.9
Women					
Controlled Hypertensives	91	39	42.8	1	0.1
Aware, Treated, Uncontr. Hypert.	42	21	50.0	---	0.0
Aware, Untreated Hypertensives	50	14	28.0	5	10.0
Unaware Hypertensives	34	5	14.7	5	14.7
Normotensives	955	165	17.5	77	8.1

physicians not only for antihypertensive treatment, but also because of other illnesses.

3.5 Self-reported health of hypertensives and normotensives

Number of chronic diseases. Table 6 shows the number of self-reported chronic diseases by age and sex. There were no significant differences between men and women. However, there were differences between age groups. About 50 percent of the male and female participants in the age group 30 to 39 reported having had at least one chronic condition detected or treated, whereas about 80 percent in the age group 60 to 69 reported having had at least one chronic condition detected and/or treated. Note that for those over 50, a large percentage reported 3 or more chronic diseases.

Table 7 shows the number of chronic diseases in relationship to BP status for the age groups under and over 50 years. In the age group 30 to 49, 29.8 percent of the "aware, untreated hypertensives," 46.3 percent of the "unaware hypertensives," and 44.0 percent of the normotensives did not report a chronic condition. However, the differences between the unaware hypertensives and the

Table 6

Number of self-reported chronic diseases by age and sex. Munich Blood Pressure Study I (MBS 1980/81)

	Total N	Number of Self-Reported Chronic Diseases		
		None %	1 - 2 %	3 and More %
Men	1042	35.1	52.9	12.0
30-39 Y.	302	49.7	48.0	2.3
40-49 Y.	326	39.6	52.1	8.3
50-59 Y.	232	22.4	55.2	22.4
60-69 Y.	182	19.2	59.3	21.4
Women	1174	31.0	56.4	12.7
30-39 Y.	298	51.7	44.0	4.4
40-49 Y.	319	31.3	60.5	8.2
50-59 Y.	315	20.2	64.8	15.2
60-69 Y.	242	19.0	55.4	25.6

Table 7

Number of self-reported chronic diseases of hypertensives and normotensives by age. Munich Blood Pressure Study I (MBS 1980/81)

| | Total N | Number of Self-Reported Chronic Diseases | | |
		None %	1 - 2 %	3 or More %
Age < 50 Y.	1245	42.8	51.3	5.9
Controlled Hypertensives	27	18.5	59.2	22.2
Aware, Treated, Uncontr. Hypert.	15	26.7	60.0	13.3
Aware, Untreated Hypertensives	47	29.8	46.8	23.4
Unaware Hypertensives	67	46.3	47.8	6.0
Normotensives	1089	44.0	51.4	4.6
Age ≥ 50 Y.	971	20.2	59.1	20.7
Controlled Hypertensives	116	9.5	54.3	36.2
Aware, Treated, Uncontr. Hypert.	63	15.9	39.7	44.4
Aware, Untreated Hypertensives	62	29.0	58.1	13.0
Unaware Hypertensives	57	31.6	57.9	10.5
Normotensives	673	20.7	62.0	17.4

normotensives are more pronounced in the age group over 50: 31.6 percent of the unaware hypertensives do not report a chronic disease compared to only 20.7 percent of the normotensives. In the age group over 50 there were pronounced differences between all treated hypertensives, controlled or not, and all others. For example, 39 percent of the treated hypertensives report three or more chronic diseases whereas only 12 percent of the untreated (aware and unaware) and 17.4 percent of the normotensives do so.

General evaluation of physical condition. The other health status measure pertains to the subjective evaluation of the individuals' present physical condition. Possible responses were: excellent, good, fair, poor. Table 8 shows the self-rated health status of hypertensives and normotensives by sex. Fewer women than men define their health as excellent or good. This tendency is found in all age groups. With regard to the BP status in men, there are only small differences in the health status between "controlled" and "aware, treated, uncontrolled" hypertensives. However, these groups seem to feel worse than the other groups. That is, about 60 percent of the treated hypertensives consider their health status as excellent or good compared to about 80 percent of those who are untreated. These results are

Table 8

**Self-reported health of hypertensives and normotensives by sex.
Munich Blood Pressure Study I (MBS 1980/81)**

	Total	Self-Reported Health Excellent/Good	
	N	N	%
Men	1042	801	76.9
Controlled Hypertensives	52	31	59.6
Aware, Treated, Uncontr. Hypert.	36	21	58.3
Aware, Untreated Hypertensives	59	46	78.0
Unaware Hypertensives	90	74	82.2
Normotensives	805	629	78.1
Women	1174	772	65.8
Controlled Hypertensives	91	51	56.0
Aware, Treated, Uncontr. Hypert.	42	21	50.0
Aware, Untreated Hypertensives	50	27	54.0
Unaware Hypertensives	34	25	73.5
Normotensives	957	648	67.7

similar to those regarding the number of self-reported chronic conditions of treated hypertensives. For women the relationship between treatment status and self-reported physical condition is less clear. Among hypertensive women who are aware of their high blood pressure, those who are treated and untreated feel the same physically. It also can be seen, that the "aware, untreated" hypertensive women repeat less often an excellent or good health status than the "aware, untreated" hypertensive men (54.0 percent compared to 78.0 percent). The data from "aware, untreated hypertensive" men suggest that the awareness of high blood pressure does not diminish the sense of physical well-being.

3.6 Results of log-linear analyses

As we have seen there are interrelationships among the variables mentioned: age, number of chronic diseases, self-reported health, number of physician visits, and treatment status. It is therefore desirable to do analyses that include all these variables simultaneously. We conducted log-linear analyses to investigate the interactions among these five variables. Only actual hypertensives were included. They were divided into those treated and untreated for hypertension. The other variables included were age, which was dichotomized at 55 years; number of chronic diseases, which was dichotomized into less than two and two or more; number of physician visits in the previous year, dichotomized into zero to two and three or more; and self-reported health dichotomized into excellent/good and fair/poor.

The χ^2 from the "best-fitting" hierarchical model had a p-value of 0.21. That is, the observed frequencies were not significantly different from the expected frequencies calculated from the model. There was one third-order interaction term containing the variables: TREATMENT STATUS - NUMBER OF CHRONIC DISEASES - FREQUENCY OF PHYSICIAN VISITS. Four second-order interaction terms were also in the model: FREQUENCY OF PHYSICIAN VISITS and AGE; TREATMENT STATUS and AGE; NUMBER OF CHRONIC DISEASES and SELF-REPORTED HEALTH; and finally SELF-REPORTED HEALTH and FREQUENCY OF PHYSICIAN VISITS. The addition of other second- or third-order terms did not significantly improve the fit of the model. In general, the results from the log-linear analyses support the findings presented earlier.

4. Discussion

The city of Munich is a community with one of the highest physician densities in the world. Given this, the control of hypertension in men and women was lower than we expected: less than one quarter of male hypertensives and less than half of female hypertensives are controlled. Even in the best treated group, that is 60 to 69 year old women, only 50 percent of hypertensives have controlled blood pressure values. As hypothesized, the chance of detecting and treating high blood pressure increased with the frequency of physician visits. This holds true for men and women and supports the results of the study by Wagner, Warner, and Slome (1980). However, frequent physician visits do not guarantee an effective treatment of hypertension. A high percentage of treated hypertensives visited their physicians more than ten times in the preceding year and failed to show normotensive or borderline BP values.

Possible explanations for the unsuccessful treatment of so many of the subjects are lack of compliance to medical regimens, alternative criteria for high blood pressure by physicians, and incorrect dosages of antihypertensive medication. The fact that treated hypertensives report more chronic diseases than untreated hypertensives and normotensives suggests that physicians are more likely to treat hypertension in patients who show further diseases or cardiovascular risk factors.

The different chronic illnesses that respondents were suffering from cannot be described here. Other analyses of the Munich Blood Pressure Study revealed that 25 percent of all treated hypertensives reported that they were diagnosed as "diabetic." Only 7 percent of the other groups reported a diagnosis of diabetes. This also suggests that the frequent physician visits of treated hypertensives is related to the treatment of other illnesses.

As expected, the number of self-reported chronic diseases is related to self-reported health status. It appeared from log-linear analyses that the observed relationship between treatment status and self-reported health could be explained through other variables, such as the number of chronic diseases. This suggests that the treatment of high blood pressure alone does not necessarily lead to a diminished sense of well-being.

This suggestion is supported by the finding of the Munich Blood Pressure Study II (1981/82) that the hypertensives generally rated their treatment positively. More than half of the treated hypertensives said that they have felt better since they began treatment, and most of the remainder reported feeling exactly the same as before. Furthermore, 80 percent reported having noticed no side-effects from their antihypertensive medication. Those findings are different from what had been suggested by others (for example, by Alderman and Madhavan, 1981).

Women do visit their physician more often than men and, therefore, may be better informed about their blood pressure. However, we cannot determine why women visit their physician more frequently than men. There were no major differences in the number of chronic diseases reported by men and women. However, women did rate their present physical condition worse than men across all age groups.

Drawing conclusions about the causal pathways among all these variables is not possible with cross-sectional data. The results do suggest that in the Munich population there is an association between the frequency of physician visits, the number of chronic diseases, and the degree of awareness, treatment, and control of hypertension. The subjects self-rated health, on the other hand, did not appear to be directly related to the treatment status of hypertensives.

References

Aday, L. U. and Andersen, R. *The development of indices of access to medical care.* Ann Arbor: Health Administration Press, 1975.

Alderman, M. H. and Madhavan, S. Management of the hypertensive patient: A continuing dilemma. *Hypertension* 3 (1981) 192-197.

Andersen, R., Kravits, J., and Andersen, O. W. *Equity in health services.* Cambridge: Empirical Analysis in Social Policy, 1975.

Everitt, B. S. *The analysis of contingency tables.* London: 1979.

Gove, W. R. and Hughes, M. Possible causes of the apparent sex differences in physical health: an empirical investigation. *American Soc. Review* 44 (1979) 126-146.

Hypertension Detection and Follow-up Program Cooperative Group. Five-year findings of the hypertension detection and follow-up program. I. Reduction in mortality of persons with high blood pressure, including mild hypertension. *Journal of the American Medical Association* 242 (1979) 2563-2571.

Hypertension Detection and Follow-up Program Cooperative Group. Five-year findings of the hypertension detection and follow-up program. II. Mortality by race, sex and age. *Journal of the American Medical Association* 242 (1979) 2572-2577.

Hypertension Detection and Follow-up Program Cooperative Group. Five-year findings of the hypertension detection and follow-up program. III. Reduction in stroke incidence among persons with high blood pressure. *Journal of the American Medical Association* 247 (1982) 633-638.

Joint National Committee on Detection, Evaluation and Treatment of High Blood Pressure. The 1980 report of the joint national committee on detection, evaluation and treatment of high blood pressure. U.S. Department of Health and Human Services, Public Health Service, National Institutes of Health. NIH Publication, 1980.

Joint National Committee on Detection, Evaluation and Treatment of High Blood Pressure. The 1984 report of the joint national committee on detection, evaluation and treatment of high blood pressure. *Archives of Internal Medicine* Vol. 144, pp. 1045-1057, May, 1984.

Keil, U. Use of national health survey systems in epidemiological research. In P. Armitage (ed.) *National health survey systems in the European economic community.* Proceedings of a conference held in Bruxelles, Luxembourg: 1977, 147-151.

Keil, U., Stieber, J., Döring, A., and Fricke, H. Ergebnisse der Münchner Blutdruckstudie und Aufbau des Münchner Blutdruckprogramms. *Öffentl. Gesundheitswesen* 44 (1982) 727-732.

Keil, U., Döring, A., Stieber, J. Community studies in the Federal Republic of Germany. In F. Gross, T. Strasser (eds.) *Mild hypertension: Recent advances.* New York: Raven Press, 1983. Pp. 63-83.

Mechanic, D. Correlates of physician utilization: why do major multivariate studies of physician utilization find trivial psychosocial and organizational effects? *Journal of Health and Social Behavior* 20 (1979) 387-396.

Wagner, E. H., Warner, J. T., and Slome, C. Medical care use and hypertension. *Medical Care* 12 (1980) 1241-1250.

PART III

STRESS

Considerable evidence from a variety of countries now suggests that the onset and duration of illness is related to mechanisms for coping with one's stress. Further, a growing body of literature argues that supportive social environments, including family and friends, can buffer the impact of stress upon the individual. Since these social support systems vary cross-nationally, we have included a section describing recent developments in research on stress, social support, and illness.

Measuring life stress and social support has been difficult. In the first paper in this section, I. Sarason and B. Sarason discuss some of these measurement problems. They present new measures for life stress and for social support. The role of psychosocial factors, including social support in illness is described with original data in the paper by G. Kaplan. Finally, Biondi presents evidence on the mechanisms relating life stress to illness.

13

LIFE CHANGE, SOCIAL SUPPORT, COPING, AND HEALTH

Irwin G. Sarason, Ph.D.
Barbara R. Sarason, Ph.D.

Epidemiology is often defined as the study of the distribution of diseases and the factors that influence their distribution. A host of epidemiologically relevant factors--among them, sanitation, environmental and cultural variables, socioeconomic status, education, migration, marital status, and the death of loved ones-- have been identified. Many of these factors exert their impacts not because of their characteristics per se but because of the way they are processed by the individual. Whether disease and maladaption result from their occurrence depends on the situational context, the personal resources of the individual, and how he or she experiences the relevant event. Within the past decade epidemiologists as well as those in other disciplines have placed increased emphasis on the idea that the impact of environment or health can only be understood when the environmental demands are viewed in the context of the resources available for coping (Cassel, 1976; Henry and Stephens, 1977; Jenkins, 1979; Kaplan, Cassel, and Gore, 1977).

It is common to speak of many of these variables in terms of stress. Recognition is growing that stress from a variety of sources can play an important role in both physical and psychological disorders. Stress as a factor in many psychiatric disorders has been recognized for some time. However, the present diagnostic system of the American Psychiatric Association, DSM III (American Psychiatric Association, 1980) also expands the range of physical disorders in which stress is a likely contributing factor far beyond the narrower range of the psychophysiological disorders described in the previous DSM II system.

Since life experiences that evoke stress responses may be related to vulnerability to both psychological and physical breakdown, researchers have sought ways of assessing them. Because recent experiences often exert powerful influences and are more easily recalled than those that occurred many years ago, efforts have been made to quantify stressful life changes for specific time periods, such as the past year. Research has shown that persons who have experienced multiple stressors in the recent past are especially susceptible to conditions in which depression, anxiety, and overreactivity of physiological systems are prominent. Undersirable experiences seem to have a more negative effect on people than do desirable ones.

For a long time, physicians had observed an association between very severe stressors (wars, concentration camps, natural disasters) and illness. Even under such extreme stress the association between the events and physical or psychological symptoms is far from perfect. Some people deteriorate rapidly under severe stress, others show minimal to moderate deterioration, and still others seem unaffected. Clinical observations have also suggested that the stressful events of everyday life might play a role in illness and researchers have inquired into the relationship to illness of less cataclysmic events such as marriage, divorce, and loss of a job. Research findings have buttressed the idea that a great variety of stressful life events may be related to decreased emotional or physical health. Yet, it would be premature to conclude that there is a simple relationship between the build-up of negative life events and vulnerability to illness. For example, animal studies have shown complex effects of stressful events upon the immunological system. Some stressors may predispose the organism to some disease states and provide a protective effect against others (Friedman, Glasgow, and Ader, 1969). There are all too many examples in the history of science of the discovery of "simple" relationships that turn out to have a variety of complications and ramifications. The study of stress-arousing life changes has followed this pattern of increasing awareness of complex interactions among a wide variety of factors.

Of special interest are a number of variables that appear to moderate or render less stressful some of these events as experienced by some persons. Tentative positive relationships between these variables and positive health measures have also been suggested. This paper describes some of these variables and shows how they can be taken into account in research on stressful life events, and then suggests a theoretical formulation concerning the relationships of certain psychosocial variables to vulnerability.

1. Assessing Life Events

Stressful life events seem to set the stage for vulnerability to health impairment. Before meaningful study of the role of life experiences as stressors can be carried out, it is necessary to describe life events in some systematic and quantifiable fashion.

1.1 The Schedule of Recent Events

Holmes and Rahes (1967) The Schedule of Recent Events (SRE) provided the first widely used tool to assess life events. It yields a score that reflects the events of the recent past weighted by previously determined values of the events likely impacts

Since its initial development, the SRE has been used in numerous studies designed to determine relationships between life stress and indices of health and adjustment. Retrospective and prospective studies have provided support for a relationship between SRE scores and a variety of health-related variables. Life stress has, for example, been related to sudden cardiac death (Rahe and Lind, 1971), myocardial infarction (Edwards, 1971; Theorell and Rahe, 1971), pregnancy and birth complications (Gorsuch and Key, 1974), chronic illness (Bedell et al., 1977; Wyler, Masuda, and Holmes, 1971), and other major health problems such as tuberculosis, multiple sclerosis, and diabetes, and a host of less serious physical conditions (Rabkin and Struening, 1976). While not presenting conclusive evidence, these studies supported the position taken by Holmes and Masuda (1974) that life stress serves to increase overall susceptibility to illness.

Although some of the studies using the SRE were motivated primarily by the desire to determine whether particular physical disorders had psychosocial antecedents, an increasing number of researchers have taken a more conceptual and methodological approach. They have dealt with topics such as the relationship between life change and stress, have devised various ways of assessing life changes, and have related life change scores to various external criteria. In the course of this work, some researchers expressed the need for an instrument that would enable subjects to characterize events beyond simply whether or not the events had occurred in the recent past. In particular the way in which the SRE lumped together both desirable and undesirable events was questioned.

1.2 The Life Experiences Survey

An example of the type of instrument that has grown out of these methodological concerns is the Life Experiences Survey (LES) (Sarason, Johnson, and Siegel, 1978). It provides both positive and negative life change scores and permits individualized ratings of the impact of events and their desirability. These individualized measures have the advantage of providing reflections of person-to-person differences in the perception of events. Evidence in support of this approach was provided by Yamamoto and Kinney (1976) who found life stress scores, based on self-ratings of degree of stress experienced, to be better predictors than scores derived by employing mean adjustment ratings similar to those used with the SRE. Other investigators have also found that individualized self-ratings of the impact of life events aid in the prediction of clinical course (Lundberg, Theorell, and Lind, 1975).

221

The LES is a 47-item, self-report measure that allows subjects to indicate events they have experienced during the past year. Subjects can also indicate the occurrence of significant events they have experienced that are not on the LES list. A special supplementary list of ten events relevant primarily to student populations is available. Other special adaptations are possible. Some LES items were chosen to represent life changes frequently experienced by individuals in the general population. Others were included because they were judged to be events that, although less frequent in occurrence, might exert a significant impact on the lives of persons experiencing them. Thirty-four of the events listed in the LES are similar in content to those found in the SRE. However, certain of these SRE items were made more specific. For example, the SRE contains the items "Pregnancy" which might be endorsed by women but perhaps not by a man whose wife or girlfriend has become pregnant. The LES allows both men and women to endorse the occurrence of pregnancy in the following manner: Female: Pregnancy; Male: Wife's/girlfriend's pregnancy. The Schedule of Recent Events includes the item "Wife begins or stops work," an item which fails to assess the impact on women whose husbands begin or cease working. The LES scale lists two items: Married male: Change in wife's work outside the home (beginning work, ceasing work, changing to a new job, etc.) and Married female: Change in husband's work (loss of job, beginning of a new job, etc.). Examples of events not listed in the SRE but included in the LES are: male and female items dealing with abortion and concerning serious injury or illness of a close friend, engagement, and breaking up with boyfriend/girlfriend. Table 1 presents excerpts from the Life Experience Survey.

Subjects respond to the LES by separately rating the desirability and impact of events they have experienced. Summing the impact ratings of evidents designated as positive by the subject provides a positive change score. A negative change score is derived by summing the impact ratings of those events experienced as negative by the subject. Scores on the LES do not seem to be influenced by the respondent's mood state at the time of filling out the questionnaire (Siegel, Johnson, and Sarason, 1979a). In addition, the LES does not seem to be appreciably correlated with the social desirability response set.

1.3 Life experience correlates

The negative change score of the LES correlates significantly with measures of anxiety, depression, and general psychological discomfort. Studies have also found that negative change scores are related to myocardial infarction (Pancheri et al., 1980), menstrual discomfort (Siegel, Johnson, and Sarason, 1979b), the attitudes of mothers of at-risk infants (Crnic et al., 1980), job satisfaction (Sarason and Johnson, 1979), and college grades (Sarason, Johnson, and Siegel, 1978; Knapp and Magee, 1979). Michaels and Deffenbacher (1980) found the LES negative change score to be related to physical (seriousness of illness), psychological (depression, anxiety), and academic (grades) variables. While some researchers have found correlates for positive life changes, the magnitude and consistency of these relationships has usually not been robust.

Table 1

Excerpts from the Life Experiences Survey

Instructions

Listed below are a number of events which may bring about changes in the lives of those who experience them.

Rate each event that occurred in your life *during the past year as good or bad* (circle which one applies).

Show how much the event affected your life by circling the appropriate statement (no effect--some effect--moderate effect-great effect).

If you have not experienced a particular event in the past year, leave it blank.

Please go through the entire list before you begin to get an idea of the

Event	Type of event	Effect of event of your life			
1. Marriage	Good Bad	No effect	Some effect	Moderate effect	Great effect
2. Death of close family member	Good Bad	No effect	Some effect	Moderate effect	Great effect
a) mother	Good Bad	No effect	Some effect	Moderate effect	Great effect
b) father	Good Bad	No effect	Some effect	Moderate effect	Great effect
c) brother	Good Bad	No effect	Some effect	Moderate effect	Great effect
d) sister	Good Bad	No effect	Some effect	Moderate effect	Great effect
e) grandmother	Good Bad	No effect	Some effect	Moderate effect	Great effect
f) grandfather	Good Bad	No effect	Some effect	Moderate effect	Great effect
g) other (specify)	Good Bad	No effect	Some effect	Moderate effect	Great effect
3. *Male* Wife's/Girlfriend's pregnancy	Good Bad	No effect	Some effect	Moderate effect	Great effect
4. *Female* Pregnancy	Good Bad	No effect	Some effect	Moderate effect	Great effect
5. Gaining a new family member (through birth, adoption, family member moving in, etc.)	Good Bad	No effect	Some effect	Moderate effect	Great effect
6. Change of residence	Good Bad	No effect	Some effect	Moderate effect	Great effect
7 Ending of formal schooling	Good Bad	No effect	Some effect	Moderate effect	Great effect

One intriguing idea that merits further study is the possibility that negative and positive life changes are differentially useful in predicting particular types of psychological and physical criteria. Negative, but not positive, life events tend to correlate with emotional malfunction, such as general psychological distress, depression, and anxiety (Johnson and Sarason, 1978), as well as with behavioral problems, such as lowered grade point average (Knapp and Magee, 1979). On the other hand, a few studies have suggested that both positive and negative life changes contribute to physical illness. Two correlational studies with introductory psychology undergraduates have shown both positive and negative life changes to be associated with self-rated illness. In one study using the LES, the number of symptoms checked was correlated with number of positive events listed, number of negative events listed, and total events (Sarason et al., 1983). The second study found similar results, with significant correlations of positive, negative, and total life changes with the medical items on the Cornell Medical Index (Coppel, 1980).

It is possible that the totality of life changes affects the body's physiological homeostasis, whereas only negative life changes are associated with personal dissatisfaction and a lowered sense of emotional well-being. Petrich and Holmes (1977) have suggested that patients should be advised to pace the occurrence of positive and negative life events wherever possible. It may be that such a maneuver would be advantageous only for patients with physical problems. Controlling the occurrence of positive events might be counterproductive for individuals experiencing emotional problems.

As this overview suggests, research on life changes is becoming more methodologically sophisticated. Scales designed to (1) assess the subjective stress associated with events (Horowitz, Wilner, and Alverez, 1979), (2) deal with the important psychometric issues (Skinner and Lei, 1980; Ross and Mirowsky, 1979), and (3) reflect the multidimensionality of life changes (Ruch, 1977) are now being developed and bode well for progress in this area.

2. Moderator Variables

The last decade has seen a proliferation of research on the relationaship between the occurrence of life events in a given period of time and a variety of indices of psychological and physical well-being. The relationships uncovered have often been significant but not large. This is not surprising given that most life stress research has neglected both (1) individual differences other than the occurrence of certain life events and (2) social-environmental factors that might influence the individual's response. Recognition of this neglect and efforts to clarify the roles of these variables are now visible in the literature.

The effects of life changes can be moderated by variables outside of or within the individual. Stressors have differential effects on people depending upon (1) individual differences in their personalities, coping skills, motivation, and past experiences and (2) differences in their environment such as the presence or absence of situational props or aids (for example, having or not having visits from family members and friends after undergoing surgery). As these mediators of life stress are identified, measured reliably, and included in research designs, increased effectiveness in prediction is likely to result.

Researchers have recently addressed the question of what variables determine which individuals are likely to be most adversely affected by life change (Jenkins, 1979; Johnson and Sarason, 1979). Lack of attention to such moderator variables constitutes a major limitation of much life events research and, therefore, the predictive value of knowledge of life changes is diluted.

2. 1 Personality variables

While a life change may be imposed on an individual, he or she determines how the change is dealt with. Identification of those personal attributes that are the most important contributors to how events are processed by people is central in that regard. Locus of control and sensation-seeking are examples of personality characteristics that have been shown to promote differential response to stress.

A personality variable that appears to be related to perception of life events as stressful is locus of control, or the degree to which people feel in control of their lives. Those who are internal in locus of control tend to perceive events as being controllable by their own actions, whereas those who are external tend to view events as being influenced by factors other than themselves. In one study, Johnson and Sarason (1978) found that negative life changes were significantly related to both anxiety and depression, but only for external subjects. The results were consistent with the view that people are more adversely affected by life stress if they perceive themselves as having little control over their environment.

Individuals vary in their desire or need for environmental stimulation, and also in their tolerance for stimulation. Some people appear to thrive on life changes; they enjoy traveling to strange places, prefer the unfamiliar to the familiar, and participate in activities such as skydiving, automobile racing, motorcycle riding, and water skiing. Other people shy away from the unfamiliar, would never think of racing cars or going skydiving, and find some everyday stiuations more arousing than they would like.

In conceptual terms, it is reasonable that sensation seeking as a personality attribute should serve as an important moderator of life stress. High sensation seekers ought to be relatively unaffected by life changes, particularly if these changes are not too extreme. These individuals may be better able to deal with the increased arousal involved in experiencing such changes. On the other hand, life change would be expected to have a negative effect on people low in sensation seeking who are less able to cope with arousing stimulus input. To the extent that stimulation seeking mediates the effects of life change, one might expect to find significant correlations between life change and problems of health and adjustment with low but not high sensation seekers. This expectation was borne out in a study by Smith, Johnson, and Sarason (1978). They examined the relationship between life events as measured by the LES, sensation seeking, and psychological distress. They found that subjects with high negative life change scores who were also low in sensation seeking reported high levels of distress. Subjects with high negative life change scores who also had high scores in sensation seeking did not describe themselves as experiencing discomfort.

The results of Smith, Johnson, and Sarason are supported by a study by Johnson, Sarason, and Siegel (1978), who found that individuals low on the sensation seeking dimension were much more likely to report that they were greatly affected by life changes than those high in sensation seeking. For those low sensation seekers, the negative change score on the LES was significantly related to measures of both anxiety and hostility. The positive change score was unrelated to dependent measures regardless of arousal seeking status.

Personality factors appear to be important in determining how an event or situation is appraised. The Smith and colleagues and Johnson and colleagues studies suggest that negative events are cognitively appraised as having different degrees of stress by those high and low in the personality characteristic of sensation seeking.

The idea that it is not the events themselves, but the cognitive appraisal of them and how it dovetails with personality that determines their stress value is also supported by data from a research program concerned with the causes of myocardial infarctions (Pancheri et al., 1980). Although their data suggest that negative life events as assessed by the LES were associated with the occurrence of heart attacks, Pancheri and his coworkers found that cognitive appraisal of these events also played a role in the stressor-infarction relationship. Two factors were especially important as moderators of the appraisal process: the individual's general tendency to react with anxiety to problematic situations and his or her coping styles.

2.2 Social support

Not only personality factors, but also environmental factors, or at least the indivudals' perception of them, have been observed to be moderators of stress. One prominent environmental factor is social support. There are theoretical and empirical reasons for believing that social support contributes to positive adjustment and personal development and also provides a buffer against the effects of stress. Cobb (1976) defined social support as the individual's perception that he or she belongs to a social network of communication and mutual obligation. Social supports include people on whom we can rely and people who let us know that they care about, value, and love us. Thus, someone who believes that he or she belongs to a social network of communication and mutual obligation experiences social support.

Observations from clinical interactions have highlighted the positive roles played by social attachments in psychological adjustment and health. Physicians daily note the salutary effects of their attention and expressed concern on patients' well-being and recovery from illness. Psychotherapists are aware of the necessity of providing their clients with the acceptance needed to overcome the anxieties of self-examination. Those who observe groups under stress have also emphasized the importance of relationships as protective factors. For example, soldiers develop strong, mutually reinforcing ties with each other that contribute to their success and survival. Research into the supportive role of social relationships for both adults and children has also been stimulated by Bowlby's (1980) theory of attachment. Experimental data have added credence to the clinical inferences and the theoretical model. After reviewing the literature, Cohen and McKay (1981) recently concluded that, while lacunae exist in current knowledge, there is increasing evidence consistent with the hypothesis that social support can function as a stress buffer.

Social Support and Health. Several studies suggest that social support functions as a moderator of the effects of stressful life events on psychological adjustment. The moderator's affect may be either preventive or rehabilitative. Lyon and Zucker (1974) found that the post-hospitalization adjustment of discharged schizophrenics was better when social support (friends, neighbors) was present. Burke and Weir (1977) found that the husband-wife helping relationship is an important moderator between experiencing stressful life events and psychological well-being. A helping spouse seems to be particularly valuable in contributintg to self-confidence and a sense of security in dealing with the demands of daily living. Brown, Bhrolchain, and Harris (1975) found that the presence of an intimate, but not necessarily sexual, relationship with a male reduced the probability of depression in women following stressful life events. Consistent with these findings, Miller and Ingham (1976) showed that social support (presence of a confidant and friends) reduced the likelihood of psychological and physical symptoms (anxiety, depression, heart palpitations, dizziness) under stress.

There is also evidence that availability of social support is facilitative to physical health and that lack of such support has a detrimental effect. Gore (1978) studied the relationship between social support and worker's health after being laid off and found that a low sense of social support exacerbated illnesses following the stress of job loss. De Araujo and associates (1972, 1973) reported that asthmatic patients with good social supports required lower levels of medication to produce clinical improvement than did asthmatics with poor social supports. There is much evidence that the health status of medical and surgical patients benefits from attention and expressions of friendliness by physicians and nurses (Auerbach and Kilmann, 1977). Sosa and colleagues (1980) found that the presence of a supportive lay person had a favorable effect on length of labor and mother-infant interaction after delivery. Other researchers have also found links between social variables, illness, and birth complications (Andrews et al., 1978; Cooley and Keesey, 1981).

In a prospective study of over 7,000 men evaluating the onset of angina pectoris (chest pain due to insufficient cardiac blood flow and associated with future myocardial infarction), Medalie and Goldbourt (1976) found that wife's love and support was an important predictor. Specifically, where patients were already high on angina, those men with low spouse support had a 68 percent increase in onset of angina with respect to those having high spouse support. A study by Berkman and Syme (1979) has provided impressive evidence that particular patterns of social interaction and levels of social support have distinctive correlations with longevity and that social disconnectedness and higher mortality are significantly associated.

Social support may be a protective factor only in instances where the individual experiences high stress. This interaction is demonstrated by Nuckolls, Cassel, and Kaplan (1972) in a study of lower-middle-class, pregnant women living in an overseas military community demonstrated how social support and stress may interact. These authors studied two factors of special interest: recent stressful life events and psychosocial assets, a major component of which was defined as the availability of social supports. Neither life changes nor psychosocial assets alone correlated significantly with complications of pregnancy. However, women high in life changes and low in psychosocial assets had many more birth complications than any other group. Only in conditions of low support and high stress was physical health adversely affected.

In another study showing this moderator effect, Sarason, Potter, Antoni, and Sarason (1982) examined the relationship of both life events and social support to illnesses among U.S. Navy Submarine School students. They computed correlations between negative and positive events (reported on the LES) and both self-reported and medically treated illnesses for subjects high or low in satisfaction with their social support and high and low in their ratings of availability of support. There were no significant results involving positive life events. For negative life events, the correlations with illness were significantly stronger for low than for high support subjects, suggesting that low availability of and satisfaction with social support may increase vulnerability to illness. The correlation between negative life events and illness was .33 for subjects high in social support satisfaction. The comparable correlations for those low in social support satisfaction was .50. In general, the results were stronger for satisfaction with supporrt than with the availability of support ratings. It is important to bear in mind that the subjects were generally young and healthy. Most of their illnesses were of the acute variety, particularly upper respiratory conditions. With subjects more at risk for illness these relation-ships would be likely to be even stronger.

Assessing Social Support. Reasonable as an emphasis on the importance of social support seems to be, the task of empirically demonstrating its effects has barely begun. One of the barriers to objective research has been the lack of a reli-able, general, and convenient index of social support. Some researchers have sim-ply gathered information about subjects' confidants and acquaintances; others have focused their attention on the availability of helpful others in coping with certain work, family, and financial problems; and still others have devised questionnaires and other techniques to assess social support. These devices range from simple paper and pencil scales (Luborsky, Todd, and Katcher, 1973) to detailed interview schedules (Henderson, 1980). The diversity of measures of social support is matched by the diversity of conceptualizations concerning its ingredients. How-ever, regardless of how it is conceptualized, social support would seem to have two basic elements: (1) number of available others to whom one can turn in times of need and (2) degree of satisfaction with the available support.

Sarason, Levine, Basham, and Sarason (1983) have recently constructed an instrument, the Social Support Questionnaire (SSQ), that reliably assesses the avai-lability and satisfaction dimensions of social support. The SSQ items require two-part answers to questions. The subjects are asked to (1) list the people to whom they could turn and on whom they could rely in given sets of circumstances (Avai-lability) and (2) rate how satisfied they are with the support that would be avail-able from that source (Satisfaction). (See Table 2). Studies with the SSQ show it to have desirable psychometric features and to correlate with reports of positive life change on the LES (Sarason et al., 1983). In general, low social support seems related to an external locus of control, relative dissatisfaction with life, and diffi-culty in persisting on a task that does not yield a ready solution.

3. The Need for a Theory of Life Change

A major problem in the study of life changes as stressors is the atheoretical character of much of the work in the field. One useful path toward a theory of life changes might be an information-processing approach. Life changes provide the individual with information that requires processing. The first step in this process-ing is attention to a stimulus configuration. Information that is attended to requires appraisal and interpretation, after which behavioral strategies evolve. Salience is a key concept in this regard. It pervades all phases of information-processing and refers to the perceptual "pull value" of a situation and its motiva-tional significance.

Table 2

The Social Support Questionnaire

The Social Support Questionnaire (SSQ) consists of 27 items, four of which are presented below. After answering each item, the test taker is asked to indicate his or her level of satisfaction with the support available by marking a six-point rating scale that ranges from "very satisfied" to "very unsatisfied."

The SSQ yields scores relating to Availability of and Satisfaction with social support.

The list taker is asked to "list all the people you know, excluding yourself, whom you can count on for help or support in the manner described. Give the person's initials and their relationship to you. If you have no support for a question, check the words "No one."

Example

Who do you know whom you can trust with information that could get you in trouble?

--- No one A) R.N. (brother)	D) T.N. (father)	G)
B) L.M. (friend)	E) L.M. (employer)	H)
C) R.S. (friend)	F)	I)

Sample Test Items

Whose lives do you feel that you are an important part of?

--- No one A)	D)	G)
B)	E)	H)
C)	F)	I)

Whom can you really count on to distract you from your worries when you feel under stress?

--- No one A)	D)	G)
B)	E)	H)
C)	F)	I)

Who helps you feel that you truly have something positive to contribute to others?

--- No one A)	D)	G)
B)	E)	H)
C)	F)	I)

3.1. Personal salience

The salience of a situation is a very personal matter and for that reason it makes sense to look at all life events and changes from an interactional perspective. No simple, standardized tally of events that happen in a given period of time can shed light on why each of the many life changes people go through is salient at a particular time, in a particular degree, and a particular way. But it does seem possible to create instruments that go beyond simply tallying which events occurred and which did not. We are now modifying the Life Experiences Survey so as to reflect how situations are processed--for example, the degree to which the person felt in control of the situation. Whether people attend to particular situations or appraise them in particular ways depends on what might be called cognitive moderators, distinctive styles of information processing. It may be that people most likely to use a maladaptive style of information processing can be identified on the basis of personal (e.g., locus of control) or situational (e.g., social support) moderator variables. It may be possible to utilize these variables to predict those individuals who are most vulnerable to the negative effects of particular stressors.

It is widely assumed that stress and anxiety have much to do with the development of both psychological and physical maladaptations. Yet there is little knowledge of the process by which stress becomes transduced into symptoms. Some forms of maladaptation may be due to inattentiveness to certain types of information and others might be due to overattentiveness, for example, the anxiety neurotic's preoccupation with danger. For still other people, stimuli might be attended to but their significance minimized or denied. From an information-processing standpoint, knowledge is needed about the causes of blockages in the processing system. Furthermore, the stressors experienced by people and the anxiety associated with them are not the same throughout the life span. Knowledge is needed on how different types of demands, constraints, and opportunities influence people at different points in their lives.

Both the salience of particular information and the coping mechanisms available are a function not only of the past history of a person but also of his or her developmental state. Life changes are important milestones in life span development. Some coping mechanisms are also age-related in their development. For example, the way in which a young child and an adolescent cognitively process the news of their parents' impending divorce differs in part because of their differing ability to understand the meaning of divorce. A young child's perception may be that he or she is personally responsible by virtue of having done something to alienate the parent who has left, "It's my fault that Daddy went away because he couldn't stand the way I whined when things went wrong." Teenagers, on the other hand, are likely to have a better understanding of the interpersonal difficulties spouses may encounter and are not as likely to see themselves as causal agents. Thus, because of the difference in the developmental level of their cognitive skills, children of those two age groups may face very different situations with which they must cope. The variety of social supports available may also be, in part, a function of developmental level. A toddler depends largely on parental figures; an adolescent has a much wider range of potential supports. Thus, how current changes are handled depends, in part, on the residues of previous changes and, in part, on the utilization of competencies in coping that occur at different stages in development. How future changes are handled depends, in part, on the outcome of the interaction of current personal and situational characteristics.

3.2. The role of coping

Stress, social support, and coping are intricately linked. If the individual possesses needed coping skills, performance in a stressful situation is likely to be adequate and emotional arousal will therefore be contained within manageable limits. When an individual's response repertory does not include coping behavior useful to meet a given set of circumstances, the perception of stress and cognitive or physiological manifestations of anxiety are likely to occur. An analysis of longitudinal data by Vaillant (1977) has indicated that coping styles apparent in youth are linked to later health status and to premature mortality.

It is becoming increasingly clear that the most valuable way in which an individual can be helped to reduce stress evoked anxiety is the acquisition of effective coping skills. Research dealing with several populations, including test-anxious college students, juvenile delinquents, military recruits, depressed patients, and presurgical patients, points to the need to first identify stress related, maladaptive behavior in a response repertory, and then to devise ways to replace it with adaptive behavior. Like the other variables discussed in this paper, coping skills enter into complex relationships with both stressful events and with other moderators. It may be necessary to set up varied programs to teach coping skills depending on the personality predispositions of the client. For example, patients who were internal in locus of control responded positively to specific preparatory information about a surgical procedure. The same information lowered the adjustment of patients external in locus of control (Auerbach et al., 1976). Another important consideration in teaching coping techniques is that no particular technique is best in all situations. Well functioning individuals use a variety of techniques in any one situation and also vary their strategies from one kind of situation to another (Folkman and Lazarus, 1980; Ilfeld, 1980). Social support may play a role in the acquisition and effectiveness of coping skills by lowering the emotional charge associated with demands, constraints, and opportunities and thus permitting the learning of appropriate skills to take place.

Studies of the effects of stressful life events must take more variables into account in their design. This complexity is needed in order to determine the role played by interactions between (a) personal characteristics (e.g., characteristic coping styles) and the demands, constraints, and opportunities that people perceive and (b) events or conditions, concurrent with stressful life events, that moderate responses to these events.

It would make sense to integrate research on life changes with theories and research concerned with how people cope with stress and the way they process potentially stressful information. Into this same package it is essential to factor the effects of moderator variables in order to describe more clearly the individual and situational differences that have been observed. A large number of research efforts have demonstrated that the number of stressful life events is related to either or both emotional adjustment and physical health. Measuring instruments such as the LES and SSQ described in this paper are designed to delineate more clearly some of these complex relationships. More emphasis on a theoretical integration of work on life events, the effect of stress, and role of individual difference variables in their effect on health should also be productive. How much measures of individual differences in personality and in perceptions of a supportive environment will add to the usefulness of measures of the cumulative effects of life changes in predicting health outcome is, of course, an empirical question. But it seems to be a question well worth asking.

References

Andrews, G., Tennant, C., Hewson, D., and Schonell, M. The relation of social factors to physical and psychiatric illness. *American Journal of Epidemiology* 108 (1978) 27-35.

Auerbach, S. M., Kendall, P. C., Cuttler, H. F., and Levitt, N. R. Anxiety, locus of control, type of preparatory information and adjustment to dental surgery. *Journal of Consulting and Clinical Psychology* 44 (1976) 809-818.

Auerbach, S. M. and Kilmann, P. R. Crisis intervention: A review of outcome research. *Psychological Bulletin* 84 (1977) 1189-1217.

Bedell, J. R., Giordani, B., Amour, J. L., Tavormina, J., and Boll, T. Life stress and the psychological and medical adjustment of chronically ill children. *Journal of Psychosomatic Research* 21 (1977) 237-242.

Berkman, L. F. and Syme, S. L. Social networks, host resistance, and mortality: A nine-year follow-up study of Alameda County residents. *American Journal of Epidemiology* 109 (1979) 186-204.

Bowlby, J. *Attachment and loss: Vol. 3. Loss.* New York: Basic Books, 1980.

Brown, G. W., Bhrolchain, M. N., and Harris, T. Social class and psychiatric disturbance among women in an urban population. *Sociology* 9 (1975) 225-254.

Burke, R. and Weir, T. Marital helping relationships: Moderators between stress and well-being. *Journal of Psychology* 95 (1977) 121-130.

Cassel, J. The contribution of the social environment to host resistance. *American Journal of Epidemiology* 104 (1976) 107-123.

Cobb, S. Social Support as a moderator of life stress. *Psychosomatic Medicine* 38(5) (1976) 300-313.

Cohen, S. and McKay, G. *Social support, stress and the buffering hypothesis: A review of naturalistic studies.* Unpublished manuscript, University of Oregon, 1981.

233

Cooley, E. J. and Keesey, J. C. Moderator variables in life stress and illness relationship. *Journal of Human Stress* 8 (1981) 35-40.

Coppel, D. B. *The relationship of perceived social support and self-efficacy to major and minor stresses.* Unpublished doctoral thesis, University of Washington, 1980.

Crnic, K. A., Greenberg, M. T., Ragozin, A. S., and Robinson, N. M. *The effects of life stress and social support on the life satisfaction and attitudes of mothers of newborn normal and at-risk infants.* Paper presented at Western Psychological Association Annual Conference, Honolulu, Hawaii, May, 1980.

de Araujo, G. D., Dudley, D. L., and van Arsdel, P. P. Psychosocial assets and severity of chronic asthma. *Journal of Allergy and Clinical Immunology* 50 (1972) 257-263.

de Araujo, G. D., van Arsdel, P. P., Holmes, T. H., and Dudley, D. L. Life change, coping ability and chronic intrinsic asthma. *Journal of Psychosomatic Research* 17 (1973) 359-363.

Edwards, M. K. *Life crisis and myocardial infarction.* Master of Nursing Thesis, University of Washington, 1971.

Folkman, S. and Lazarus, R. S. An analysis of coping in a middle aged community sample. *Journal of Health and Social Behavior* 21 (1980) 219-239.

Friedman, S. B., Glasgow, L. A., and Ader, R. Psychosocial factors modifying host resistance to experimental infections. *Annals of the New York Academy of Sciences* 164 (2) (1969) 381-392.

Gore, S. The effect of social support in moderating the health consequences of unemployment. *Journal of Health and Social Behavior* 19 (1978) 157-165.

Gorsuch, R. L. and Key, M. K. Abnormalities of pregnancy as a function of anxiety and life stress. *Psychosomatic Medicine* 36 (1974) 352.

Henderson, S. A development in social psychiatry: The systematic study of social bonds. *Journal of Nervous and Mental Disease* 168 (1980) 62-69.

Henry, J. and Stephens, P. *Stress, health and the social environment: A sociobiologic approach to medicine.* New York: Springer-Verlag, 1977.

Holmes, T. H. and Rahe, R. H. The social readjustment rating scale. *Journal of Psychosomatic Research* 11 (1967) 213-218.

Holmes, T. H. and Masuda, M. Life change and illness susceptibility. In B. S. Dohrenwend and B. P. Dohrenwend (eds.) *Stressful life events: Their nature and effects.* New York: John Wiley and Sons, 1974.

Horowitz, M., Wilner, N., and Alverez, W. Impact of Event Scale: A measure of subjective stress. *Psychosomatic Medicine* 41 (1979) 203-218.

Ilfeld, F. W. Coping styles of Chicago adults: Description. *Journal of Human Stress* 6 (1980) 2-10.

Jenkins, C. D. Psychosocial modifiers of response to stress. In J. E. Barrett et al. (eds.) *Stress and mental disorder.* New York: Raven Press. 1979.

Johnson, J. H. and Sarason, I. G. Life stress, depression and anxiety: Internal-external control as a moderator variable. *Journal of Psychosomatic Research* 22 (1978) 205-208.

Johnson, J. H. and Sarason, I. G. Moderator variables in life stress research. In I. G. Sarason and C. D. Spielberger (eds.) *Stress and anxiety (Vol. 6).* Washington: Hemisphere, 1979.

Johnson, J. H., and Sarason, I. G., and Siegel, J. M. Arousal seeking as a moderator of life stress. Unpublished manuscript, University of Washington, 1978.

Kaplan, H. B., Cassel, J. C., and Gore, S. Social support and health. *Medical Care* 15 (1977) 47-58.

Knapp, S. J. and Magee, R. D. The relationship of life events to grade point average of college students. *Journal of College Student Personnel* (November, 1979) 497-502.

Lazarus, R. S. and Launier, R. Stress-related transactions between person and environment. In L. A. Pervin and M. Lewis (eds.) *Perspectives in interactional psychology.* New York: Plenum, 1978.

Luborsky, L., Todd, T. C., and Katcher, A. H. A self-administered social assets scale for predicting physical and psychological illness and health. *Journal of Psychosomatic Research* 17 (1973)(109-120.

Lundberg, V., Theorell, T., and Lind, E. Life changes and myocardial infarction: Individual differences in life change scaling. *Journal of Psychosomatic Research* 19 (1975) 27-32.

Lyon, K. and Zucker, R. Environmental supports and post-hospital adjustment. *Journal of Clinical Psychology* 30 (1974) 460-465.

Medalie, J. H. and Goldbourt, U. Angina pectoris among 10,000 men: II. Psychosocial and other risk factors as evidenced by a multivariate analysis of a five year incidence study. *American Journal of Medicine* 60 (1976) 910-921.

Michaels, A. D. and Deffenbacher, J. L. *Comparison of three life change assessment methodologies.* Unpublished manuscript, Colorado State University. 1980.

Miller, P. and Ingham, J. G. Friends, confidants, and symptoms. *Social Psychiatry* 11 (1976) 51-58.

Nuckolls, K. B., Cassel, J., and Kaplan, B. H. Psychosocial assets, life crisis, and the prognosis of pregnancy. *American Journal of Epidemiology* 95(5) (1972) 431-441.

Pancheri, P., Bellaterra, M., Reda, G., Matteoli, S., Santarelli, E., Publiese, M., and Mosticoni, S. *Psycho-neural-endocrinological correlates of myocardial infarction.* Paper presented at the NIAS International Conference on Stress and Anxiety, Wassenaar, Netherlands, June, 1980.

Petrich, J. and Holmes, T. H. Life change and onset of illness. *Medical Clinics of North America* 61 (1977) 825-838.

Rabkin, J. G. and Struening, E. L. Life events, stress, and illness. *Science* 194 (1976) 1013-1020.

Rahe, R. H. and Lind, E. Psychosocial factors and sudden cardiac death: A pilot study. *Journal of Psychosomatic Research* 15 (1971) 19-24.

Ross, C. E. and Mirowsky, J., II. A comparison of life-event-weighting schemes: Change, undesirability, and effect-proportional indices. *Journal of Health and Social Behavior* 20 (1979) 166-177.

Ruch, L. O. A multidimensional analysis of the concept of life change. *Journal of Health and Social Behavior* 18 (1977) 71-83.

Sarason, I. G. Test anxiety, stress, and social support. *Journal of Personality* 49 (1981) 101-114.

Sarason, I. G. and Johnson, J. H. Life stress, organizational stress, and job satisfaction. *Psychological Reports* 44 (1979) 75-79.

Sarason, I. G., Johnson, J. H., and Siegel, J. M. Assessing the impact of life changes. Development of the Life Experiences Survey. *Journal of Consulting and Clinical Psychology* 46 (1978) 932-946.

Sarason, I. G., Levine, H. M., Basham, R. B., and Sarason, B. R. Assessing social support: The Social Support Questionnaire. *Journal of Personality and Social Psychology* 44 (1983) 127-139.

Sarason, I. G., Potter, E. H., Antoni, M. H., and Sarason. B. R. *Life events, social support, and illness.* Seattle, Washington: Office of Naval Research Technical Report, 1982.

Siegel, J. M., Johnson, J. H., and Sarason, I. G. Mood states and the reporting of life changes. *Journal of Psychosomatic Research* 23 (1979a) 103-108.

Siegel, J. M., Johnson, J. H., and Sarason, I. G. Life changes and menstrual discomfort. *Journal of Human Stress* 5 (1979b) 41-46.

Skinner, H. A. and Lei, H. The multidimensional assessment of stressful life events. *Journal of Nervous and Mental Disease* 168 (1980) 535-541.

Smith, R. E., Johnson. J. H., and Sarason, I. G. Life change, the sensation seeking motive, and psychological distress. *Journal of Consulting and Clinical Psychology* 46 (1978) 348-349.

Sosa. R., Kennell, J., Klaus, M., Robertson, S., and Urritia, J. The effect of a supportive companion on perinatal problems, length of labor, and mother-infant interaction. *New England Journal of Medicine* 303 (1980) 597-600.

Theorell, T. and Rahe, R. H. Psychosocial factors and myocardial infarction: 1. An inpatient study in Sweden. *Journal of Psychosomatic Research* 15 (1971) 25-31.

Vaillant, G. E. *Adaptation to life.* Boston: Little, Brown, 1977.

Whitcher, S. J. and Fisher, J. D. Multidimensional reaction to therapeutic touch in a hospital setting. *Journal of Personality and Social Psychology* 37 (1979) 87-96.

Wyler, A. R., Masuda, M., and Holmes, T. H. Magnitude of life events and seriousness of illness. *Psychosomatic Medicine* 33 (1971) 115-122.

Yamamoto, K. J. and Kinney, O. K. Pregnant women's ratings of different factors influencing psychological stress during pregnancy. *Psychological Reports* 39 (1976) 203-214.

14

PSYCHOSOCIAL ASPECTS OF CHRONIC ILLNESS: DIRECT AND INDIRECT ASSOCIATIONS WITH ISCHEMIC HEART DISEASE MORTALITY

George A. Kaplan, Ph.D.

1. Introduction

Every aspect of human health and disease bears the imprint of psychosocial processes. Although this is a rather strongly worded statement, it can, I believe, stand up to the hardest scrutiny at both empirical and conceptual levels. The facts leading to this conclusion have been stated by many. Psychosocial factors are related to exposure to harmful agents or environments, to the practice of risk-increasing or decreasing behaviors, to increased susceptibility to particular conditions, progression of disease, help-seeking behavior, access to medical care, adherence to and effect of medical treatments, recovery from serious conditions, and a wide range of other outcomes. Although the quality of the evidence varies widely, the variety of methodologies and endpoints considered in the literature lends confidence to my first statement.

But, perhaps, this is painting with too broad a brush. Isn't it possible that our emphasis on psychosocial factors merely reflects the "unfinished business" of the biomedical sciences? From this perspective, advances in clinical and laboratory science will account progressively for the leftover variance we now attribute to psychosocial factors. There are a number of important conceptual, methodological, and philosophical issues to be raised in discussion of such a view. For the present, however, I would like to leave those aside and instead concentrate on issues raised by current psychosocial research on chronic illness with particular emphasis on coronary heart disease.

To convince ourselves of the importance of psychosocial factors, we would presumably want to be able to point to a coherent body of literature that underscores the importance of a relatively small number of psychosocial variables. This, however, is no easy task. With the expansion of research interest in psychosocial aspects of health has come an almost geometric increase in the number of factors which are thought to be important. To some extent, the task can be simplified by considering only a portion of the health trajectory.

Consider, for example, health behaviors. A recent attempt (Cummings et al., 1980) to integrate the various factors thought to be related to health behaviors brought together the judgments of nine theorists involved in research on the initiation, maintenance, and practice of these behaviors. Altogether, 109 variables were identified via an examination of the publications of these nine experts. (One blanches at the number which might have been generated by 19 instead of nine experts!) Using a process of multidimensional scaling of similarity judgments made by the nine judges, the authors were able to end up with six domains of variables: accessibility to health care, evaluation of health care, perception of symptoms and threat of disease, social network characteristics, knowledge about disease, and demographic characteristics. In a sense, this represents a kind of consensus conference, and similar efforts could be carried out with respect to etiologic and treatment issues. However, such a process does not clarify the underlying relationships so much as it presents us with the current level of agreement among experts.

When we turn to psychosocial influences of etiologic significance, we are faced with a similarly bewildering task. Again, some limits on the discussion make the task easier. Consider, as we shall for the remainder of this chapter, psychosocial influences on coronary heart disease morbidity and mortality. Jenkins (1971, 1976) reviewed much of the then current literature on "psychologic and social precursors of coronary disease." His 1971 and 1976 reviews covered a total of 250 articles. Were such an attempt to be made in 1983, the number of articles would be in the 500 to 1,000 range! Jenkins identified a number of factors that appeared to have some relationship to coronary disease etiology. These ranged from status incongruity to the Type A behavior pattern. To the list of factors considered by Jenkins in those reviews, we would now have to add decision latitude and job demands at work (Karasek et al., 1981, 1982), amount of self-referent speech (Scherwitz et al., 1983), perceived health (Kaplan and Camacho, 1983), hostility (Shekelle et al., 1983; Williams et al., 1980), acculturation (Marmot and Syme, 1976), social network participation (Berkman and Syme, 1979), and perceived control and predictability (Glass, 1977) to name only a few.

Thus we are left with an undifferentiated list of variables, all of which are associated with the development of coronary heart disease. In my view, progress in psychosocial epidemiology will depend to a great extent on our ability to convert long lists of variables to coherent theory or models. Such models of the impact of psychosocial variables on disease incidence, progression, and mortality, if properly constructed, will allow us to see the common themes, the interrelationships between variables, and the direct and indirect pathways of influence, and will provide some focus in our search for pathophysiological mechanisms. Without such unifying attempts, we will be likely to see an increasing accumulation of isolated and disconnected findings--a state of affairs which is not likely to lead to significant advances.

There have been some attempts at integration. Cassel's (1976) observations in "The Contribution of the Social Environment to Host Resistance," were one such attempt. He was able to show how findings from a number of studies could be interpreted as demonstrating the joint deleterious effect on health of poor social support and lack of feedback from the environment concerning one's actions. Syme and his colleagues have similarly made important contributions of this sort. Recently, they (Satariano and Syme, 1981) have argued that a common feature of the Type A behavior pattern, stressful life events such as retirement and bereavement, acculturation, intra- and inter-generational mobility, and other variables of psychosocial importance is social isolation or the disruption of social ties. Similarly, we have recently presented evidence (Kaplan and Camacho, 1983) that perception of one's level of health may focus the effect of a number of psychosocial variables that have associations with coronary heart disease mortality.

These attempts are important because they can lead to considerable simplification of a large body of evidence. Important as the attempts are, however, they leave many significant questions. For example, we need to understand the linkages between social connections or support and other psychosocial factors that influence health. We need to know if there are interactions of etiologic significance between variables that measure health practices, personality, stress and coping, perceived health, social connections and support, life satisfaction, positive and negative affect, and other variables of psychosocial significance. Such an examination and the resultant simplification would add to our understanding of causal pathways and lead to both intervention methods and further research.

2. The Human Population Laboratory

We have been examining such questions in our work at the Human Population Laboratory, a unit of the State of California's Department of Health Services. The Human Population Laboratory was established in 1959, its purpose being threefold:

1.   to assess the health, including physical, mental, and social dimensions, of persons living in Alameda County, California

2.   to ascertain whether particular levels in one dimension of health tend to be associated with comparable levels in other dimensions

3.   to determine relationships of various demographic characteristics and ways of living (including personal habits, familial, cultural, economic, and environmental factors) to levels of health.

One of the major efforts of the Human Population Laboratory (HPL) has been a now 18-year-old prospective study of a sample of Alameda Country residents (Berkman and Breslow, 1983; Hochstim, 1970). Figure 1 shows the design of this study. It is a three-wave prospective study with measurement of self-reported physical health status, health habits or practices, psychological functioning, and social functioning during each wave. In addition, two mortality clearances have been executed to identify the deaths occurring between the first and second waves (1965-1974) and the second and third waves (1975-1983).

The participants selected in 1965 were chosen on the basis of a three-stage, stratified random sample of household units in Alameda County. This procedure identified approximately 8,300 eligible adults in 4,735 households. Eligibility was defined as non-institutionalized adults, 20 years of age or older (16 if married). Eighty-six percent of the eligible adults returned completed questionnaires, and these 6,928 individuals constitute the cohort that has been followed since 1965. Differences between respondents and non-respondents were examined via an examination of the results of a household enumeration which was done before placing the questionnaire. Non-respondents were generally older, white, male, and single. However, since the non-respondents were such a small portion of the total sample, and the differences were quite small, the respondents were judged to be an adequate representation of the non-institutionalized adults in Alameda County.

This cohort was intensively traced in 1974, vital status was determined for all 1965 respondents who could be found, and an attempt was made to interview all survivors. A computerized death clearance procedure coupled with the tracing efforts identified 717 deaths in the nine-year period. Of the remaining 6,211 respondents from 1965, 98 percent were successfully located in 1974, and completed questionnaires were obtained from 85 percent of those with whom they were placed.

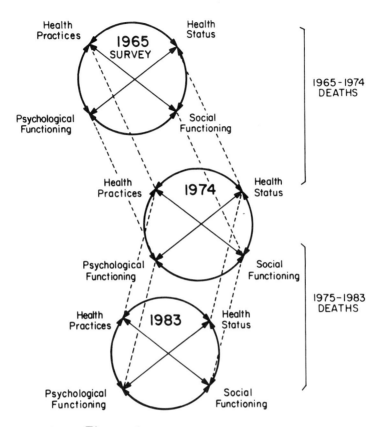

Figure 1.   Alameda County Study

Through a similar procedure, we have recently ascertained the current vital status of the entire cohort. For the period through 1980, an additional 426 deaths have been discovered, bringing the total deaths for the 15-year period, 1965-1980, to 1,143. We are currently in the process of tracing and reinterviewing a 50 percent random sample of the survivors of the 1974 survey. Our current results indicate a response rate of nearly 90 percent for those with whom questionnaires were placed and approximately 5 percent loss to follow-up.

3. Factors Related to Ischemic Heart Disease Mortality

With this in mind, let me briefly review some findings which have come from the HPL studies of this cohort. I will present data for the 2,352 respondents 50 years or over in 1965, the first nine years of follow-up, and deaths from ischemic heart disease (410-414, International Classification of Diseases, 8th Revision).

Health Practices. Previous analyses (Belloc and Breslow, 1972; Belloc, 1973; Breslow and Enstrom, 1980; Wiley and Camacho, 1980; Wingard, Berkman, and Brand, 1982) of this data have shown that not smoking, moderate alcohol consumption, average levels of relative weight for height, moderate leisure-time physical activity, and seven to eight hours of sleep usually are significantly related to both current and future health status. Using univariate and multivariate techniques, it has been shown that people who practiced higher numbers of these health practices had lower levels of both morbidity and mortality than those who did not. Figure 2 demonstrates the strong association between a simple index of the number of health practices engaged in and mortality from ischemic heart disease over the nine-year follow-up period.

Social Functioning. Early HPL studies (Berkman, 1971; Renne, 1974) had shown the cross-sectional relationships between social well-being and health. It remained for Berkman and Syme (1979) to demonstrate that an aspect of social functioning--social network participation--was prospectively associated with mortality, even when controls were instituted for health practices, current physical

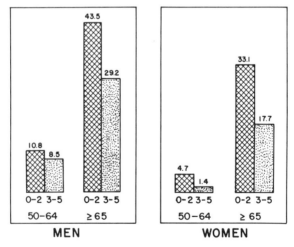

Figure 2.   Association between Health Practices (0-2 vs. 3-5) and IHD Mortality/100, 1965-1974.

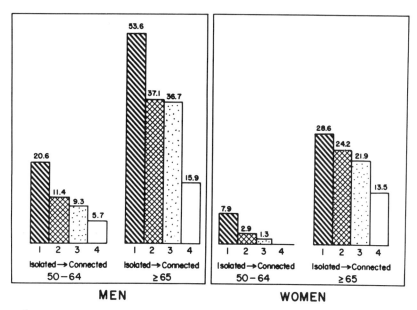

Figure 3. Association between Social Network Index and IHD Mortality/100, 1965-1974.

health status, and a variety of other confounders. This finding, which has had a major impact on social epidemiology, has now been replicated in the Tecumseh Study and others (Blazer, 1982; House, Robbins, and Metzner, 1982). Figure 3 shows that this relationship is strong and consistent for ischemic heart disease mortality.

Perceived Health. We have recently shown that an individual's level of perceived health (i.e., whether he perceives his health as "excellent," "good," "fair," or "poor") is also associated with future health (Kaplan and Camacho, 1983). Individuals who perceived their health as "poor" in 1965 when compared with those who perceived their health as "excellent" were at approximately twice the risk of death between 1965 and 1974. This is true even when there were controls for 1965 physical health status, health practices, social network participation, depression, life satisfaction, and a number of other potential confounders. Figure 4 shows rates of ischemic heart disease death for different levels of perceived health. This finding has been replicated in other studies (Mossey and Shapiro, 1982; Salonen, personal communication).

243

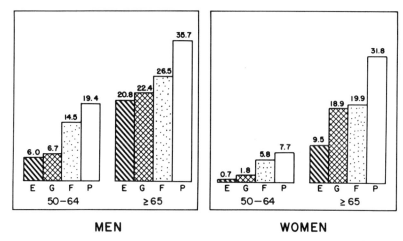

MEN   WOMEN

Figure 4.   Association between Perceived Health and IHD Mortality/100, 1965-1974 (E=excellent, G=good, F=Fair, P=poor)

Depression. A measure of depression based on the items shown in Table 1 shows a strong gradient of mortality. This measure, derived by Roberts (1981) from the questions available on the HPL questionnaire, has been validated against several other depression scales and gives good reliability and validity. As you can see in Figure 5, those who scored as depressed (one standard deviation or more above the mean) were at significantly increased risk of death from ischemic heart disease over the next nine years.

Life Satisfaction. A measure developed by Berkman (1977) and by Wingard (1980), measuring the extent to which respondents reported being satisfied with various domains of their lives, is also related to risk of death from ischemic heart disease. Individuals who reported low life satisfaction were at increased risk of death from ischemic heart disease (Figure 6).

Helplessness. Table 2 presents the items appearing on another scale. This scale was derived from a factor analysis of psychosocial items on the HPL questionnaire by Berkman (1977) and can be described in a variety of ways. It seems to measure a mixture of helplessness and personal uncertainty. In some ways, it might represent Antonovsky's (1979) concept of coherence. For reasons of simplicity, I'll refer to this as a measure of helplessness. As you can see in Figure 7, it, too, is quite strongly related to mortality from ischemic heart disease.

Socioeconomic Status. Finally, Figure 8 shows that a measure of socioeconomic status based on education (0-8 years versus more) and family income

Table 1

Items in Depression Index (Roberts, 1981)

Felt depressed or unhappy
Appetite poor
Lonely or remote from others
Never felt on top of the world
Too tired to do enjoyable things
Little enjoyment from leisure time
Less energy than others
Never feel pleased about accomplishments
Felt bored
Restless
Felt left out, even in a group
Never felt excited or interested
Hard to feel close to others
Never satisfied with performance
Cannot relax easily
Bothered by getting tired in short time
Felt vaguely uneasy

(inadequate or marginal versus more) is also associated with ischemic heart disease mortality.

4. The "Kitchen Sink" Model

At this point, we are in the customary state of having a list of interesting measures, all of which are associated with ischemic heart disease mortality. Before proceeding further, let me assure you that these associations persist as significant when multivariate analyses are performed with adjustment for age, sex, and 1965 physical health status. The usual strategy at this point is to put all the variables into a single multivariate analysis and examine the independent effect of each variable when there is adjustment for all others. Table 3 presents the results of such an analysis. Again, we are considering only 50+-year-olds and the 217 ischemic heart disease deaths that occurred during the nine-year follow-up. As you can see, health practices, social isolation, perceived health, and helplessness are all significantly associated with ischemic heart disease death. One is tempted to stop here; however, there are a number of conceptual and statistical problems with such an "everything but the kitchen sink" approach. In the real world, variables are

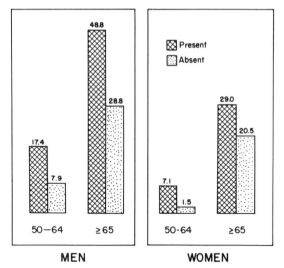

Figure 5. Association between Depression and IHD Mortality/100, 1965-1974.

correlated, and this collinearity exacts a substantial cost in multivariate analyses.

Figure 9 shows the extent to which this is true. Practice of a low number of health practices is associated with "poor" perceived health, social isolation, presence of depression and helplessness, low life satisfaction, and low socioeconomic status. Similarly, those who report "poor" perceived health report higher rates of helplessness, depression, social isolation, life satisfaction, low health practices, and low socioeconomic status. Those who are classified as "helpless" report higher rates of low health practices, low life satisfaction, social isolation, "poor" perceived health, low socioeconomic status, and depression. That these variables are correlated should not come as a great surprise; after all, one would expect that various psychosocial measures on the same individual would be correlated, but such patterns of correlation do not signal our defeat. Although they do make the interpretation of cross-sectional data problematic, analyses of prospective data can still be quite valuable.

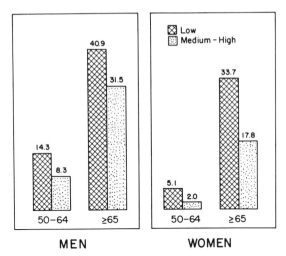

Figure 6. Association between Life Satisfaction (Low vs. Med./High) and IHD Mortality/100, 1965-1974.

5. Analyses of Direction and Extent of Confounding

When we are examining prospective data, we are able to look in some detail at the nature of the correlations between variables and the extent to which one variable confounds the association between another variable and the outcome of interest. Consider the following. A given risk variable (E) may be related to a disease outcome (D):

$$\text{a) } E \longrightarrow D$$

However, a confounder variable (C) may be related to both the disease outcome:

b)

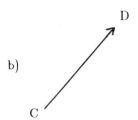

and to the risk variable*

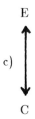

Thus we have the following situation*

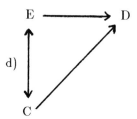

An examination of the extent and symmetry or asymmetry of confounding influences between E and C can tell us something about presumed causal relationships. For example, consider first the unadjusted association between E and D. Now compare this to the association between E and D when adjusted for C. At the same time, consider the change in the association between C and D when there is adjustment for E. If the adjustment for C has little or no effect on the association between E and D, while the adjustment for E has a great effect on the association between C and D, then we can postulate the following model:

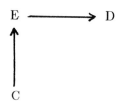

The interpretation of this model is that E has a *direct* association with D while C has an *indirect* association, its influence being felt only because of its association with E.

We have used this logic to examine the associations with ischemic heart disease mortality just discussed. To do this, we performed a series of 28 multiple logistic analyses in which each of the seven variables is considered both by itself and in conjunction with each of the other six variables. All analyses include simultaneous adjustment for age, sex, and 1965 self-reported physical health status.

Inspection of Table 4 shows that there is considerable variability in the extent to which one variable confounds the association between another variable and mortality. In this table, we show the percentage reduction in approximate relative risk associated with adjustment for another variable. In this case, the baseline value is the relative risk associated with the particular variable of interest when there are controls for age, sex, and physical health status. For example, the approximate relative risk associated with "many" versus "few" health practices decreases 32 percent when there is adjustment for perceived health but only 9.4 percent when there is adjustment for life satisfaction. In some cases, there is considerable asymmetry in these confounding effects. For example, the association between social network functioning and ischemic heart disease mortality is reduced only 4.8 percent when there is adjustment for life satisfaction. On the other hand,

Table 2

Items in Helplessness Index (Berkman, 1977)

I am easily sidetracked from things I start to do.

I have a hard time making up my mind about things I should do.

I keep putting things off, and I don't get as much done as others do.

Much of the time I'm not sure what I really want.

I have periods of days, weeks, or months when I can't get going.

I often do things on the spur of the moment without stopping to think.

It seems to me that other people find it easier to decide what is right than I do.

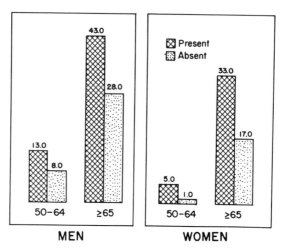

Figure 7.   Association between Helplessness and IHD Mortality/100, 1965-1974.

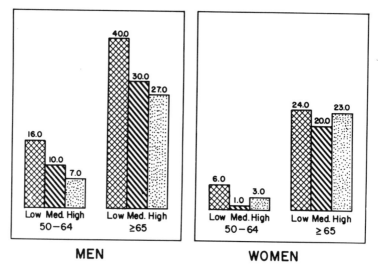

Figure 8.   Association between Socioeconomic Status and IHD Mortality/100, 1965-1974.

250

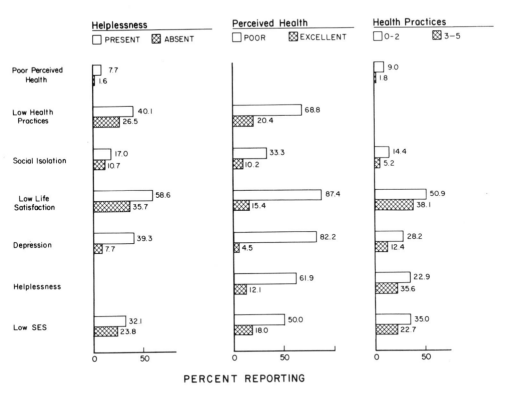

Figure 9. Interrelationships between Psychosocial Measures.

the association between life satisfaction and ischemic heart disease mortality is reduced 19.5 percent when there is adjustment for social network functioning. Inspection of these patterns of mutual influence allows us to partition the total association with mortality for a particular variable into components representing a direct association and components representing the influence of other variables--an indirect association.

In the next seven figures, I will demonstrate the results of this type of analysis for each variable we've seen to be related to ischemic heart disease mortality. Let me repeat that the intent is to examine the patterns of confounding between variables and their relationship to ischemic heart disease mortality and, through an examination of the directionality and extent of confounding, to arrive at a model of the way these variables are interrelated in their impact on ischemic heart disease mortality. After we've examined each variable individually, I'll then put all of them together in a way that, I hope, will result in some simplification.

Table 3

Relative Risk of Death from Ischemic Heart Disease, 1965-1974

	Relative Risk*	95% C.I.
Health Practices (0 vs. 5)	3.11	(2.68-3.61)
Social Network Index (Isolated vs. Most Connected)	2.86	(1.65-4.96)
Perceived Health (Poor vs. Excellent)	1.48	(1.14-1.93)
Helplessness (Present vs. Absent)	1.53	(1.08-2.18)
Life Satisfaction (Dissatisfied vs. Else)	1.07	(.64-1.80)
Depression (Present vs. Absent)	.95	(.67-1.35)
Socioeconomic Status (Low vs. Else)	1.03	(.83-1.28)

*Adjusted for age, sex, and 1965 physical health status and all variables above.

Table 4

Relative Risk of Ischemic Heart Disease Death with Adjustment for Age, Sex, and 1965 Physical Health Status

Variable	RR	%ΔRR adjusted for:						
		HP	SNI	PH	LS	DEP	HLP	SES
Health Practices (HP)	4.24		-19.8	-31.6	-9.4	-12.3	-9.9	-1.4
Social Network Index (SNI)	3.55	-11.8		-5.4	-4.8	-3.9	-7.3	-8.4
Perceived Health (PH)	2.44	-22.1	-9.0		-11.1	-9.8	-21.7	-20.1
Life Satisfaction (LS)	1.49	-7.4	-19.5	-10.1		-10.1	+.7	-2.0
Depression (DEP)	1.49	-12.1	-10.7	-13.4	-9.4		-22.1	-4.7
Helplessness (HLP)	1.77	-7.9	-7.9	-5.1	-4.0	-4.5		-2.2
Socioeconomic Status (SES)	1.40	-9.3	-11.4	-7.9	-2.9	-5.0	-7.1	

In these analyses, we have represented a path as direct if it remains significant after adjustment for other variables. Indirect paths are indicated where the adjusted relative risk is reduced by approximately 20 percent or more when there is adjustment for a second variable or where the association of the first variable with the mortality outcome becomes non-significant with adjustment for the second variable. Also, the area of each circle is proportional to the relative risk associated with a particular variable when adjusted for age, sex, and physical health status.

With this in mind, consider first the pathways associated with health practices. Figure 10 shows that the impact of health practices occurs in three ways. One is a direct path not involving any other variables in the analysis. The other two pathways are indirect pathways involving perceived health and social network participation. That is, levels of perceived health and social network participation account for some of the association between the practice of certain discretionary behaviors and ischemic heart disease mortality. Another way of saying this is that people who practice low numbers of health practices are at increased risk, partially because they are also socially isolated and perceive their health as "poor."

Figure 11 shows the results for social network participation. No other variable has a significant effect on the relationship between social network participation and ischemic heart disease mortality. Thus there is only a direct effect.

The situation for perceived health is somewhat more complicated as seen in Figure 12. Here we see that perceived health has a direct pathway and three indirect paths, one through health practices, one through socioeconomic status, and one through helplessness. Thus the increased risk of ischemic heart disease death for people who perceive their health as "poor" rather than "excellent" is due to their level of perceived health as well as the fact that they tend to be lower on the health practices index, belong to lower socioeconomic strata, and feel "helpless." Recall for a moment that health practices also had a path through perceived health. This is the only example we will see of a symmetric confounding relationship between two variables. The fact that it is between perceived health and health practices raises a number of interesting questions which we will discuss later.

Figure 13 shows that the association between socioeconomic status and ischemic heart disease mortality is unaffected by any of the other variables considered, a finding which is in line with other reports (Rose and Marmot, 1981; Salonen, 1982).

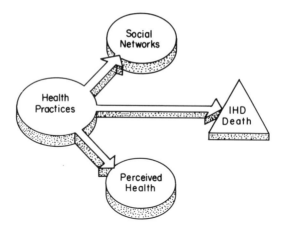

Figure 10. Direct and Indirect Associations between Health Practices and Death

Figure 11. Direct Associations between Social Networks and IHD Death.

A similar finding obtains, as shown in Figure 14, when we consider the helplessness variable. Note that no other variable has a significant effect on the association between helplessness and ischemic heart disease mortality. This is particularly striking when you consider that some of the other variables included in this analysis reflect presence of depression, low life satisfaction, and low health practices.

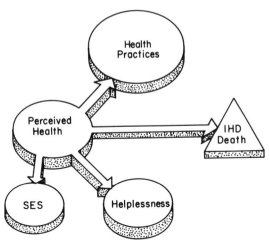

Figure 12.   Direct and Indirect Associations between Perceived Health and IHD
Death.

Figure 13.   Direct Associations between Socioeconomic Status and IHD Death.

Different patterns obtain when we consider depression and life satisfaction.
In both cases, the association with ischemic heart disease mortality is entirely due
to other variables. Thus, in Figure 15, depression exerts its influence indirectly
through its association with health practices, perceived health, social networks,
socioeconomic status, and helplessness. That is, people who are depressed are at
increased risk because they have higher rates of "poor" perceived health, low
health practices, social isolation, low socioeconomic status, and helplessness.

Finally, Figure 16 shows the results for life satisfaction. Here again, the
associations are all indirect, entirely reflecting the influence of health practices,
social networks, perceived health, and helplessness.

Figure 14.   Direct Associations between Helplessness and IHD Death.

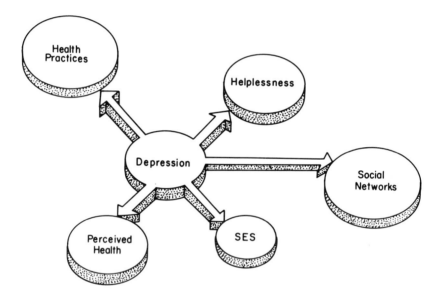

Figure 15.   Indirect Associations between Depression and IHD Death.

Figure 17 shows all of these relationships with ischemic heart disease mortality portrayed simultaneously. Let me first assure you that although this appears to be rather complex, there *has* been a substantial decrease in complexity from what could have resulted. This model shows five direct pathways with ischemic heart disease mortality, and each of these pathways bears some additional comment.

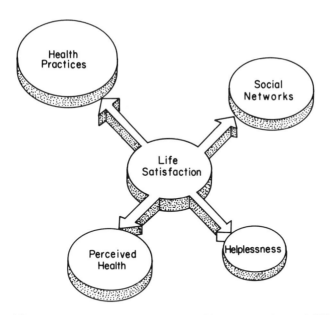

Figure 16. Indirect Associations between Life Satisfaction and IHD Death.

Socioeconomic Status. Measures of education and income have long been known to be associated with varying rates of coronary heart disease, although the particular directionality of the association is not necessarily consistent over time (Antonovsky, 1968; Morgenstern, 1980). One of the explanations offered to account for this inconsistency is the diffusion of various protective or high risk behaviors to different socioeconomic strata at various points in time (Cassel, Heyden, and Bartel, 1971; Inkeles and Smith, 1974; Morgenstern, 1980). Although a proper analysis of this topic would require a sequential cohort design with information on secular trends in the practice of these high risk behaviors and other variables, our analyses and those of others suggest that such an explanation will not be adequate to account for socioeconomic status-ischemic heart disease mortality associations and changes in these associations over time (Holme et al., 1980; Salonen, 1982). We may need to look at other factors, for example, work related ones. Low socioeconomic status individuals are often exposed to demanding jobs with little control, a situation which Karasek and his colleagues (1981) and others have shown to be associated with increased risk of coronary disease.

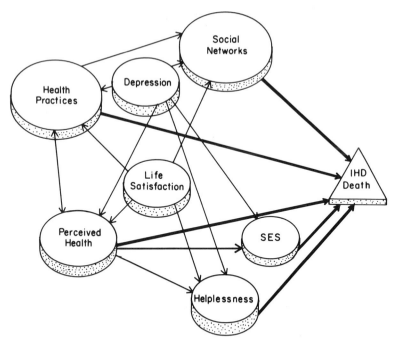

Figure 17.   Direct and Indirect Associations between Psychosocial Measures and IHD Death.

Individuals in lower socioeconomic strata also have differential exposure to various harmful substances. A recent publication (Mahaffey et al., 1982) using HANES II data collected by the National Center for Health Statistics demonstrated a substantial relationship between blood lead levels and family income, race, and degree of urbanization. Differential exposure to coronary disease relevant risks associated with different socioeconomic strata should be studied further.

Social Networks. We have demonstrated, as did Berkman and Syme (1979), that in this data set there is a strong relationship between social network participation and mortality. There are several findings here which add to the previous findings. Our results show that this relationship holds beyond the age Berkman and Syme examined (they looked only at individuals aged 30-69; our analyses covered 50-94). Furthermore, the association is strong for mortality from ischemic heart disease. In addition, and more importantly, social network participation is shown to be independent of other risk factors in multivariate analyses and actually accounts for part of the association between health practices, life satisfaction, and depression and coronary mortality. There are still many unanswered questions about this relationship between social connections and mortality. Our analyses are based on mortality data only and therefore do not allow us to distinguish between factors associated with the development versus progression of coronary heart disease. There is, however, some information that bears upon this issue. Marital status, which is an important component of the measure of social network participation used in these analyses, has been shown in other studies to be related to a wide range of endpoints. Recently, Chandra and colleagues (1982) demonstrated that marital status was related to both hospital case fatality rates and long-term survival in a population of 1,401 myocardial infarction patients, even when there were controls for severity of disease. It seems likely that social network participation is protective both because of internal psychophysiological disregulation present in the hospital environment in those who are isolated, leading to increased hospital case fatality rates and also because of external pathways reflecting the availability of coping resources once released from the hospital, with poorer resources leading to poorer outcome post-discharge.

Some data concerning the role of social connections in the incidence of myocardial infarction are available. An analysis of Finnish hospital discharge records (Koskenvuo et al., 1981) demonstrated lower incidence of coronary disease for the married. In preliminary analyses in collaboration with Syme and Salonen of data collection in North Karelia, Finland, on over 10,000 people followed for seven years, we are seeing indications of an association between measures of social connection and the incidence of myocardial infarction. We hope our analyses will clarify some of the issues concerning the role of psychosocial factors in incidence and prevalence of coronary heart disease which have recently been raised (Reed et al., 1983).

Health Practices. The role of health practices continues to be strong in relation to ischemic heart disease mortality. The pathways associated with social network participation and perceived health are particularly interesting. Individuals who practice fewer health practices also tend to be more socially isolated, and this social isolation accounts for some of the association between low health practices and increased risk of death. There is some evidence of a synergistic relationship between the two variables, but further work needs to be done to clarify its nature.

The indirect pathway from health practices through perceived health is an interesting one to speculate about. The bi-directionality of this pathway is particularly interesting. It implies that there is a dynamic relationship between the two variables in their association with heart disease mortality. This may indicate substitutability between the two variables, or it may simply mean that our examination at one point in time of these two variables is masking some temporal process. Perhaps, people who perceive their health as excellent or good over time tend to increase their practice of measures which are protective and decrease their practice of those which are deleterious to their health. Or, perhaps, high levels of perceived health are a response to increases in health practices. We will be examining these issues in more detail in future analyses of our three-wave data.

Perceived Health. The association between level of perceived health and subsequent mortality is intriguing. This result has now been replicated in several other studies, and I am convinced that it does not simply reflect poor measurement of physical health status. In our analyses, the relationship is strong and consistent across all levels of self-reported physical health status. Analyses by Mossey and Shapiro (1982) of the mortality experience of an over-65-year-old cohort followed for five years arrived at similar conclusions. They were able to control for physical health status using physical exams and medical and hospitalization utilization data. In analyses of the mortality experience of the North Karelia cohort (Salonen, personal communication), perceived level of health appears to make a strong and independent contribution to ischemic heart disease mortality, even when there are controls for family history, serum cholesterol level, mean arterial pressure, and smoking. Thus the finding seems to be a robust one.

In other unpublished analyses, we have found that many other variables related to coronary heart disease rates are also highly associated with lower levels of perceived health. Upon closer examination, they seem to exert their influence indirectly via perceived health. In this category are measures of job strain, family disruption, negative childhood experiences, and geographical mobility. Perhaps then, perceived health is a variable that focuses the effects of other psychosocial variables over time. We will return to this idea in a moment.

Also of interest are the indirect pathways mediating the impact of perceived health--socioeconomic status and helplessness. It is important to point out that although part of the association between perceived health and ischemic heart disease mortality is an indirect one involving helplessness, when there is control for that measure, the association between perceived health and mortality is strong. Cohort members who perceived their health as "poor" rather than "excellent" are at 1.91 the risk of death in the follow-up period. Similarly, when there is control for socioeconomic status, the increased risk is 1.95. Thus, the association between perceived health and mortality does not appear to be due to factors such as a response set associated with poorer education and income or feelings of helplessness.

Helplessness. Finally, we come to the last variable that has a direct association with ischemic heart disease mortality--that which I've called helplessness. The association between this variable and ischemic heart disease mortality is striking both in its magnitude and in its consistency. Those who are scored as "helpless" have a relative risk of ischemic heart disease death 1.80 that of those who aren't helpless, even when there are controls for age, sex, and physical health status. No other variable is so little influenced by other confounders; the maximum reduction in relative risk being 7.9 percent with adjustment for social network participation or health practices.

Now, I have to admit that I am not really sure what to call this measure. Berkman (1977) called it "personal uncertainty," but to me there are a variety of other constructs involved. We will have to do some careful work examining this variable in relationship to other psychological measures, but, in the meantime, I think it is important to point out the extent to which it appears to involve other constructs which have figured in the literature. There certainly is the element of helplessness, but there are also strong elements of what Antonovsky (1979) calls coherence and what Kobasa (1982) calls hardiness. In addition, there is a sense of things being out of control and unpredictable (Glass, 1977). We intend to explore the meaning of this variable in much more depth.

Life Satisfaction and Depression. When other controls are instituted, life satisfaction and depression are only indirectly associated with ischemic heart disease mortality. What is particularly interesting is the large number of indirect pathways accounting for the association between these variables and ischemic heart disease mortality. Depression is associated with higher rates via its link with health practices, social network functioning, perceived health, socioeconomic status, and helplessness. Another way of summarizing this is to say that depressed people are at increased risk in a number of ways, and it is these patterns of higher risk levels on other variables which account for their overall increased risk. Of course, we cannot tell from these analyses what the causal relationships are between these variables. For example, does social isolation cause depression, or vice versa, or are they causally unrelated?

Of particular note is the connection with helplessness. This latter variable is one of the pathways by which depression is associated with higher rates of death, but the reverse is not true. That is, depression does not mediate the association between helplessness and ischemic heart disease mortality. This asymmetry lends some confidence to our feelings that we are dealing with two substantively different scales. The helplessness pathway which partially accounts for the effect of depression on ischemic heart disease death is particularly provocative given current thinking about depression (Garber and Seligman, 1980).

Finally, there is life satisfaction. Again, of particular note is that the association between this measure and ischemic heart disease mortality is via indirect pathways involving health practices, perceived health, helplessness, and social network participation. Of note is that social network participation appears to have by far the greatest confounding effect on life satisfaction. It may be that those who are dissatisfied with their functioning in social domains are more isolated, and it is this isolation that primarily increases their risk of death.

261

6. Discussion and Conclusions

The analyses just presented demonstrate again the significance of psychosocial factors in health. Using data collected on a large community sample which has been followed for nine years, it has been possible to demonstrate that wide range of psychosocial factors is associated with mortality from ischemic heart disease. Furthermore, the analysis of the patterns of confounding between variables has allowed us to gain some insight into the pathways by which such measures are associated with increased mortality risk.

In the introduction to this chapter, we pointed out the need for synthesis in psychosocial epidemiology and proposed an attempt in that direction. It behooves us then to step back and assess to what extent this has been successful. We have shown how it is possible to represent the indirect and direct pathways by which particular measures are associated with risk of ischemic heart disease death, and this has resulted in some simplification; however, the overall picture is still complex. Perhaps, this simply reflects the overall complexity of the subject matter, and we shouldn't expect any great deal of synthesis.

However, certain patterns emerge in Figure 17. The interpretation of some of these is clearer if we return first to Cassel's (1976) seminal paper. He suggested that lack of feedback to the individual concerning consequences of her actions might be associated with deleterious health outcomes. For our purposes, the critical insight has to do not so much with the notion of control over consequences (Antonovsky, 1979; Bandura, 1977; Langer, 1975) as much as it has to do with the notion of feedback. Feedback is, by definition, the flow of information back to the individual from the environment following some action, and it is this element, with its bi-directional and recursive character, which may help us to understand Figure 17. In this view, individuals act both "on" and "in" their environment, and their actions lead to changes in the environment, which affect them as "in" the environment! What is perhaps unique to psychosocial epidemiology is the recognition that these transactional processes with their dense interdependence make the distinction between host and environment problematic.

With this in mind, let us return to Figure 17. Note that health practices and social network participation both involve day-to-day processes in which the individual acts on and gets feedback from the environment, which may or may not lead to changes in the individual and/or environment. In the former case, activities like smoking or jogging lead to the perception of symptoms or bodily states that tell individuals something about their physical condition. In the extreme case, changes in physical activity and the subsequent changes in feeling states have been reported to be one of the ways in which early myocardial infarction patients assess the meaning of anginal pain or "indigestion" (Cowie, 1976; Kaplan, 1981). Similarly, the absence of negative symptoms in smokers and the apparent message that "all is well" may contribute to their difficulty in stopping smoking. Thus discretionary behaviors may be one of the ways in which individuals get information about their underlying physical state, which is used by them to maintain or modify their actions or environment. In this respect, the bi-directional pathway between perceived health, which we will argue later is a measure of underlying physical health status, and health practices is particularly important.

Social network participation also involves day-to-day activities which provide one with ongoing feedback regarding the structure, meaning, and value of one's social relationships. The data presented here and elsewhere (Berkman and Syme, 1979) as well as that concerning bereavement and other major social losses (Cottington et al., 1980; Helsing and Szklo, 1981) surely point to the importance of maintenance of these day-to-day social connections. Furthermore, the fact that measures of depression and helplessness do not account at all for the association between social isolation and increased risk suggest that there is something important about the absence of ongoing social feedback which is important far beyond any affective dysfunction. Also of interest is the finding that some of the increased risk associated with a lower health practices score is related to higher rates of social isolation. It may be that those who are unable to use feedback in the social domain to regulate and maintain their social ties are also those who suppress or deny symptoms associated with heavy drinking, smoking, and/or inactivity. Whether this reflects state, trait, or contextual factors such as occupation and stress remains to be seen.

The situation is different when we now turn to helplessness, life satisfaction, and depression. Depression and life satisfaction do seem to carry with them increased risk of ischemic heart disease death but only because of their association with virtually all other measures we have considered. On the other hand, helplessness retains a strong and independent association. It seems reasonable to speculate that whereas health practices and social network participation reflect day-to-day influences, helplessness is more proactively oriented. Individuals who see the world as incoherent and unpredictable and feel helpless and uncertain are likely to appraise new situations as stressful and threatening and also to see themselves as unable to cope. Thus, in this view, the association between this measure and increased risk of ischemic heart disease death may reflect the absence of coping resources with which to meet new environmental or personal demands. The fact that this association is independent of social network participation suggests that it reflects the relative lack of internal resources as opposed to resources associated with other people. Again, it is intriguing to note the association between perceived health and helplessness. Apparently, people who perceive their health as "poor" also feel helpless, and these feelings of helplessness account for some of the increased risk. This finding is consistent with the interpretation of helplessness as an absence of internal coping resources. Individuals who perceive their health as "poor" and who feel helpless are less likely to be able to engage in activities that might improve their health.

263

Understanding the role of socioeconomic factors in coronary disease continues to be problematic. As pointed out earlier, we agree with other investigators that the increased risk associated with being relatively poor and uneducated is not explainable by any other factors we have measured. Again, some focus on the time dimension may be useful. In our cohort of over-50-year-olds, those who reported being poor and uneducated in 1965 presumably shared certain common features with respect to their past and future. Those who had less than eight years of education had more than 30 years of exposure to the consequences of this, and those who reported "inadequate" or "marginal" income are likely to have also suffered economic privation for some time. Those who reported both, the low socioeconomic strata in our analyses, have thus had considerable exposure to a wide variety of undesirable factors and, what's more, presumably could look forward to similar difficulties in the future. In this view, membership in a low socioeconomic strata exerts a profound influence on past, present, and future. It thus represents a temporal context covering a large part of the individual's history. This type of "contextual" view and analyses of the life experiences, behaviors, and responses following from it will be necessary before we can really understand the role of socioeconomic factors in health. Of interest is that both depression and perceived health were associated with increased risk partially because those who were depressed or perceived their health as "poor" were more highly represented in the low socioeconomic strata.

Finally, we come to perceived health and its pathways. To be sure, judgments of perceived health reflect something about the physical condition of the individual. Individuals who report the presence of chronic conditions or symptoms also report lower levels of perceived health, with the correlation averaging around .40 (Kaplan and Camacho, 1983). However, we have found that measures of negative life events, psychological dysfunction, unhappy childhood experiences, and work stress are also associated with lower levels of perceived health, and it is these lower levels of perceived health which appear to account for the poorer health experience of individuals who report such problems. These findings and others have led us to speculate that levels of perceived health reflect an underlying level of physiologic functioning which is the result of various physical, social, and psychological insults to the individual. In this view, the level of perceived health an individual reports may represent a barometer of host resistance or susceptibility. Recent work in psychoneuroendocrinology (Henry, 1982) and psychoneuroimmunology (Ader, 1981) has pointed to mechanisms which would allow experience at a variety of levels to be focused in such a common way. Our analyses have shown how the association between perceived health and ischemic heart disease mortality includes indirect pathways involving health practices, helplessness, and socioeconomic status. The possibility that an individual's report of perceived health might index a level of physiologic functioning integrating these three levels of functioning seems to us to be quite exciting.

We have tried to show how analyses of the patterns of association between psychosocial measures and ischemic heart disease mortality can help us to understand the pathways by which these factors are associated with increased risk. In the process of these analyses, a number of methodological issues have arisen which bear some comment. Our analyses have been restricted to mortality outcomes, and although there is some evidence that psychosocial factors also exert an influence on the incidence of ischemic heart disease, much more work needs to be done utilizing a variety of health outcomes. It would be valuable to be able to examine in the same data set outcomes reflecting incidence, prevalence, and mortality. Analyses of this sort would go far toward clarifying the biological and social pathways leading to the associations which we have presented.

We also need to explore more fully procedures for examining multiple, correlated measures. In traditional, biomedically oriented, epidemiologic research, we are usually primarily interested in a single or small number of exposure factors, and other measures are considered only insofar as they confound or modify the association between those factors and some health outcome. However, in psychosocial epidemiology, and perhaps in other areas of epidemiology as well, it is becoming clear that there is something about the structure of the *relationships* between variables which is worthy of interest. We have presented one technique for examining these structural relationships and their association with ischemic heart disease mortality in the present chapter, but much more work needs to be done in this area.

But by far the biggest problem lies within the design of one-wave prospective studies. When we examine a cohort and measure a number of variables at one point in time and then assess the health experience of that cohort at some later point, we have progressed far beyond what would be available in a cross-sectional study. However, we have not gone far enough. Consistently, in the discussion of the patterns of confounding between variables, we have asked questions about the temporal relationships between these variables. For example, do people who practice low numbers of health practices have higher rates of "poor" perceived health because of their low levels of health practices, or vice versa? Do changes in depression or physical health status lead to changes in social network functioning, or vice versa? We could go on and on generating numerous hypotheses about the causal orderings of psychosocial measures and their relationship with health outcomes. Thoits (1982) has pointed to similar problems in the analyses of the "buffering" effect of social support on the consequences of negative life events. What is needed are cohort studies in which the cohort members are assessed at numerous points in time. Such studies would allow us to examine the trajectories of change and influence between psychosocial variables and how these trajectories and patterns of influence are associated with changes in health. Future work at the Human Population Laboratory will utilize our three-wave data covering 18 years in just such a way. We hope our results and others' using this approach will help us to unravel the web of psychosocial causation.

References

Ader, R. A. (ed.) *Psychoneuroimmunology.* New York: Academic Press, 1981.

Antonovsky, A. *Health, stress, and coping.* San Francisco: Josey-Bass Publishers, 1979.

Antonovsky, A. Social class and the major cardiovascular diseases. *Journal of Chronic Diseases* 21 (1968) 65-106.

Bandura, A. Self-efficacy mechanism in human agency. *American Psychologist* 37 (1982) 122-147.

Belloc, N. B. Relationship of health practices and mortality. *Preventive Medicine* 2 (1973) 67-81.

Belloc, N. B. and Breslow, L. Relationship of physical health status and health practices. *Preventive Medicine* 1 (1972) 409-421.

Berkman, L. F. *Social networks, host resistance, and mortality: A follow-up study of Alameda County residents.* Ph.D. Dissertation, University of California, Berkeley, 1977.

Berkman, P. L. Spouseless motherhood, psychological stress, and physical morbidity. *Journal of Health and Social Behavior* 10 (1969) 323-334.

Berkman, L. F. and Breslow, L. *Health and ways of living.* New York: Oxford, 1983.

Berkman, L. F. and Syme, S. L. Social networks, host resistance, and mortality: A nine-year follow-up study of Alameda County residents. *American Journal of Epidemiology* 109 (1979) 186-204.

Blazer, D. G. Social support and mortality in an elderly community population. *American Journal of Epidemiology* 115 (1982) 684-694.

Breslow, L. and Enstrom, J. E. Persistence of health habits and their relationship to mortality. *Preventive Medicine* 9 (1980) 469-483.

Cassel, J. The contribution of the social environment to host resistance. *American Journal of Epidemiology* 104 (1976) 107-123.

Cassel, J., Heyden, S., and Bartel, G. Incidence of coronary heart disease by ethnic group, social class and sex. *Archives of Internal Medicine* 128 (1971) 901-906.

Chandra, V., Szklo, M., Goldberg, R., and Tonascia, J. The impact of marital status on survival after an acute myocardial infarction: A population-based study. *American Journal of Epidemiology* 117 (1983) 320-325.

Cottington, E. M., Matthews, K. A., Talbott, E., and Kuller, L. H. Environmental events preceding sudden death in women. *Psychosomatic Medicine* 4 (1980) 567-574.

Cowie, B. The cardiac patient's perception of his heart attack. *Social Science and Medicine* 10 (1976) 87-96.

Cummings, K. M., Becker, M. H., and Maile, M. C. Bringing the models together: An empirical approach to combining variables used to explain health actions. *Journal of Behavioral Medicine* 3 (1980) 13-145.

Garber, J. and Seligman, M. E. P. *Human helplessness.* New York: Academic Press, 1980.

Glass, D. C. *Behavior patterns, stress, and coronary disease.* New York: Wiley, 1977.

Helsing, K. J. and Szklo, M. Mortality after bereavement. *American Journal of Epidemiology* 114 (1981) 41-52.

Henry, J. P. The relation of social to biological processes in disease. *Social Science and Medicine* 16 (1982) 369-380.

Hochstim, J. R. Health and ways of living--the Alameda County, California, population laboratory. In Kesler, I. I., and Levin, M. L. (eds.) *The community as an epidemiologic laboratory.* Baltimore: Johns Hopkins University Press, 1970.

Holme, I., Helgeland, A., Hjermann, I., Leren, P., and Lund-Larsen, P. G. Four-year mortality by some socioeconomic indicators: The Oslo Study. *Journal of Epidemiology and Community health* 34 (1980) 48-52.

House, J. S., Robbins, C., and Metzner, H. L. The association of social relationships and activities with mortality: Prospective evidence from the Tecumseh Community Health Study. *American Journal of Epidemiology* 116 (1982) 123-140.

Inkeles, A. and Smith, D. H. *Becoming modern: Individual change in six developing countries.* Cambridge: Harvard University Press, 1974.

Jenkins, C. D. Recent evidence supporting psychologic and social risk factors for coronary disease. *New England Journal of Medicine* 294 (1976) 987-994, 1034-1038.

Jenkins, C. D. Psychologic and social precursors of coronary disease. *New England Journal of Medicine* 284 (1971) 244-255, 207-317.

Kaplan, G. A. Understanding the understanding of heart attack: Patient and physician perspectives. Paper presented at the American Psychological Association Meetings, Los Angeles, September, 1981.

Kaplan, G. A. and Camacho, T. Perceived health and mortality: Nine-year follow-up of the Human Population Laboratory cohort. *American Journal of Epidemiology* 117 (1983) 292-304.

Karasek, R. A., Baker, D., Marxer, F., Ahlbom, A., and Theorell, T. Job decision latitude, job demands and cardiovascular disease: A prospective study of Swedish men. *American Journal of Public Health* 71 (1981) 694-705.

Karasek, R. A., Theorell, T., Schwartz, J., Peiper, C., and Alfredsson, L. Job, psychological factors and coronary heart disease. *Advances in Cardiology* 29 (1982) 62-67.

Kobasa, S. C. The hardy personality. In Sanders, G. S., and Suls, J. (eds.) *Social psychology of health and illness.* Hillsdale, N.J.: Lawrence Erlbaum Associates, 1982.

Koskenvuo, M., Kaprio, J., Romo, M., and Langinvaino, H. Incidence and prognosis of ischemic heart disease with respect to marital status and social class: A national record linkage study. *Journal of Epidemiology and Community Health* 35 (1981) 192-196.

Langer, E. J. The illusion of control. *Journa of Personality and Psychology* 32 (1975) 311-328.

Mahaffey, K. R., Annest, J. L., Roberts, J., and Murphy, R. S. National estimates of blood lead levels: United States, 1976-1980. Association with selected demographic and socioeconomic factors. *New England Journal of Medicine* 307 (1982) 573-579.

Marmot, M. G. and Syme, S. L. Acculturation and coronary heart disease in Japanese-Americans. *American Journal of Epidemiology* 104 (1976) 225-247.

Morgenstern, H. The changing association between social status and coronary heart disease in a rural population. *Social Science and Medicine* 14 (1980) 191-201.

Mossey, J. M. and Shapiro, E. Self-rated health: A predictor of mortality among the elderly. *American Journal of Public Health* 72 (1982) 800-808.

Reed, D., McGee, D., Yano, K., and Feinleib, M. Social networks and coronary heart disease among Japanese men in Hawaii. *American Journal of Epidemiology* 117 (1983) 384-396.

Renne, K. S. Measurement of social health in a general population survey. *Social Science Research* 3 (1974) 25-44.

Roberts, . Prevalence of depressive symptoms among Mexican Americans. *Journal of Mental and Nervous Disorders* 169 (1981) 213-219.

Rose, G. and Marmot, M. G. Social class and coronary heart disease. *British Heart Journal* 45 (1981) 13-19.

Salonen, J. T. Socioeconomic status and risk of cancer, cerebral stroke, and death due to coronary heart disease and any disease: A longitudinal study in eastern Finland. *Journal of Epidemiology and Community Health* 36 (1982) 294-297.

Satariano, W. and Syme, S. L. Life change and illness in the elderly: Coping with change. Paper presented at the National Academy of Sciences Conference on Biology and Behavior, Woods Holes, Mass., June 22-24, 1979.

Scherwitz, L., McKelvain, R., Laman, C., Patterson, J., Dutton, L., Yusim, S., Lester, J., Kraft, I., Rochelle, D., and Leachman, R. Type A behavior, self-involvement, and coronary atheroslcerosis. *Psychosomatic Medicine* 45 (1983) 47-57.

Shekelle, R. B., Gale, M., Ostfeld, A., and Paul, O. Hostility, risk of coronary heart disease, and mortality. *Psychosomatic Medicine* 45 (1983) 109-114.

Thoits, P. Conceptual, methodological, and theoretical problems in studying support as a buffer against life stress. *Journal of Health and Social Behavior* 23 (1982) 145-159.

Wiley, J. A. and Camacho, T. C. Life-style and future health: Evidence from the Alameda county Study. *Preventive Medicine* 9 (1980) 1-21.

Williams, R. B., Haney, T. L., Lee, K. L., Kong, Y., Blumenthal, J. A., and Whalen, R. E. Type A behavior, hostility and coronary atherosclerosis. *Psychosomatic Medicine* 42 (1980) 539-549.

Wingard, D. L. *The sex differential in mortality rates: Biological and social factors.* Ph.D. Dissertation, University of California, Berkeley, 1980.

Wingard, D. L., Berkman, L. F., and Brand, R. J. A multivariate analysis of health-related practices: A nine-year mortality follow-up of the Alameda County Study. *American Journal of Epidemiology* 116 (1982) 767-775.

15

STRESS, PERSONALITY, IMMUNITY, AND CANCER: A CHALLENGE FOR PSYCHOSOMATIC MEDICINE

Massimo Biondi, M.D.
Paolo Pancheri, M.D.

Many clinical observations seem to suggest that emotional factors, at least in certain cases, can play a role in the genesis and course of neoplastic disease (Bahnson, 1980; Baltrusch, Austarheim, and Baltrusch, 1964; Brown et al., 1974; Crisp, 1970; Gengerelli and Kirkner, 1954; Holden, 1978; Kissen, 1969; Kissen and LeShan, 1964; Pancheri and Biondi, 1979; Stoll, 1979; Surawicz et al., 1976). Experimental studies with animals support this hypothesis, demonstrating how several emotionally stressful stimuli can increase or modify the susceptibility to the development of cancer (Dechambre and Gosse, 1973; Henry et al., 1975; La Barba, 1970; Newberry et al., 1972; Pavlidis and Chirigos, 1980; Pradhan and Ray, 1974; Riley, 1974; Sklar and Anisman, 1979, 1980). Other studies show that immune reactivity in both animal and man can be significantly impaired by stressful emotional stimuli, thereby sustaining from a physiopathological standpoint the possibility of a connection among stress, emotions, and neoplasia (Ader, 1980; Biondi, 1980, 1983; Rasmussen, 1969; Rogers et al., 1979; Stein et al., 1976).

Studies of the psychobiology of cancer, however, do not minimize the role of environmental and genetic risk factors. The aim of psychosomatic cancer research is, rather, to contribute to a more comprehensive understanding of the pathogenesis of the disease and to assess if and to what extent emotional stress could increase the individual risk of cancer in a given population.

1. Studies on Personality, Stress, and Cancer

Research shows that patients with tumors of several sites often exhibit some typical personality characteristics, relative to control group subjects, such as a reduced capacity or "discharge" for emotional expression (Blumberg, 1954; Greer and Morris, 1975; Grissom et al., 1975; Kissen, 1966; Kissen et al., 1969), a tendency to use psychological defense mechanisms as repression and denial (Abse et al., 1974; Bahnson, 1969; Bahnson and Bahnson, 1964, 1966, 1969; Pancheri et al., 1979), a reduced insight (Abse et al., 1974; Bahnson and Bahnson, 1964, 1966), and conventional and rigid behavior (Abse et al., 1974; Bahnson and Bahnson, 1964, 1966; Kissen, 1966; Pancheri et al., 1979). Moreover, other studies have found that in the anamesis of cancer patients there is a greater frequency of life stress events prior to the onset of the disease, as, for example, significant emotional losses (for instance, the loss by death or other cause of a loved one) (Booth, 1969; Conti et al., 1981a; LeShan, 1966; LeShan and Worthington, 1956), followed by the development of hopelessness and helplessness states (Greene, 1966; LeShan, 1966; Renneker et al., 1963; Schmale and Iker, 1966, 1971; Spence, 1975).

In order to reduce methodological bias and the variance related to the heterogeneity of cancer groups, especially in the early psychosomatic cancer research, recent studies have been carried out on selected samples suffering from disease in the same organ site and in the same general stage of progression. Among other kinds of cancer, breast carcinoma seems to be one of the most suitable clinical models for psychosomatic investigations on cancer, due to sex homogeneity, similarity of predisposing and risk factors, the organ site, subjective complaints, and the possibility of a precise histopathological staging of disease by TNM classification.

There have only been a few systematic and controlled studies relating breast cancer to personality and life stress events (Giovacchini and Muslin, 1965; Renneker et al., 1963; Tarlau and Smalheiser, 1951). Reznikoff (1955), for example, found that in a group of patients with mammary nodules, the women with cancer more frequently reported semotional losses in infancy, a history of marital dissatisfaction, and major difficulties in identification with the female role than the women with benign nodules. Schonfield (1975) examining 112 patients with suspected mammary nodules on the day prior to a biopsy, found that women with malignant tumors scored higher on the L scale of the Minnesota Multiphasic Personality Inventory. He interpreted this finding as an expression of the patient's need for "denial" of having a malignant tumor. He also found a higher than average score for life stress events in the three years before disease diagnosis in benign nodule patients. For a sample of 160 subjects, Greer and Morris (1975) found that breast cancer patients, as compared with a control group with benign nodules, showed a behavior pattern characterized by incapacity to express hostile and aggressive feelings. They also found in patients over forty years of age an inability to express other emotional states. Subsequently, on a longitudinal study, a significant correlation was found between suppression of hostile feelings and the raising of serum IgA levels in cancer patients (Morris et al., 1981; Pettingale et al., 1977).

However, other research on personality and life stress events has given different results. Snell and Graham (1971), in a study based on standardized interviews with more than 300 breast cancer patients, did not find a higher frequency of life stress events in the previous five years, as compared to a control group. Unfortunately, the control group was not homogeneous and it included cancer patients with other localizations, a fact that could have altered the findings.

2. Studies on Stress, Emotions, and Immunity

In the last fifteen years, a considerable amount of research has shown that immune humoral and cellular reactivity can be significantly altered by stressful stimuli (Ader, 1980; 1981; Biondi, 1980, 1983; Rasmussen, 1969; Rogers et al., 1979; Stein et al., 1976). For example, distressing emotional stimuli can reduce and delay antibody synthesis and response (Edwards and Dean, 1977; Hill et al., 1967; Solomon, 1969; Vessey, 1964), inhibit the recirculation of T lymphocytes (Spry, 1972), reduce or modulate the specific and aspecific reactivity of both T cells (Joasoo and McKenzie, 1976; Monjan and Collector, 1977; Pitkin, 1966) and B cells (Gisler et al., 1971; Monjan and Collector, 1977).

Other studies have more specifically investigated the biological mediators between stress, emotional arousal, and immune modifications. Central nervous system processes (i.e., emotional arousal) could affect peripheral immune reactivity in several ways. Neuroendocrine systems and pathways, such as the hypothalamus-pituitary-adrenal axis, the hypothalamus-adrenal medullary axis, the hypothalamus-pituitary-growth hormone axis, and the adrenergic and cholinergic agents of the autonomic nervous system, seem to be the main biological mediators of the interaction between brain and the immune system (Ader, 1980; Biondi, 1977, 1980; Gisler, 1974; Gisler et al., 1971; Bourne et al., 1974; Hadden et al., 1970; Stein et al., 1976; Strom et al., 1977). Having systematically examined recent findings on psychoneuroimmunology research, Biondi has proposed a two-level model of the psychobiology of the immune system, consisting of a first level of

273

basic regulation and a second level of *psychoneuroendocrine modulation* of immune functioning. The first level of immune regulation is under strict genetic control, and determines the basic biological rules and functioning of the system (antigen recognition, processing and efferent responses, feedback regulating mechanisms, etc.), quite independently from influences of other organism systems. At the second regulation level, the immune system is influenced by several neuroendocrine mediators (ACTH, cortisol, GH, catecholamines, cholinergic agents, enkephalins, sexual hormones, etc.) each of which modulates cell responses and functions to a different degree. At this second level of modulation, the immune system is sensitive to central nervous system activity and to psychosocial stimuli, including emotional arousal and stress (Biondi, 1983).

In human psychoneuroimmunology, however, just a few studies relating stressful and emotional stimuli to immune processes have been carried out. Bartrop and his colleagues have shown a reduction of lymphocyte reactivity to mitogens (PHA, ConA) following a highly distressing event such as the death of a spouse in a sample of healthy subjects. Greene and his colleagues (1978) reported a significant correlation between life change scores and impairment of lymphocyte citotoxicity. during and after the stress of sleep deprivation and continued task performance, Palmblad and colleagues demonstrated significant alteration of granulocyte and lymphocyte reactions (Palmblad et al., 1979) and of human interferon production (Palmblad et al., 1976).

Taken together, these findings suggest a relationship between stress and disease, such as infectious processes, autoimmune diseases, and cancer (Riley, 1981). Up to now, however, the main topic in psychoimmunology has been the relationship between stressful stimuli and immune modifications, with little attention paid to assessing to what extent individual emotional reactions and coping styles affect the alterations of the immune system. As in other fields, such as psychobiology of the autonomic nervous system and psychoendocrinology, the role of personality traits and of individual cognitive evaluation of stimuli could be keys to understanding stress, emotions, and immune modifications.

3. Aims of the Research

This research in cancer psychobiology consisted of two different studies. The aim of the first study is to compare breast cancer patients with a control group to specifically assess the following hypotheses about cancer:

a. the presence in cancer patients of a more "normal" personality profile

b. the tendency in cancer patients to use predominantly defense mechanisms of the "repression-denial" type, with greater suppression of conflictural states generating anxiety

c. a level of state and trait anxiety in cancer patients less than, or at least not in excess of, those in the control group, in a similar distressing situation

d.   the presence in the anamneses of cancer patients of greater scores of life stress events, as assessed by specific methods, at various time intervals prior to the diagnosis of illness

The aim of the second study is to evaluate the relationship between immune reactivity and emotional reactivity in the examined subjects. The pre-surgical situation is viewed as a "real life stressor" capable of generating differential emotional arousal according to the different individual coping styles.

4. Methodology

The experimental design envisaged a transversal study based on the assessment of personality, anxiety, defense mechanisms, incidence of life stress events, and cellular immune reactivity in a group of breast carcinoma patients (the experimental group) and in a control group with fibrocystic mastopathic disease, a disease closely similar for organ site localization and subjective symptomatology, concern and complaints. The two groups were homogeneous according to age, sex, social status, and level of education. A statistical analysis of data compared variables on personality, anxiety, defense mechanisms, and life stress events scores for:

a.   the cancer group vs. the fibrocystic disease group

b.   the immune-hyporeactive vs. the normo-immunoreactive group

4.1. Subjects

The sample consisted of 43 inpatients of the Regina Elena Institute for the Study and Therapy of Tumors, in Rome, of whom 23 were breast carcinoma patients (the experimental group) and 20 were fibrocystic disease patients (the control group). Criteria for inclusion in the study were as follows: under 65 years of age, middle or lower middle class, a minimum of five years of formal education, and the absence of any other type of somatic pathology known at the moment of evaluation.

Breast carcinoma group: consisted of 23 patients (mean age 41.8, SD 8.7) with clinical and confirmed histopathological diagnosis of "breast carcinoma," according to common anatomopathological criteria, classifiable as T1a, T2a, without fixation to underlying pectoral fascia or muscle, or skin compromission, with or without positive axillary lymph nodes (N0, N1), and without distant metastases (M0). Patients with cancer other than carcinoma were excluded from the study, as were those with renal, hepatic, cardiovascular, or metabolic concomitant pathology. All of the patients were surgically treated and, at the time of the interview and of the administration of psychometric tests, were not informed about their exact diagnosis, but were told about the possibility of an imminent surgical intervention.

Fibrocystic disease group: consisted of 20 patients (mean age 40.3, SD 9.2) with clinical and postsurgical histopathological diagnoses of fibrocystic breast disease. The psychological and environmental conditions of this group were similar to that of the experimental group: they were inpatients of the same ward, had therapeutic relationships with the same medical staff, and received the same or similar diagnostic and clinical routine procedures. Also, these patients were not informed at the time of interview and of psychometric testing about their exact diagnoses, but were informed about the possibility of an imminent surgical intervention. This control group was selected since they suffered from a disease with the same anatomical site, and with initial clinical and subjective symptomatology most similar to that of the experimental group. Their disease was thus able to produce a similar level of emotional reaction and somatic concern.

4.2. Interview

Within three days of admission to the hospital and before the exact diagnosis might be noted, all of the patients participated in a standardized interview with one of the two authors. In the course of the interview, after having been informed of the activity of a psychological team within their hospital division, they were able to choose whether or not to participate in the psychometric evaluation. Only two patients did not accept.

4.3. Psychometric testing

After the interview, each patient completed the following tests:

a. *Minnesota Multiphasic Personality Inventory* (MMPI) (Dahlstrom et al., 1972): the test was chosen for its extensive clinical validation, for the possibility of a multidimensional personality evaluation, and for its wide application in psychosomatic research.

b. *State and Trait Anxiety Inventory* (STAI) (Spielberger et al., 1970): the test was chosen for its extensive clinical and research validation and for the possibility of giving a measure of state and trait anxiety, of particular interest in the setting of a pre-surgical evaluation.

c. *Reaction Scheme Test* (TSR) (Pancheri and Biondi, 1979): the test, designed at Psychiatric Clinic 5a of the University of Rome, consists of description of 38 actual stressful situations which could generate a state of anxiety and conflict in the subject. Each subject is requested to indicate his or her most probable reaction to each one of the situations among five available alternatives. These alternatives represent five different behavioral responses to cope with stressful situations.

The first scale, P (projection) scale, assesses the tendency toward an aggressive or extrapunitive response with, for example, projection of guilt. The second scale, C (colpevolizzazione, i.e., self blaming) evaluates the tendency toward introjection of aggressiveness and assumption of guilt. The next scale, N, (negazione) describes the subject's tendency to use defense mechanisms of the "repression-denial" type, thereby minimizing or annulling the impact or the significance of the stressful situation. The fourth, R (razionalizzazione, i.e., rational coping) assesses the tendency to confront the situation on the basis, for example, of identifying its determinants and actively searching for an adequate response. The last scale, PB, (proiezione bloccata, i.e., blocked projection) measures to what extent the subject is inhibited from openly reacting, in spite of his wish to be emotionally expressive or extrapunitive.

4.4. Assessment of life stress events

For the assessment and evaluation of life events the Schedule of Recent Experiences (Holmes and Rahe, 1967) and the Life Experience Survey (Sarason et al., 1978) were utilized. The Schedule of Recent Experiences (SRE) and the Life Experience Survey (LES), in Italian translation by Pancheri and Biondi, were already in use at Psychiatric Clinic 5a in numerous studies in psychosomatic and psychiatry (Conti et al, 1981a, 1981b; Pancheri and Biondi, 1979; Pancheri et al., 1979a, 1979b, 1980b). The SRE and LES consist of a list of events for the most part identical, but they provide different information since they are based on two different evaluation systems: a "predetermined" or "social weight" of any event, quite independent from the subjective evaluation, expressed in Life Change Units (LCU), adopted in the SRE; a "subjective," graded positive or negative score, based on the individual evaluation of each event, according to the impact in his life adopted in the LES. Thus, the SRE gives a measure of stress in a month or year period, in terms of psychosocial weights, while the LES gives a precise subjective evaluation, as reported by the subject himself.

Instructions to the patient and completion of life event list. For each item in the list the patient was requested to write on her chart both the date of the event and her subjective evaluation of the event impact at the time it occurred. The subjective evaluation of events was done according to Sarason et al. (1978) methodology, with a 7 point scale, ranging from -3 (impact extremely negative) to +3 (impact extremely positive). Written instructions were as follows: "This is a list of events that can happen more or less frequently in the lives of people. Each one of these events can produce at the moment it occurs different, negative or positive, impact on life. Reading this list, try to remember if some of these events have happened in your life and write on the page the date and the impact the event had on your life at the time. You can evaluate the impact of events from very positive to very negative, with various intermediate possibilities."

Calculation of life stress events scores. From the protocols, four life event scores were calculated:

a. Life Change Units, according to the Holmes and Rahe criteria (1967). A standardization of LCU values from the American population, because a standardization for the Italian population was not established, since it has been demonstrated that the LCU weights are relatively constant in western countries (Komaroff et al., 1968; Rahe, 1969)

b. Negative Change Score, according to Sarason et al. criteria (1978)

c. Positive Change Score, as above

d. Total Change Score, as above

4.5. Clinical diagnostic evaluation

After admission to the hospital, all the patients underwent the following routine diagnostic assessments: a general physical check-up; x-ray examination of the cranium, thorax, pelvis, and colon; ECG and hematochemical evaluation; TNM assessment; and a post-surgical histopathological diagnosis.

4.6. Immune evaluation

Before surgery or any kind of therapy, immune cellular status was assessed both through in vivo and in vitro tests. *Test in vivo: skin test* by means of ml. 0.1 intradermic PPD (Sclavo), 50 U Varidase (Lederle), solution 1 : 1000 Candida (Pasteur). The skin tests were considered positive if erythema and induration appeared from 24 to 48 hours after. The following response degrees were established, according to Sega and Sega (1978):

negative: no reaction or erythema only 0 - 5 mm.;

positive +: erythema and induration 5 - 9 mm.;

positive ++: erythema and induration 10 - 20 mm.;

positive +++: erythema and induration greater than 40 mm., or vesicle formation

The overall skin test reactivity was classified on a scale as follows, on the basis of the number of (+) for the three antigens:

high reactivity: from 7+ to 9+;

medium reactivity: from 4+ to 6+;

low reactivity: from 0+ to 3+.

Test in vitro: a) *E rosette formation test*, assessed on the basis of the % count of E rosette forming cells, according to Sega et al. (1980) (normal values 55 ± 10 %), b) *PHA lymphocyte transformation test*, with assessment of the response through the % increase of glycolisis (delta I value) of PHA stimulated lymphocytes, compared to unstimulated lymphocytes, according to Cordiali Fei et al. (1980) (normal values 37 ± 7).

4.7. Data elaboration

Psychometric tests: Mean value was calculated fro the cancer and for the fibrocystic disease groups for each one of the psychometric test parameters.

Statistical analysis was performed by means of the 't' test (independent sample, two tails), with a significance level of p< .05.

Life stress events: The mean value of the four life stress events scores, relative to the 10, 5, and 3-year period before diagnosis, was calculated for each subject, and for the cancer and the fibrocystic disease group. Statistical analysis was performed as above for each one of the life events scores and for the 10, 5, 3-year time intervals.

Immune reactivity and psychosocial variables: according to the aim of the second study to investigate the relationship among personality, stress, and immunity, in the situation of awaiting surgery. Two groups of patients, one immuno-hyperactive to all the three immune tests, and a second normo-reactive to the same three tests, were distinguished, without considering the diagnosis of cancer or fibrocystic disease. The criteria for inclusion into the two respective groups were:

a. *immune hyporeactive group*: only subjects who manifested at the same time low reactivity to the skin tests (from 0 to 3 + for all three antigens), a value of less than 30% at the PHA transformation test, and a value less than 45% at the E rosette formation test were included in this group

b. *immune normoreactive group*: only subjects who manifested medium or high reactivity to the skin tests (totaling from 4 to 9+ for all three antigens), a value between 30 and 44% at the PHA transformation test, and a value between 45 and 65% at the E rosette formation test were included in this second group

On the basis of these criteria, all patients were included in the "normoreactive" group (5 with cancer and 6 with fibrocystic disease), while the "hyporeactive" group had 14 patients (8 with cancer and 6 with fibrocystic disease). The normoreactive and the hyporeactive groups were then compared in terms of psychometric tests scores (MMPI, STAI, and TSR) and for life stress events. The significance of differences between the two groups was tested by means of 't' test for independent samples (two tails). It was not possible to carry out a similar analysis within the cancer group or the fibrocystic disease group, due to the limited number of subjects assessed as totally normoreactive or totally hyporeactive (by excluding mixed subjects) in each separate group.

5. Results

5.1. First study: personality, stress, and breast cancer

Minnesota Multiphasic Personality inventory. The mean profile of breast cancer and fibrocystic disease groups are within the normal range but, as Figure 1 shows, the mean profile of the fibrocystic disease group is clearly higher and, on various scales, evaluations are in the neurotic area (Hs, D, Hy scales). Statistical analysis shows that the fibrocystic group scores significantly higher on F, Hs, D, Pd, and Sc scales.

279

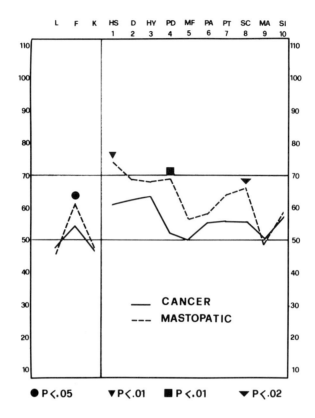

Figure 1:   M.M.P.I. group profiles of breast carcinoma and fibrocystic breast disease patients.

State and Trait Anxiety Inventory. State and trait anxiety levels turned out to be moderately higher in the fibrocystic disease group, without, however, reaching statistical significance. The cancer group, state anxiety : $\bar{x}$ = 50.91, SD 9.48, trait anxiety : $\bar{x}$ = 41.97, SD 10.61; the fibrocystic group, state anxiety : $\bar{x}$ = 54.05, SD 9.44, trait anxiety : $\bar{x}$ = 45.55, SD 8.90.

Reaction Scheme Test. The cancer group showed a significantly greater use of denial-repression defense mechanisms and self blaming. The fibrocystic disease group scored significantly higher on the blocked projection scale. (See Figure 2 and Table 2).

Table 1

Means, standard deviations, Student T, and significance for the 13 MMPI basic scales

| | Cancer | Mastopatic Disease | | |
	$\bar{x}$ SD	$\bar{x}$ SD	t	p
L	47.9± 8.9	46.1± 8.7	0.6	n.s.
F	54.1± 10.8	61.8± 10.9	2.2	< 0.05
K	48.0± 9.9	47.9± 9.1	.04	n.s.
HS	60.8± 11.7	72.3± 8.7	3.5	< 0.01
D	61.8± 11.3	68.9± 5.8	2.4	< 0.01
HY	62.3± 9.5	67.2± 8.1	1.7	n.s.
PD	53.3± 9.2	61.8± 6.9	3.2	< 0.01
MF	51.4± 11.3	56.9± 8.0	1.7	n.s.
PA	55.1± 11.9	59.2± 11.9	1.0	n.s.
PT	57.0± 12.3	63.4± 11.1	1.7	n.s.
SC	56.4± 13.5	66.1± 11.5	2.4	< 0.02
MA	50.1± 8.7	49.5± 10.2	0.2	n.s.
SI	58.0± 11.6	59.8± 7.0	0.5	n.s.

Life stress events. The mean temporal profile of life stress events for the cancer and fibrocystic groups, for the ten years preceding disease diagnosis, were practically identical for the three LES scores (negative, positive, and total change scores) and for LCU scores. Further, statistical analysis of the four life events scores relative to time intervals of 3, 5, and 10 years preceding the diagnosis did not indicate significant differences between cancer and fibrocystic disease groups. (Figures 4 and 5 and Table 3).

Immune evaluations in cancer and fibrocystic groups. There were no significant differences in cellular immune reactivity between breast and fibrocystic disease groups. In the *skin tests*, the cancer group had 13 hyporeactive and 10 normoreactive subjects. The fibrocystic group had 13 hyporeactive and 7 normoreactive subjects. In the *PHA transformation test*, 11 cancer subjects were hyporeactive and 12 normoreactive. In the fibrocystic group, 9 subjects were hyyporeactive and 11 normoreactive. In the *E rosette test*, 11 cancer subjects were classified as hyporeactive and 12 as normoreactive. In the fibrocystic group, 6

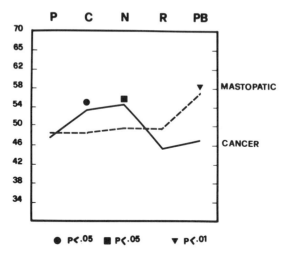

Figure 2: Mean group Reaction Scheme Test profiles of breast carcinoma and fibrocystic disease patients: P, projection; C, colpevolization, i.e. self-blaming; N, denial-repression; R, rational coping; Pb, blocked projection.

Table 2

Means, standard deviations, and significance levels for
Reaction Scheme Test

	Cancer	Mastopatic Disease		
	$\bar{x}$ SD	$\bar{x}$ SD	t	p
Scale P	6.0± 3.5	6.5± 4.4	0.3	n.s.
Scale C	10.0± 3.8	7.8± 2.4	2.1	< 0.05
Scale N	10.2± 3.7	7.8± 2.9	2.2	< 0.05
Scale R	8.0± 3.5	9.5± 5.9	0.9	n.s.
Scale PB	2.9± 2.7	6.5± 3.8	3.5	< 0.01

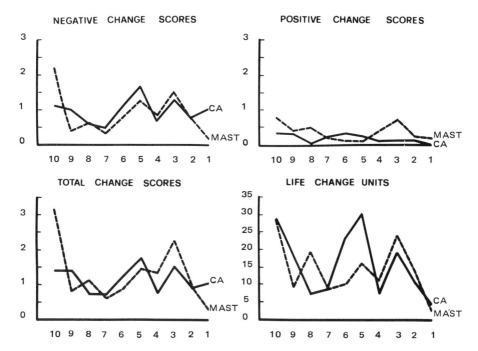

Figure 3: Mean temporal life stress events profiles (Life Experience Survey and Schedule of Recent Experiences) for the ten-year period preceding disease diagnosis.

subjects were hyporeactive and 14 normoreactive. Overall, 8 subjects in the cancer group and 6 in the fibrocystic group manifested hyporeactivity to all the tests carried out. (Table 4).

5.2. Second study: stress, emotions and immunity

The comparison between the hyporeactive group and the normoreactive group resulted in the following findings.

Minnesota Multiphasic Personality Inventory. The MMPI profile of the normoreactive group come out more psychologically "disturbed" (i.e., higher for the most part on the clinical scales) than the hyporeactive group, which exhibited scores more near the normal range (Figure 5). In addition, the normoreactive group exhibited scores significantly higher on the L scale and on the Si scale (Figure 5 and Table 5).

283

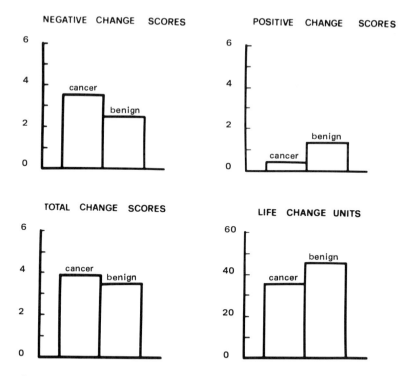

Figure 4: Mean group life stress events scores (LES and SRE) for the three-year period before disease diagnosis, for breast cancer and fibrocystic groups.

State and Trait Anxiety Inventory. The level of state anxiety was distinctly higher (p< .01) in the normoreactive group ($\bar{x}$ = 59.72, SD 9.20) than in the hyporeactive group ($\bar{x}$ = 45.71, SD 6.32). Trait anxiety levels showed a similar result, significantly higher in the normoreactive group ($\bar{x}$ = 45.18, SD 7.30) compared with the hyporeactive group ($\bar{x}$ = 37.21, SD 7.26) (p< .01) (Figure 6).

Reaction Scheme Test. Immune hyporeactive subjects were distinguished from normoreactive subjects at a statistically significant level by a higher score for "denial-repression" reaction scheme (p< .02) and by a lesser score of projective reaction scheme (p< .01) (Figure 7 and Table 6).

Life stress events scores. Normoreactive subjects displayed higher scores for all three time intervals (3, 5, 10 years prior to the diagnosis) and in all four life events indexes. The differences were statistically significant for the previous 3-year period on the total change score (p< .05), for the 5-year period on the negative change score (p< .01) and on the LCU score (p< .01) (Figure 8 and Table 7).

Table 3

Means, standard deviation, and significance for the four life
stress events measurement indexes for the 3, 5, and 10-year
period preceding illness onset

	Cancer		Mastopatic Disease		t	p
3 YEARS						
	$\bar{x}$ SD		$\bar{x}$ SD			
Negative Change Scores	3.9±	4.1	2.6±	2.7	1.4	n.s.
Positive Change Scores	0.3±	0.6	1.1±	1.8	1.5	n.s.
Total Change Scores	3.5±	4.1	3.7±	4.4	0.1	n.s.
Life Change Units	36.1±	54.4	44.0±	57.9	0.5	n.s.
5 YEARS						
	$\bar{x}$ SD		$\bar{x}$ SD			
Negative Change Scores	5.5±	4.8	4.8±	4.3	0.4	n.s.
Positive Change Scores	0.5±	0.8	1.8±	3.9	1.7	n.s.
Total Change Scores	6.0±	4.8	6.7±	6.7	0.6	n.s.
Life Change Units	74.0±	62.2	74.3±	68.5	0.01	n.s.
10 YEARS						
	$\bar{x}$ SD		$\bar{x}$ SD			
Negative Change Scores	9.8±	7.3	9.6±	8.0	0.07	n.s.
Positive Change Scores	1.8±	1.8	4.1±	5.0	1.9	n.s.
Total Change Scores	11.6±	7.8	13.7±	10.0	0.8	n.s.
Life Change Units	161.9±	111.0	156.2±	133.0	0.1	n.s.

6. Discussion

The research findings indicate that breast cancer patients in the first stage of disease development differ at statistically significant levels from a control group of fibrocystic breast disease by 1) a more "normal" personality profile (MMPI), with fewer neurotic characteristics, 2) a tendency to lower levels of state and trait anxiety (STAI), 3) a greater tendency to use defense mechanisms of the "denial-repression" type, but also by 4) a greater tendency toward self-blaming (Reacting Scheme Test). The groups do not differ in the number of life stress events as assessed by the Schedule of Recent Experiences and by the Life Experience Survey, for various periods preceding disease diagnosis.

Second, the research findings demonstrate that in the emotionally stressful pre-surgical situation significant differences can be found between two groups of

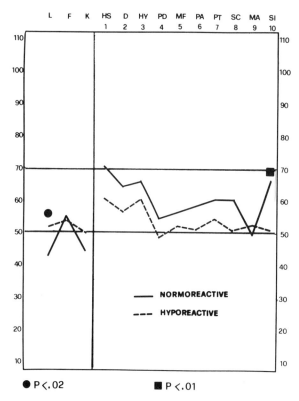

Figure 5: M.M.P.I. profiles in immuno-hyporeactive and normoreactive subjects, in the stress setting of awaiting surgery.

patients with hyporeactive and normoreactive responses to tests for cellular immune function.

Before evaluating the findings, it is necessary to point out how the research methodology promoted strictly controlled experimental conditions. The two groups of patients were not informed of their diagnosis at the time of the interview. They suffered from a pathology with identical anatomical localization. They were in the same emotionally stressful setting of waiting for imminent surgery. They were thus susceptible to a similar distress and concern about their physical health and life. The differences which resulted from psychometric testing could, therefore, be interpreted in the light of the main psychosomatic hypotheses on cancer psychobiology studies.

In accord with previous research, a less "disturbed" and more "normal" personality profile was found (Abse et al., 1974; Bahnson and Bahnson, 1964, 1966; Bahnson, 1969; Kissen, 1966; Kissen et al., 1969): lower scores on the F MMPI

Table 4

Immunological Characterization
of the Experimental Groups

	ROSETTE E		PHA		SKIN TESTS	
	N	H	N	H	N	H
Cancer (n=23)	12	11	12	11	10	13
Mastopatic Disease (n=20)	14	6	11	9	7	13

N = NORMOREACTIVE
H = HYPOREACTIVE

scale, lower scores on the Hs scale, a lower score on D scale, a greater social conformity and morality (lower Pd scale), less tendency to fantasize and a more concrete attitude, realistic, syntonic with environmental reality (lower score on the Sc scale), and a tendency to lower levels in state and trait anxiety (STAI).

Further, in accord with other studies of cancer patients' prevalent psychological defense mechanisms (Bahnson, 1969; Bahnson and Bahnson, 1966, 1969; Pancheri et al., 1979a), breast cancer patients were found to have behavioral reaction patterns based on denial-repression, i.e., trying to minimize or exclude from the conscious both the conflictural situation and the emotional reaction aroused by it. This result is interesting since in a previous study using the Reaction Scheme Test we found that patients with cervical uterine carcinoma in early stage (in situ, I and II) showed higher scores on the denial-repression scale compared to a control group of uterine fibroma patients (Pancheri et al., 1979a).

On the other hand, it should be emphasized that the differences of personality profile and defense mechanisms reported in a transversal design study cannot be certainly deemed as "pre-existent" to the illness, nor for that matter, in an etiopathogenic relationship with it.

One interesting alternative explanation is that the differences between breast cancer and fibrocystic disease patients are different emotional reactions to the pre-surgical stress situation. This interpretation allows for interesting inferences on personality characteristics and coping styles of cancer and control patients. The common pre-surgical situation for the two groups is an intense, stressful stimulus, which should elicit similar concern and emotional arousal. Thus, the differences

Table 5

Means, standard deviations, Student T, and Significance for MMPI
(Normoreactive vs. Hyporeactive)

	Normoreactive (n=11) $\bar{x}$ SD	Hyporeactive (n=14) $\bar{x}$ SD	t	p
L	43.5± 6.8	52.0± 8.6	2.5	< 0.02
F	55.0± 10.4	54.3± 10.8	0.1	n.s.
K	45.5± 7.1	50.2± 9.5	1.2	n.s.
HS	70.8± 9.9	61.7± 11.9	1.9	n.s.
D	63.0± 10.0	57.4± 10.3	1.3	n.s.
HY	62.9± 8.4	60.4± 15.0	0.4	n.s.
PD	54.0± 7.6	49.2± 9.0	1.3	n.s.
MF	55.6± 13.6	51.8± 6.5	0.8	n.s.
PA	56.0± 8.8	51.1± 7.2	1.4	n.s.
PT	60.0± 9.2	55.5± 14.3	0.8	n.s.
SC	60.0± 9.8	51.5± 13.5	1.6	n.s.
MA	49.6± 11.1	52.5± 8.5	0.7	n.s.
SI	66.0± 9.6	50.5± 6.9	4.4	< 0.01

Table 6

Means, standard deviations, and significance levels for Reaction
Scheme Test (Normoreactive vs. Hyporeactive)

	Normoreactive (n=11) $\bar{x}$ SD	Hyporeactive (n=14) $\bar{x}$ SD	t	p
P	8.6± 5.2	3.8± 2.4	2.7	< 0.01
C	10.0± 3.9	8.0± 3.5	1.2	n.s.
N	6.3± 3.8	12.0± 5.7	2.6	< 0.02
R	9.5± 4.8	9.9± 4.1	0.2	n.s.
PB	3.3± 3.7	3.6± 4.5	0.03	n.s.

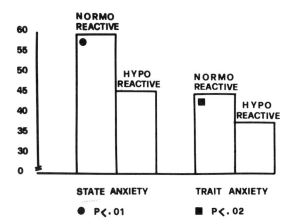

Figure 6: Mean state and trait anxiety scores for immuno-hyporeactive and normoreactive subjects in the stress setting of awaiting for surgery.

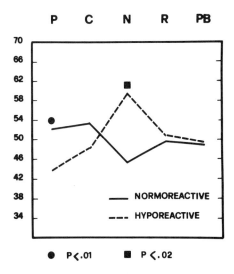

Figure 7: Reaction Scheme Test profiles of immuno-hyporeactive and normoreactive subjects.

289

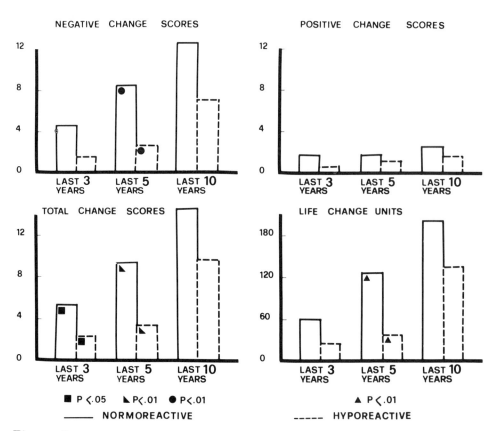

Figure 8: Mean group life stress events scores (LES and SRE) for the 3, 5, and 10-year period preceding disease diagnosis in immuno-hyporeactive and normoreactive subjects.

between cancer and fibrocystic patients allows us to hypothesize different personality characteristics, and above all, different individual coping styles to the emotionally stressful situation, probably pre-existent to the condition of disease.

In contrast to previous clinical reports (LeShan, 1966; Rennecker et al., 1963), a greater number of life stress events (as assessed by SRE and LES scores) before disease was not found. Our findings support previous negative results on the relationship between life events scores and breast cancer (Grissom et al., 1975; Schonfield, 1975; Snell and Graham, 1971). The present study, however, has not investigated the frequency of emotional loss events, due to the limited number of subjects. The separate analysis of loss events items of the LES in cancer and control groups in another study led to interesting and suggestive findings, despite the similarity of group life events scores, such as the LCU and the negative change score (Conti et al., 1981a).

Table 7

The four life stress events measurement indexes for the 3, 5, and 10-year period preceding illness onset in normoreactive and hyporeactive patients

	Normoreactive (n=11)	Hyporeactive (n=14)	t	p
3 YEARS				
	$\bar{x}$ SD	$\bar{x}$ SD		
Negative Change Scores	4.4± 4.5	1.9± 1.8	1.7	n.s.
Positive Change Scores	1.8± 0.3	0.2± 0.5	1.6	n.s.
Total Change Scores	5.6± 4.4	2.2± 1.7	2.0	< 0.05
Life Change Units	58.9± 40.0	21.7± 13.4	1.7	n.s.
5 YEARS				
	$\bar{x}$ SD	$\bar{x}$ SD		
Negative Change Scores	8.3± 4.0	2.4± 1.9	4.5	< 0.01
Positive Change Scores	1.8± 1.5	1.3± 1.7	0.2	n.s.
Total Change Scores	9.5± 3.5	3.7± 1.7	4.4	< 0.01
Life Change Units	126.0± 55.0	39.1± 17.0	4.9	< 0.01
10 YEARS				
	$\bar{x}$ SD	$\bar{x}$ SD		
Negative Change Scores	12.0± 5.9	7.0± 6.1	1.7	n.s.
Positive Change Scores	2.6± 2.0	1.7± 1.5	1.0	n.s.
Total Change Scores	14.7± 5.0	9.7± 7.2	1.8	n.s.
Life Change Units	198.2± 86.6	132.8± 88.0	1.7	n.s.

Finally, results on *stress and immunity* through comparing immune hyporeactive and normoreactive subjects, although preliminary and in a small sample, are interesting. Subjects with normal cellular immune reactivity in the pre-surgical stress situation showed more open emotional reactions, with greater anxiety, more concern for their own somatic state, greater capacity for the projection of aggressiveness, and less tendency to react with denial-repression mechanisms. On the contrary, subjects with impaired cellular immune reactivity showed a behavior pattern characterized by a tendency to use denial-repression defense mechanisms, lower anxiety, and less somatic preoccupation. It is possible that immune reactivity and emotional reaction dependent on a third variable, for example, the degree of somatic compromission or the disease process. But the criteria of patient selection (excluding anyone with concurrent somatic disease; including subjects free from pharmacological intervention at the time of psychometric and immune evaluations; only carcinoma patients at an early stage of disease) and the presence in each of the two groups, hyporeactive and normoreactive, of a similar proportion of patients with either cancer or fibrocystic disease, deny this possibility.

From the perspective of psychoimmunology, these findings could be interpreted in terms of the psychobiological response to the pre-surgical situation. Many studies in animals and humans have demonstrated that, in fact, immune cellular reactivity can be significantly impaired by stressful emotional stimuli (Bartrop et al, 1977; Greene et al., 1978; Joasoo and McKenzie, 1976; Monjan and Collector, 1977; Palmblad et al., 1979; Spry, 1972). Our findings, based not on a single immune cellular response, but on the presence of contemporary hyporeactivity to three different cellular tests (PHA, E rosette, and skin tests) suggest that individual emotional reaction and coping styles to stressful stimuli can affect the modifications of immune functions under stress differently, as has been shown for autonomic and neuroendocrine responses. Other psychobiological research, in fact, has suggested that the severity and the persistence of hormonal and psychophysiological alterations varies according to whether the subject reacts to stressful stimuli with an open emotional reaction on the behavioral level or, instead, suppresses his emotional reaction (Barrell and Price, 1977; Freeman et al., 1967; Hofer et al., 1972; Hokanson and Burgess, 1962; Hokanson and Shetler, 1961). The "balance" model proposed in psychosomatics by Pancheri (1978, 1980) has provided, through research on psychogenic amenorrhea (Ermini et al., 1978) and on myocardial infarction (Pancheri et al., 1980), some clinical evidence of such a relationship between behavioral and biological reactions.

From this standpoint, the comparison between immune hyporeactivity and normoreactivity in subjects who show a different emotional response to the pre-surgical stress setting seems extremely suggestive. Such an hypothesis, however, should be confirmed by further studies.

Acknowledgements

For the clinical and psychometric assessment, Dr. ssa C. Conti; for the assessment and laboratory procedure of immune reactivity, Dr. ssa P. Cordiali Fei, Dr. ssa M. C. Apollonj, Dr. F. M. Sega of the Immune Laboratory (Dir. Prof. E. Sega), of the Regina Elena Institute for the Study and Therapy of Tumors, Rome.

References

Abse, D. W. et al. Personality and behavioral characteristics of lung cancer patients. *Journal of Psychosomatic Research* 18 (1974) 101-113.
Ader, R. Psychosomatic and psychoimmunologic research. *Psychosomatic Medicine* 42 (1980) 307-321.
Ader, R. (ed.) *Psychoneuroimmunology.* New York; Academic Press, 1981.

Bahnson, C. B. Stress and cancer: The state of the art. *Psychosomatics* 21 (1980) 975-981.

Bahnson, C. B. Psychophysiological complementarity in malignancies: Past work and future vistas. *Annals of the New York Academy of Sciences* 164 (1969) 319-330.

Bahnson, C. B. and Bahnson, M. B. Ego defences in cancer patients. *Annals of the New York Academy of Sciences* 164 (1969) 547-577.

Bahnson, C. B. and Bahnson, M. B. Role of the ego defenses: Denial and repression in the etiology of malignant neoplasms. *Annals of the New York Academy of Sciences* 125 (1966) 827-845.

Bahnson, C. B. and Bahnson, M. B. Denial and repression of primitive impulses and of disturbing emotions in patients with malignant neoplasms. In Kissen, D. M., and LeShan, L. (eds.) *Psychosomatic Aspects of Neoplastic Disease.* London: Pitman, 1964.

Baltrusch, H. J., Austarheim, K., and Baltrusch, E. Psyche - nervous system - neoplastic process: An old problem with new interest. Part III.: Clinical observations and investigations on psychosomatics of cancer, particularly by the pychoanalytical or psychosomatic method. *Zeitschrifts fur Psychosomatische Medizin und Psychoanalisis* 10 (1964) 157-169.

Barrell, J. J. and Price, D. D. Two experiential orientations toward a stressful situation and their related somatic and visceral responses. *Psychophysiology* 14 (1977) 517-521.

Bartrop, R. W., Lazarus, L., Luchurst, E., and Kiloh, L. G. Depressed lymphocyte function after bereavement. *Lancet* 1 (1977) 834-837.

Biondi, M. Psicoimmunologia. In Pancheri, P. (ed.) *Trattato di Medicina Psicosomatica.* Firenze: USES, 1984.

Biondi, M. Stress, immunita' e malattia. *Medicina Psicosomatica* 25 (1980) 183-200.

Biondi, M. Stress, sistema nervoso centrale e sistema immunitario. Medicina Psicosomatica. 22 (1977) 53-70.

Blumberg, E. M. Results of the psychological testing of cancer patients. In Gengerelli, J. A., and Kirkner, F. J. (eds.) *Psychological Variables in Human Cancer.* Berkeley: University of California Press, 1954.

Booth, G. General and organic specific object relationship in cancer. *Annals of the New York Academy of Sciences* 164 (1964/69?) 568-576.

Bourne, H. R. et al. Modulation of inflammation and immunity by cyclic AMP. *Science* 184 (1974) 19-28.

Brown, J. H., Varsamis, J., Toews, J., and Shane, M. Psychiatry and oncology: A review. *Canadian Psychiatric Association Journal* 19 (1974) 219-222.

Conti, C., Biondi, M., and Pancheri, P. Valutazione statistica degli eventi stressanti in 144 pazienti psichiatrici e neoplastici. *Rivista di Psichiatria* 16 (1981a) 357-377.

Conti, C., Biondi, M., and Pancheri, P. Stress e depressione: Studio controllato in 243 pazienti psicosomatici e psichiatrici. *Rivista di Psichiatria* 16 (1981b) 342-355.

Cordiali Fei, P., Floridi, A., Apollonj, M. C., and Natali, P. G. Estimation of PHA induced transformation in peripheral blood lymphocytes through the measurement of their increased glycolysis. *Immunology Communication* 9 (1980) 210.

Crisp, A. H. Some psychosomatic aspects of neoplasia. *British Journal of Medical Psychology* 43 (1970) 313-331.

Dahlstrom, W. G., Welsh, S. G., and Dahlstrom, L. E. *An MMPI Handbook.* Minneapolis: University of Minnesota Press, 1972.

Dechambre, R. P., and Gosse, C. Individual versus group caging of mice with grafted tumors. *Cancer Research* 33 (1973) 140-144.

Edwards, E. A., and Dean, L. M. Effects of crowding of mice on humoral antibody formation and protection to lethal antigenic challenge. *Psychosomatic Medicine* 39 (1977) 19-24.

Ermini, M. et al. Psychoneuroendocrine aspects of secondary amenhorrhea. In Carenza, L., Pancheri, P., and Zichella, L. (eds.) *Clinical Psychoneuroendocrinology in Reproduction.* London: Academic Press, 1978.

Freeman, E. H. et al. Personality variables and allergic skin reactivity. *Psychosomatic Medicine* 29 (1967) 312-321.

Gengerelli, A. J., and Kirkner, F. J. (eds.) Psychological Variables in Human Cancer. Berkeley: University of California Press, 1954.

Giovacchini, P. L. and Muslin, H. Ego equilibrium and cancer of the breast. *Psychosomatic Medicine* 27 (1965) 524-532.

Gisler, R. H. Stress and the hormonal regulation of the immune response in mice. *Psychotherapy and Psychosomatics* 23 (1974) 197-208.

Gisler, R. H., Bussard, A. E., Mazie, J. C., and Hess, R. Hormonal regulation of the immune response. I. Induction of an immune response in vitro with lymphoid cells from mice exposed to acute systemic stress. *Cellular Immunology* 2 (1971) 634-645.

Gisler, R. H. and Schenkel-Hulliger, L. Hormonal regulation of the immune response. II. Influence of pituitary and adrenal activity on immune responsiveness in vitro. *Cellular Immunology* 2 (1971) 646-657.

Greene, W. A. Psychosocial setting of the development of leukemia and lymphoma. *Annals of the New York Academy of Sciences* 125 (1966) 794-801.

Greene, W. A. et al. Psychosocial factors and immunity: Preliminary report. *Psychosomatic Medicine* 40 (1978) 87 (Abstr.)

Greer, S. and Morris, T. Psychological attributes of women who develop breast cancer: A controlled study. *Journal of Psychosomatic Research* 19 (1975) 147-153.

Grissom, J. J., Weiner, B. J., and Weiner, A. Psychological correlates of cancer. *Journal of Consulting and Clinical Psychology* 43 (1975) 119.

Hadden, J. W., Hadden, E. M., and Middleton, E. Lymphocyte blast transformation. I. Demonstration of adrenergic receptors in human peripheral lymphocytres. *Cellular Immunology* 1 (1970) 583-595.

Henry, J. P., Stephens, P. M., and Watson, F. M. C. Forced breeding, social disorder and mammary tumor formation in CBA/USC mouse colonies: A pilot study. *Psychosomatic Medicine* 33 (1975) 277-283.

Hill, C. W., Greer, W. E., and Felsenfeld, O. Psychological stress, early response to foreign protein and blood cortisol levels in vervets. *Psychosomatic Medicine* 29 (1967) 279-283.

Hofer, M. A., Wolff, C. T. Friedman, S. B., and Mason, J. W. A psychoendocrine study of bereavement. I. 17 OHCS excretion rates of parents following death of their children from leukemia. *Psychosomatic Medicine* 34 (1972) 481-504.

Hokanson, J. E. and Burgess, M. The effects of three type of aggression on vascular processes. *Journal of Abnormal and Social Psychology* 64 (1962) 446-449.

Hokanson, J. E. and Shetler, S. The effect of overt aggression on physiological arousal level. *Journal of Abnormal and Social Psychology* 63 (1961) 446-448.

Holden, C. Cancer and the mind: How are they connected? *Science* 200 (1978) 1363-1369.

Holmes, T. H. and Rahe, R. H. The social readjustment rating scale. *Journal of Psychosomatic Research* 11 (1967) 213-218.

Joasco, A. and Mc Kenzie, J. M. Stress and the immune response in rats. *International Archives of Allergy and Applied Immunology* 50 (1976) 659-663.

Kissen, D. M. The present status of psychosomatic cancer research. Geriatrics 24 (1969) 129-137.

Kissen, D. M. The significance of personality in lung cancer in men. *Annals of the New York Academy of Sciences* 125 (1966) 820-826.

Kissen, D. M., Brown, R. I. F., and Kissen, M. A further report on personality and lung cancer: Psychosocial factors in lung cancer. *Annals of the New York Academy of Sciences* 164 (1969) 535-545.

Kissen, D. M. and LeShan, L. (eds.) *Psychosomatic Aspects of Neoplastic Disease.* London: Pitman, 1964.

Komaroff, A. L., Masuda, M., and Holmes, T. H. The social readjustment rating scale: A comparative study of Negro, Mexicans and White Americans. *Journal of Psychosomatic Research* 12 (1968) 121-128.

LaBarba, R. C. Experiential and environmental factors in cancer. A review of research with animals. *Psychosomatic Medicine* 32 (1970) 259-275.

LeShan, L. An emotional life history pattern associated with neoplastic disease. *Annals of the New York Academy of Sciences* 125 (1966) 780-793.

LeShan, L. Psychological states as factors in the development of malignant disease: A critical review. *Journal of the National Cancer Institute* 22 (1959) 1-18.

LeShan, L. and Worthington, R. E. Some recurrent life history patterns observed in patients with malignant disease. *Journal of Nervous and Mental Disease* 124 (1956) 460-465.

Monjan, A. A. and Collector, M. I. Stress-induced modulation of the immune response. *Science* 197 (1977) 307-310.

Morris, T., Greer, T., Pettingale, K. W., and Watson, M. Patterns of expression of anger and their psychological correlates in women with breast cancer. *Journal of Psychosomatic Research* 25 (1981) 111-117.

Newberry, B. H. et al. Shock stress and DMBA induced mammary tumors. *Psychosomatic Medicine* 34 (1972) 295-303.

Palmblad, J. et al. Stressor exposure and immunological response in man: Interferon producing capacity and phagocytosis. *Journal of Psychosomatic Research* 20 (1976) 193-199.

Palmblad, J., Petrini, B., Wasserman, J., and Akerstedt, T. Lymphocyte and granulocyte reactions during sleep deprivation. *Psychosomatic Medicine* 41 (1979) 273-278.

Pancheri, P. *Stress, Emozioni, Malattia.* 2nd ed. Milano: Mondadori EST, 1983.

Pancheri, P. Stress, personality and interacting variables: An interpretive model for psychoneuroendocrine disorders. In Carenza, L., Pancheri, P., and Zichella, L. (eds.) *Clinical Psychoneuroendocrinology in Reproduction.* London: Academic Press, 1978.

Pancheri, P. and Biondi, M. *Psicologia e Psicosomatica dei Tumori.* Roma: La Goliardica, 1979.

Pancheri, P. et al. Infarto del miocardio, reazioni emozionali, psiconeuroendocrine e complicazioni a breve termine. *Rivista di Psichiatria* 15 (1980a) 405-433.

Pancheri, P. et al. Valutazione quantitativa degli eventi stressanti ed insorgenza di malattie psicosomatiche e psichiatriche. *Rivista di Psichiatria* 15 (1980b) 291-316.

Pancheri, P. et al. Studio controllato sulle caratteristiche di personalità, meccanismi di difesa ed eventi stressanti nel carcinoma del collo dell' utero. *Rivista di Psichiatria* 114 (1979a) 210-221.

Pancheri, P. et al. Life stress events and state-trait anxiety in psychiatric and psychosomatic patients. In Sarason, I. G. and Spielberger, C. D. (eds.) *Stress and Anxiety.* Washington: Hemisphere, 1979b.

Pavlidis, N. and Chirigos, M. Stress-induced impairment of macrophage tumoricidal function. *Psychosomatic Medicine* 42 (1980) 47-54.

Perrin, G. M. and Pierce, L. Psychosomatic aspects of cancer. *Psychosomatic Medicine* 21 (1959) 397-421.

Pettingale, K. W., Greer, S., and Tee, D. E. H. Serum IgA and emotional expression in breast cancer patients. *Journal of Psychosomatic Research* 21 (1977) 395-399.

Pitkin, D. H. Effects of physiological stress on the delayed hypersensitivity reactions. *Proceedings of the Society for Experimental Biology and Medicine* 120 (1966) 350-352.

Pradhan, S. N. and Ray, P. Effects of stress on growth of transplanted and DMBA induced tumors and their modification by psychotropic drugs. *Journal of the National Cancer Institute* 53 (1974) 1241-1245.

Rahe, R. H. Multi-cultured correlation of life changes scaling: America, Japan, Denmark and Sweden. *Journal of Psychosomatic Research* 13 (1969) 191-195.

Ray, P. and Pradhan, S. N. Growth of transplanted and induced tumors in rats under a schedule of punished behavior. *Journal of the National Cancer Institute* 52 (1974) 575-577.

Rassmussen, A. F. Emotions and immunity. *Annals of the New York Academy of Sciences* 164 (1969) 458-461.

Renneker, R. E. et al. Psychoanalytical explorations of emotional correlates of cancer of the breast. *Psychosomatic Medicine* 25 (1963) 106-124.

Reznikoff, M. Psychological factors in breast cancer: A preliminary study of some personality trends in patients with cancer of the breast. *Psychosomatic Medicine* 17 (1955) 96-110.

Riley, V. Psychoneuroendocrine influences on immunocompetence and neoplasia. *Science* 242 (1981) 1100-1109.

Riley, V. Mouse mammary tumors: Alteration of incidence as apparent function of stress. *Science*

Rogers, M. P., Dubey, D., and Reich, P. The influence of the psyche and the brain on immunity and disease susceptibility: A critical review. *Psychosomatic Medicine* 41 (1979) 147-164.

Sarason, I. G., Johnson, J. H., and Siegel, J. M. Assessing the impact of life changes: Development of the life experience survey. *Journal of Consulting and Clinical Psychology* 46 (1978) 432-445.

Schmale, A. and Iker, H. Hopelessness as a predictor of cervical cancer. *Social Science and Medicine* 5 (1971) 95-100.

Schmale, A. and Iker, H. The effect of hopelessness and the development of cancer. 1. Identification of uterine cervical cancer in women with atypical cytology. *Psychosomatic Medicine* 28 (1966) 714-721.

Schonfield, J. Psychological and life experience differences betweeen Israeli women with benign and cancerous breast lesions. *Journal of Psychosomatic Research* 19 (1975) 229-234.

Sega, E. et al. Immunological monitoring during combined radiochemoimmunotherapy in inoperative lung cancer. *Oncology* 37 (1980) 390-396.

Sega, E. and Sega, F. M. Il ruolo dell' immunologo nella caratterizzazione clinica-biologica, statica e dinamica, dei tumori. *Corsi Nazionali di Terapia Antiblastica*. Bologna: Editrice Universitaria Bolognese, 1978.

Sklar, L. S. and Anisman, H. Social stress influences tumor growth. *Psychosomatic Medicine* 42 (1980) 347-365.

Sklar, L. S., and Anisman, H. Stress and coping factors influence tumor growth. *Science* 205 (513-515.

Snell, L. and Graham, S. Social trauma as related to cancer of the breast. *British Journal of Cancer* 25 (1971) 721-734.

Solomon, G. F. Stress and antibody response in rats. *International Archives of Allergy* 35 (1969) 97-104.

Spence, D. P. Language correlates of cervical cancer. *Psychosomatic Medicine* 37 (1975) 95.

Spielberger, C. D., Gorsuch, R. L., and Lushene, R. E. *Manual for the state-trait anxiety inventory.* Palo Alto: Consulting Psychologist Press, 1970.

Spry, C. Inhibition of lymphocyte recirculation by stress and corticotropin. *Cellular Immunology* 4 (1972) 86-92.

Stein, M., Schiavi, R. C., and Camerino, M. Influence of brain and behavior on the immune system. *Science* 191 (1976) 435-440.

Stoll, B. A. (ed.) *Mind and cancer prognosis.* New York: Wiley, 1979.

Strom, T. B., Lundin, A. P., and Carpenter, C. B. The role of cyclic nucleotides in lymphocyte activation and function. *Progress in Clinical Immunology* 3 (1977) 115-153.

Surawicz, F. G., Brightwell, D. R., Wietzel, W. D., and Othmer, E. Cancer, emotions and mental illness: The present status of understanding. *American Journal of Psychiatry* 133 (1976) 1306-1309.

Tarlau, M. and Smalheiser, I. Personality patterns in patients with malignant tumors of the breast and cervix. *Psychosomatic Medicine* 13 (1951) 117-121.

Vessey, S. H. Effects of grouping on levels of circulating antibodies in mice. *Proceedings of the Society for Experimental Biology and Medicine* 115 (1964) 252-256.

PART IV

BEHAVIORAL INTERVENTIONS

The final portion of this book focuses on the adequacy of behavioral methods for the prevention and treatment of chronic disease. Psychologists, drawing largely on epidemiologic data, have concluded that behavioral methods may be used to modify behaviors which are associated with development of disease. Unfortunately, most investigators in these fields have rather limited knowledge of the epidemiological data. As a result, they are often overly optimistic in their expectations. This section reviews behavioral interventions in light of the expected effectiveness as estimated through current epidemiologic studies on illness and life stress. Each of these papers presents evidence from a current program. The paper by Coates and Ewart describes the evidence for controlling hypertension in adolescents with behavioral interventions. Nader and coworkers offer pilot data from a study in Galveston, Texas, U.S.A., suggesting that risk factors for heart disease can be modified through family interventions. The modification of behaviors which may result in control of Type II diabetes is discussed in a paper by R. Kaplan and Atkins. Maccoby and colleagues present an overview of the Stanford Heart Disease Prevention Program. This program is currently being evaluated in a five city study in California, U.S.A. In the final paper, Shephard considers the potential benefits of exercise and exercise programs.

16

PRIMARY PREVENTION Of HYPERTENSION A PROGRAM WITH ADOLESCENTS[*]

Thomas J. Coates, Ph.D.
Craig K. Ewart, Ph.D.

1. Introduction

Most discussions of blood pressure in adolescents begin with the proviso that hypertension is not a prevalent problem in adolescence and, therefore, does not require significant attention. We will argue the opposite, namely, that because blood pressure tracks and is associated with specific status (e.g., race, sex, socioeconomic class) and modifiable variables (e.g., diet, weight), primary prevention programs should be developed for adolescents. We will present a brief review of the data supporting the tracking hypothesis. Following this will be a discussion of (1) those immutable characteristics that place a person at relatively high risk for hypertension and (2) potentially modifiable behaviors which, when changed, might be useful in the primary prevention of hypertension. Finally, we

[*]Preparation of this manuscript was supported in part by Grant # 5-R01-HL29431 from the National Heart, Lung, and Blood Institute

will present available empirical evidence demonstrating the utility of specific programs for lowering blood pressure in adolescents.

2. Basal Blood Pressure and Reactivity

Arterial blood pressure is among the most important factors in increased risk for morbidity and mortality due to cardiovascular disease (Stamler et al., 1976). Hypertension left untreated damages arterial walls, promotes kidney damage, and strains the heart, which is forced to work harder to move the blood against higher-than-optimal pressures.

Blood pressure *reactivity* in young persons also may be a significant risk factor for the development of hypertension. It is postulated that hypertension develops in susceptible young persons who respond to environmental or emotional stress by high cardiac output, increased peripheral resistance, or both. When faced with a challenging stimulus, the organism responds with increased output of epinephrine and norepinephrine. The catecholamines cause a rise in peripheral resistance and cardiac output, and acute blood pressure increases result. Fixed hypertension may occur over time as acute increases in blood pressure cause progressive changes in anatomical structures and changes in blood pressure regulation mechanisms.

3. Blood Pressure Tracking

Tracking refers to the phenomenon that a person's relative status on a physiological index is relatively invariant; once elevated, blood pressure tends to remain elevated relative to others in the population.

Level of elevated blood pressure during adolescence and young adulthood is predictive of late hypertension (Miall and Lovell, 1967; Paffenbarger, Thorne, and Wing, 1968).

Remarkable tracking of blood pressure has been observed in the Bogalusa study. Data from 3,524 children, aged 5, 8, 11, and 14 at initial examination, were recollected one year later. The correlation between examination and reexamination was 0.70/0.50 (systolic/diastolic). Observations from a group of 35 fifth graders examined monthly were pooled to examine intrachild blood pressure and to estimate regression toward the mean. In a multiple-regression analysis, the previous year's blood pressure and an index of present body size accounted for 39 to 55 percent of systolic blood pressure variability, whereas the previous year's systolic blood pressure contributed a partial correlation coefficient of 0.60 to 0.70 for each age cohort.

4.1. Family history

Family history of hypertension is one the major predictors of hypertension. Paffenbarger and his colleagues (1968), in their longitudinal study of hypertension in college students, reported that history of parental blood pressure contributed to the equation for predicting hypertension in later life. Zinner, Levy, and Kass (1971) extended these observations to a population ranging in age from 2 to 14 years, with a mean of 8.3 years. The sample included 721 children from 190 natural families. Maternal-child correlation coefficients were .16 for systolic and .17 for diastolic pressures; sib-sib systolic/diastolic correlations were .34 and .32, respectively. Within-family variance, of children's blood pressure was significantly lower than between-family variance for both systolic ($F=3.08$) and diastolic ($F=3.68$) pressures.

Larger epidemiological studies generally have confirmed with-family concordance. Holland and Beresford (1975), for example, studied 501 families selected at random but stratified by family size and social class. The major determinants of blood pressures in children five to eight years of age were parental weight and blood pressure. Children's blood pressures also were correlated highly with those of their siblings. Kass and his colleagues (1975) extended the findings downward to a sample ranging from 2 to 14 years of age. They also studied the sample four years later and found familial aggregations that again were significant. Klein and colleagues (1975) reported parallel results with black and white families. Langford and Watson (1973) reported similar correlations for diastolic blood pressure among full siblings aged 14 to 20 (.379 and among half-siblings (.354).

Feinleib and his colleagues (1975) estimated that as much as 60 percent of the variance in blood pressure is due to genetic factors. These estimates were drawn by studying 248 monozygous twins and 264 dizygous twins from five study centers across the country. Correlations among monozygous twins' blood pressures (.55/.58) were higher than the correlations found among dizygous twins (.25/.77). Using data from other studies to show the generally lower correlation among siblings, these investigators estimated relative genetic and environmental contributions to blood pressure using a simple additive model.

That genetic differences alone account for blood pressure variance may be questioned on two grounds. First, as Feinleib and his colleagues (1975) pointed out, their results are derived from a relatively homogeneous population. Genetic variance might be inflated because environmental variance has been suppressed. Second, there is a striking similarity in sib-sib and parent-sib relationships found in many different ethnic groups and with different levels of blood pressure and potentially difference physical and social environments.

In summary, there is little question that genetic factors are operative in elevated blood pressure and that controversies about the exact contribution of genetic factors will continue. The question does need to be reframed, however, so that the more important question is not obscured:

What interactions of environmental and genetic factors produce high blood pressure? The complexity of the phenomenon must not be obscured by the myopia of investigators with favorite but perhaps too narrow hypotheses.

4.2 Sex

Differences between the sexes in average blood pressure presumably emerge in late adolescence. Average blood pressures among adult males are typically higher than among adult females; this relationship generally holds true across racial groups as well (Stamler et al., 1976). The trends are less clear cut among young persons. The Task Force on Blood Pressure Control (1977) reported no blood pressure differences between males and females from 2 to 14 years of age. After the age of 14, however, average blood pressures and the prevalence of hypertension among males increased above the levels reported for females. Voors, Foster, Fredricks. Webber, and Berenson (1976) also found quite similar pressures among males and females aged 1 to 15 years. Other studies with older adolescents have reported characteristic sex differences both among blacks (Dube et al., 1975; and whites (Miller and Shekelle, 1976).

4.3. Race

Blacks have an unusually high prevalence of essential hypertension and related disorders in the United States (Chenoweth, 1973; Finnerty, Shaw, and Himmelsback, 1973). Average blood pressure readings for black males and black females exceed those of white males and females; the prevalence of hypertension in black males is two times greater than that in white males and is associated with higher morbidity and mortality (Stamler et al, 1976).

Voors and his colleagues (1976), in the Bogalusa Study, using six blood pressure observations, found that black children had significantly higher blood pressure than white children. This difference became obvious beginning at age ten. Comstock (1957) found significant elevations among blacks aged 15 to 24. Kotchen and his colleagues (1974) found significant racial differences among 18 to 19-year olds. With sensitive measurements in large samples, characteristic racial differences might be detected among preadolescent children as well. The National Center for Health Statistics (1973, 1977) also reported small but consistent differences in mean diastolic pressures between black and white children aged 5 to 11 years.

Other studies using casual measure found no differences. The Task Force on Blood Pressure Control (1977) reported no racial differences among two to five-year olds. Differences may not emerge until middle to late adolescence.

The distribution of pressures among blacks is displaced to the upper end of values in comparison to whites; more blacks than white may show sustained elevations in blood pressure when rescreened a second time (Kilcoyne et al., 1974; Miller and Shekelle, 1976). Voors and his colleagues (1976) reported that a significantly greater percentage of black children than would be expected by chance have blood pressures above the 95th percentile.

4.4. Socioeconomic class

Lower socioeconomic class is associated with elevated blood pressure. Langford, Watson, and Douglas (1968), in a study of 5,000 black students and 5,500 white students reported higher blood pressures in rural than in city students and an inverse relationship between socioeconomic status and blood pressure among urban students. The usual black-white blood pressure differences were abolished when urban, black, upper income girls were compared with rural whites, and this relationship also was reversed significantly in males when the same comparison was made. Kotchen and colleagues (1974) replicated these results among black students only. Inner-city blacks had higher blood pressures than blacks attending a racially integrated school in a middle-class, residential area. Among blacks, higher blood pressures were found in children whose parents worked as laborers or were unemployed than in children of parents in professional occupations. Holland and Beresford (1975) reported inverse relationships between blood pressure and socioeconomic class in 501 London families.

4.5. Hemodynamics of hypertension

In some young subjects there appears to be an increased cardiac output and tachycardia indicating a hyperkinetic circulation. Such hyperkinetic or "high output types" also may be hypervolemic. Recent echocardiographic studies in children by Davignon, Rey, Payot, Biron, and Mongeau (1977) have confirmed prior invasive hemodynamic studies of others showing that hypertensive children can be either "hyperkinetic" or "hyperresistant."

Left ventricular hypertrophy characteristically is present at autopsy in patients with established hypertension. Recently, myocardial hypertrophy has been described in the spontaneously hypertensive rate before the appearance of elevated pressures and when hypertension is prevented by early immunosympathectomy (Cutilletta et al., 1978). These studies have suggested that myocardia hypertrophy may have a role in the pathogenesis of hypertension.

The echocardiogram has added a new dimension to the study of cardiovascular performance in hypertension (Sannerstedt, Bjure, and Varnaushas, 1970; Davignon et al., 1977). Measurements of stroke volume, systolic time intervals, and left ventricular wall thickness have shown subtle signs of diminished left ventricular function in some established hypertensives. However, normal or increased cardiac performance in some young or borderline hypertensive subjects has shown them to be associated with faster heart rates and shorter preejection fractions and lower ratios of reejection period (PEP) and left ventricular ejection time (LVET).

Zahka. Neill, Kidd. Cutilletta. and Cutilletta (1981) examined 38 normotensive subjects of normotensive parents and 44 hypertensive subjects who were referred to a pediatric hypertension center. The average blood pressure of the hypertensive patients was 144 + 1.5/95 - 1.3 mm Hg, compared with 114 + 1.6/61.7 + 1.4 mm Hg (p = 0.001) in the normotensive group. While the difference is certainly significant, the elevated pressures would be considered only mildly elevated. Various indices of left ventricular mass, however, were significantly greater in the hypertensive patients than in the normotensive controls. Eight percent of the hypertensive patients met criteria of left ventricular hypertrophy. Furthermore, the degree of cardiac hypertrophy correlated poorly with the systolic, diastolic, and mean blood pressures. A similar lack of correlation between hypertrophy and blood pressure has been reported in adult hypertensive patients. Left ventricular hypertrophy seems to persist despite good therapeutic control and is present even in borderline hypertension. It could be argued that a degree of hypertrophy could be expected in patients with pressure overload; however, it appears that the degree of hypertrophy found by Zahka and colleagues (1981) is out of proportion to the level of hypertension. Based upon the hypothesis, myocardial hypertrophy, at least in part, may be independent of the degree of hypertension and could have a role in the pathogenesis of the syndrome. The presence of hypertrophy, therefore, may be useful in identifying patients who will go on to develop hypertension.

4.6. Ponderosity: weight and overweight

Weight and weight status are correlated consistently with blood pressure across the complete distribution of blood pressures. Londe, Bourgoigne, Robson, and Goldring (1971) reported that the prevalence of obesity was higher in hypertensive persons (53%) than in normotensive controls (14%).

Blood pressures are higher among samples of obese persons than among samples of normal weight persons. De Castro and colleagues (1976) studied 320 male high-school students. Using 20 pounds above mean weight for height as a criterion for obesity, the obese had average blood pressures of 124/80 mm Hg versus 116/73 mm Hg for the non-obese. Court, Hill, Dunlop, and Boulton (1974) studied 109 obese persons 1.1 to 17.8 years of age who ranged from 3 to 113 percent overweight. The correlation between measures of subscapular skinfold and pressures were robust (systolic: males = .88, females = .78; diastolic; males = .80, females = .70). Coates and colleagues (1982) reported significant relationships in 36 overweight adolescents (13 to 17 years of age; 15 to 100 percent overweight for sex, age, and height) among the factors of weight, percent overweight, and systolic blood pressure. The same relationships were found before and after the students participated in a program designed to reduce their obesity. Epidemiological studies have reported consistently that increases in blood pressure are correlated with increases in weight. These relationships hold true in black and white children and across the entire range of blood pressures and age groups (Dube et al, 1975;

Holland and Beresford, 1975; Miller and Shekelle, 1976; Stine, Hepner, and Greenstreet, 1975; Voors et al., 1976). Voors and his colleagues (1976) found that the ponderosity index consistently entered first in stepwise multiple-regression equations in predicting systolic and diastolic blood pressures among all age groups. Bivariate Pearson correlation coefficients between body weight and systolic/diastolic blood pressures were .54/.48.

4.7. Diet: salt intake

Epidemiological studies have been used to assert that the prevalence of hypertension in populations is related to salt intake (Dahl and Love, 1957). While the populations studied (e.g., Polynesia, Micronesia, Africa) differ in many ways, salt intake has been consistently low in these societies. Animal and physiological studies have been used to support the hypothesis that sodium intake contributes to the determination of arterial pressure over a long period of time within the constraints imposed by an individual genetic endowment (Dahl, 1972).

In general, few topics have stimulated more controversial and inconsistent data than the hypothesis that salt intake elevates blood pressure (Willett, 1981). Many have asserted that because salt intake in the United States is high (2.8 to 6.0 grams Na per day), greatly in excess of the 0.4 to 0.8 grams of sodium per day needed, a high prevalence of hypertension has resulted.

Studies elucidating the salt-hypertension hypothesis among children and adolescents are few. Langford, Watson, and Douglas (1968) found higher blood pressures among rural than among city students, and inverse relationships between blood pressure and socioeconomic class. Langford and Watson (1973) selected 100 black female sibling pairs for study. Diastolic blood pressures were taken three times per day over eight days in the subjects' homes. Each of the girls also collected a urine specimen every 24 hours for six consecutive days. The Na/Ca ratio was lower ($\bar{x} = 20.4$) among those with lower pressures (less than 105 mm Hg systolic) than among those with higher pressures ($\bar{x} = 34.3$; greater than 125 mm Hg systolic). The ratio also was higher among rural females than among urban (females ($\bar{x} = 33.6$ versus 24.7, respectively). An inverse relationship also was found between socioeconomic status and Na/Ca excretion. However, blood pressures and sodium excretion of Na/Ca ratio were not correlated. An earlier study also failed to confirm correlations between sodium excretion or salt taste threshold and blood pressure, but not according to a direct linear function. Langford and Watson (1975) studied 108 black girls ages 19 to 21 using blood pressures collected over eight days and urine samples collected over six days. One significant correlation emerged: the correlation between diastolic blood pressure and Na/K ratio was population, blood pressure may be a direct function of salt intake and an indirect function of potassium and perhaps calcium intake.

Tuthill and Calabrese (1979) demonstrated a statistically significant upshift of 3 to 5 mm Hg in mean blood pressure between high-school sophomores in two communities with water containing vastly different amounts of sodium. In a second study Tuthill and Calabrese (1981) found identical results among third graders in the same two communities. A confounding factor in the second study was higher sodium intake in the community with higher-sodium water. Moreover, the data supported the hypothesis of the sodium-blood pressure relationships at the aggregate community level but not at the individual level. For these reasons, Willett (1981) suggested caution in accepting these promising results as definitive.

Data from the Bogalusa Heart Study suggest that a proportion of the population may be sensitive to salt (Berenson, 1980). Blacks with the highest blood pressures had lower levels than whites for similar 24-hour sodium levels. The black males of the high stratum showed a cluster of correlations pointing toward sensitivity of the sodium/potassium intake ratio. Blacks also have positive correlation between blood pressure and sodium intake (Frank et al., 1982)

While the role of salt in the genesis and maintenance of elevated blood pressure among children and adolescents is not certain, clinical studies among adults support the utility of reducing and controlling mild hypertension by restricting sodium intake (Corcoran, Taylor, and Page, 1951; Dole et al., 1950). The relative potency of many antihypertensive medications parallels the potency of these drugs in promoting sustained sodium depletion. Finally, several recent clinical studies have supported the utility of reducing mild hypertension by restricting sodium intake. The Stanford three-community study (Farquhar et al., 1977) also reported a significant longitudinal correlation between change in urinary sodium/potassium ratio and a change in blood pressure.

The Task Force on Blood Pressure Control (1977) concluded that the exact significance of salt intake in the genesis of hypertension has not been determined. However, there is general agreement that salt intake should be reduced by individuals with hypertension and by those at risk of developing it. The hypothesis that reduction of salt intake in children and adolescents will result in lower blood pressure remains to be proven.

4.8. Behavior factors

It has been noted for some time that individuals differ in their pattern and degree of autonomic response to environmental stimulation An early individual difference variable correlated with exaggerated blood pressure response was the presence of cardiovascular disease.

Some data support the hypothesized risk-factor status of cardiovascular reactivity. Hines (1937) measured blood pressures of 400 normotensive children while they were subjected to a cold pressor test. The children were subdivided into "normal reactors" and "hyperreactors" (those whose pressures rose above 25/20 mm Hg; 18 percent of the sample) on the basis of blood pressure increases during the test. Barnett, Hines, Schirger, and Gage (1963) followed 207 of the sample 27 years later. A significant proportion of the hyperreactors became hypertensive, while none of the normal reactors became hypertensive.

Two recent studies have assessed the relation of family variables to exaggerated blood pressure reactivity in adolescents. Falkner and colleagues (1979) studied blood pressure response during mental arithmetic in adolescents with varying risk for developing essential hypertension. Three groups were studied: genetic hypertensives (N=33) were those with normal basal blood pressure but at least one parent with essential hypertension; labile genetic hypertensives (n=17) were those who already showed elevated basal blood pressure and also had at least one parent with essential hypertension; controls (N=25) were those who had normal basal blood pressure and no family history of hypertension. (Based on recent studies and reconceptualizations, we would prefer the term "borderline" to the term "labile" used by those investigators; see Horan, Kennedy, and Padgett (1981) and Insel and Chadwick (1982). Subjects were male and female, black and white, ages 14 to 15 years. Labile genetic hypertensives and normotensive genetic subjects showed greater sustained increases than controls in systolic and diastolic blood pressure during mental arithmetic. Post-stress plasma catecholamines were higher in the labile and genetic hypertensives than in the control subjects. The investigators concluded that these findings demonstrated increased central nervous system adrenergic activity and cardiovascular responses in labile hypertensives and in normotensive subjects with a genetic risk for developing hypertension.

Baer and colleagues (1980) studied family interactions and posed some interesting questions regarding transmission of hypertension in families. Three-member families (father, mother, and boy or girl aged 8 to 13 years) of hypertensive (n=16) or normotensive (N=15) fathers were videotaped as they interacted under standardized conditions calling for disagreement. Families with hypertensive fathers showed more negative interactions than families with normotensive fathers. Following these interactions, blood pressures of children in hypertensive families rose, whereas blood pressures in normotensive families fell. However, there was a difference between hypertensive and normal families on *one* negative code only: "Not Tracking" (i.e., looking away to avoid eye contact). Observers were not blind to subjects' diagnostic status. Patients may have been trying to avoid conflict out of the awareness that it affects blood pressure.

Type A (coronary prone) Behavior. The Type A behavior pattern has been associated repeatedly with increased acute blood pressure response to stress. This responsivity is one hypothesized pathway by which Type A might express itself in disease. According to Brand, Rosenman, Sholtz, and Friedman (1976), "Type A behavior is characterized by enhanced aggressiveness and competitive drive, preoccupation with deadlines, and chronic impatience and sense of time urgency in contrast to the more relaxed and less hurried Type B behavior pattern." The Western Collaborative Group study (Brand et al., 1976) and the Framingham study (Haynes, Feinleib, and Kannel, 1980) both provided prospective verification that Type A is a risk factor for cardiovascular disease and that this effect is independent of such traditional risk factors as smoking, blood pressure, and cholesterol. Haynes and colleagues (1978) developed a 300-item questionnaire for use in the Framingham prospective study. The questionnaire contains 14 subscales (reliability from .64 to .86) in three general areas: (1) behavior types: Type A men, Type A women, emotional ambitious lability, noneasygoing; (2) situational stress: nonsupport from boss, marital disagreement, marital disatisfaction, aging worries, personal worries; (3) somatic strain: tension state, daily stress, anxiety symptoms, and anger symptoms. There was moderate to good (67% to 80%) concordance with the structured interview in discriminating Type A from Type B subjects.

In a cross-sectional study of 1822 persons 45 to 47 years of age, Haynes and colleagues (1978) found that women (45 to 64 years of age) with coronary heart disease (CHD), mostly angina, scored significantly higher on Type A, emotional lability, aging worries, tension, and anger symptoms scales than did women free of CHD. Among men under 65, Type A, aging worries, daily stress, and tension were associated with myocardial infarction.

Haynes, Feinleib, and Kannel (1980) followed 1674 subjects in the Framingham study for eight years. Women who developed CHD scored significantly higher in Framingham Type A, suppressed hostility, tension, and anger than did women remaining free of CHD. The prospective study demonstrates the utility of the Framingham Type A scale in predicting disease.

Type A behavior also has been related to the incidence and prevalence of clinical CHD in men and women (Haynes et al., 1980), angiographically determined severity of atherosclerosis (Blumenthal et al., 1978; Friedman et al., 1968), and the progression of atherosclerosis in men (Krantz et al., 1981).

Type A Behavior and Reactivity in Children. The relationship between Type A behavior and blood pressure reactivity in children and adolescents has been documented in recent research. Siegel and Leitch (1981) found a positive correlation between elevated systolic blood pressure and Type A behavior in adolescents. Children and adolescents in the Bogalusa Heart Study who reported that they felt an exaggerated sense of time urgency had higher mean arterial blood pressure than students who responded negatively to this item (Voors et al., 1982). Spiga and Peterson (1981) studied fourth and fifth grade children in a Catholic school. The Mathews Youth Test for Health (MYTH) was used; the 18 highest and 18 lowest scoring males were selected to participate. These students were matched in dyads by Type A and B behaviors so that there were six AA, six AB, and six BB dyads. The dyads played a mixed-motive game in which each player could choose to compete or cooperate on each trial; rewards for individual players were contingent upon both players' choices. Type As in AA dyads showed more competitiveness than Type As in AB dyads and Type Bs in BB dyads. Type As in AA dyads also exhibited greater fluctuations in blood pressure during the task than

310

other subjects. Other investigators who have found a relationship between Type A and cardiovascular over-responding in children and adolescent are Lawler, Allen, Critcher. and Standard (1981), Bergman and Magnusson (1979).

5. A Multivariate Look

In our laboratory research, we have made an effort to examine in combination the relative contribution of family history, ponderosity. age, and the Type A behavior pattern to basal blood pressure and to blood pressure response in adolescents (Coates, Parker, and Kolodner. 1982). Subjects (21 black and 21 white males ranging in age from 14 to 17 years) were recruited from the Pediatric Blood Pressure Center at The Johns Hopkins Hospital and from local high schools. The characteristics of the subjects are presented in Table 1.

Table 1

Participants in the study of Blood Pressure Reactivity (N = 42) (from Coates et al., 1982)

	Mean	S.D.
Age	15.60	1.40
Weight (lbs.)	153.00	28.50
Average percent above ideal weight	1.10	0.22
Systolic blood pressure (resting)	122.80	11.60
Diastolic blood pressure (resting)	64.20	13.40
Heart rate (resting)	70.00	9.80
Type A (Jenkins)[1]	-2.19	8.33
Job involvement (Jenkins)[1]	-8.83	8.80
Speed and impatience	-3.54	8.89
Hard-driving and competitive (Jenkins)[1]	-8.30	14.24
Bortner Rating Scale	165.34	34.32
Percent with hypertensive parents	54	
Points on Alluisi Task during 15 minutes on CRT	78.46	22.58

[1]These are standard scores, normalized on a version of the Jenkins Activity Survey developed for adolescents (Spiga and Petersen. 1981).

Subjects were enrolled in the study after at least three successive blood pressure determinations using suitable cuff sizes had established their basal blood pressure. Approximately one week after the third clinic-based assessment, subjects returned to the Small Group Programmed Environment Laboratory at the Phipps Psychiatric Clinic at The Johns Hopkins Hospital.

Each subject was escorted to the recreation area, where he was seated at a desk. An interview took place to collect additional data on family history. The subject was weighed and height was measured. The subject then completed two measures of Type A behavior, the Jenkins Activity Survey (JAS) adapted for adolescents and the Bortner Rating Scale. The adult version of the JAS is a self-administered questionnaire (60 items) with "Type A," "Hard-driving," "Speed and Impatience," and "Job Involvement" subscales (Demobroski et al., 1978). Spiga and Peterson (1981) adapted a shortened form (47 items) of this instrument for adolescents, and the adapted version was used on our study.

Following completion of the written tests, the subject was escorted into the small workshop area, where he was seated before the cathode-ray terminal (CRT). He was instructed about the sequence of events to follow and was fitted with a remotely monitored and controlled blood pressure cuff (Vita Stat Model 900-5), a digital thermistor for monitoring skin temperature, and an optically transduced plethysmograph for monitoring peripheral blood flow. Each subject was instructed in how to complete the Alluisi Performance Battery and was given 15 minutes of adaptation to the room and to the monitoring devices. Each subject then was exposed to alternating 15 minute periods of (1) baseline, quiet resting; (2) task performance on the Alluisi Performance Battery on the CRT; (3) baseline, quiet rest; and (4) metronome conditioned relaxation.

The Alluisi Performance Battery required subjects to complete five tasks simultaneously: probability monitoring, horizontal addition and subtraction of three-digit numbers, matching to sample histograms, detecting when a stationary signal moves, and detecting when a moving signal becomes stationary (Emurian, Emurian, and Brady, 1978).

Stepwise multiple-regression equations with free entry of variables were computed to assess independent variables related to basal blood pressure and to blood pressure change from baseline to Alluisi. The number of variables permitted to enter each equation was limited to five so that models would not be overdetermined. Tables 2 and 3 present multiple-regression results showing predictors of baseline systolic and diastolic blood pressure. Our findings were similar to those from the Bogalusa Heart Study (Berenson, 1980) in that we were able to account for 40 percent of the variance in systolic blood pressure. Ponderosity, age, and parental hypertension were significant predictor variables. Most important, measures of Type A behavior entered into the multiple-regression equations and added significantly to the prediction.

Table 2

Dependent Variable: Baseline Systolic Blood Pressure
(from Coates et al., 1982)

Independent Variable	Beta	R
Weight/height^2	.410	.47
Age	.228	.55
Parental hypertension	.298	.58
Job Involvement (JAS)	.212	.62
Type A (Bortner)	.149	.64

$R^2 + .41$
$F = 4.62 \ (p < .01)$

Table 3

Dependent Variable: Baseline Diastolic Blood Pressure
(from Coates et al., 1982)

Independent Variable	Beta	R
Weight/height^2	.608	.56
Type A (Bortner)	.286	.64
Hard-driving (JAS)	.307	.67
Age	.275	.69
Type A (JAS)	-.295	.71

$R^2 = .50$
$F = 7.05 \ (p < .001)$

We are able to account for 51 percent of the variance in baseline diastolic blood pressure using both ponderosity and Type A behavior as independent variables. This may represent one of the strongest documentations of the relation between behavioral factors and baseline blood pressures in adolescents. Certainly, the study deserves replication.

313

We were primarily interested in accounting for blood pressure response during performance on the Alluisi Task. Tables 4, 5, and 6 present multiple-regression results showing predictors of the absolute and the relative (percent increase above baseline) changes from baseline to Alluisi in systolic and diastolic blood pressure. Behavioral factors entered most strongly in predicting absolute and relative systolic blood pressure change. Performance on the Alluisi task was negatively related to systolic blood pressure increases (that is. students who performed better showed smaller relative and absolute increases in systolic

Table 4

Dependent Variable: Absolute Change in Systolic Blood Pressure (from Coates et al., 1982b)

Independent Variable	Beta	B
Alluisi points	-.442	.38
Bortner Type A	.239	.48
Mean baseline systolic pressure	.195	.54
Parental hypertension	-.273	.58

$R^2 = .34$
$F = 4.57$ (p< .01)

Table 5

Dependent Variable: Relative Change in Systolic Blood Pressure (from Coates et al., 1982b)

Independent Variable	Beta	B
Bortner Type A	.372	.366
Alluisi points	-.404	.531
Weight/height^2	.254	.575
Parental hypertension	-.214	.606
Age	.078	.617

$R^2 = .381$
$F = 4.06$ (p< .01)

Table 6

Dependent Variable: Relative Change in Diastolic
Blood Pressure (from Coates et al., 1982b)

Independent Variable	Beta	R
Weight/height^2	-.294	.32
Age	.161	.38
Race	.093	.39

$R^2 = .16$
$F = 3.62$ $(p< .05)$

pressure), and performance was positively related to Type A as measured by the Bortner inventory. Baseline systolic pressure, parental hypertension, and relative obesity were other variables entering into these questions.

6. Modifying Blood Pressure in Adolescents

6.1. Weight loss

Weight loss can promote blood pressure reduction among adults (Chiang, Perlman, and Epstein, 1969; Tyroler et al., 1975). It has been suggested, however, that drops in blood pressure with weight loss are due entirely to the concomitant reduction in salt intake (Dahl, 1972). Reison and colleagues (1978), however, reported reductions in blood pressure concomitant with weight loss and independent of restriction in salt intake among adults. Patients, overweight and hypertensive, fell into three groups: those not receiving antihypertensive drug therapy (Group 1) and those on regular drug therapy but with adequate control of hypertension (Groups 2a and 2b). Group 1 and 2a received dietary counseling for weight loss, while Group 2b did not. All patients in Groups 1 and 2a lost at least 3 kg.: mean loss was 13.5 kg. (+6.3) and standard deviation was 14.9 (+5.3 kg.) during the same period. Seventy-five percent of Group 1 and 61 percent of Group 2a returned to a normal blood pressure, and correlations between weight loss and reductions in systolic and diastolic blood pressure were significant in both groups (Group 1 – .42/.56; Group 2a = .24/.30). Mean urinary sodium excretion from a 24-hour sample was similar among all three groups following treatment.

Coates and colleagues (1982) examined the relationship between weight loss and changes in blood pressures among normotensive, overweight adolescents. Subjects were 36 adolescents participating in a study of the efficacy of various treatment variables to facilitate weight loss. Prior to their participation in weight-loss classes, subjects reported to the laboratory on two separate mornings.

315

Table 7

Changes in Weight and Blood Pressures during Weight Loss

	Pre	Post	Change
Weight	X = 179.06	169.98	-9.08**
	S.D. = 41.06	40.37	
	R = 136-269	122-182	
Percent overweight	X = 40.62	33.78	-6.84**
	S.D. = 23.13	24.74	
	R = 9-100	2.5-110	
Systolic blood pressure	X = 114.54	107.30	-7.25**
	S.D. = 10.34	14.28	
	R = 98-131	80-129	
Diastolic blood pressure	X = 72.84	66.27	-6.57**
	S.D. = 8.21	9.27	
	R = 75-93	51-85	

*p< .05
**p< .01

Adapted from T. J. Coates, R. W. Jeffery, L. A. Slinkard, J. D. Killen, and B. G. Danaher. Frequency of contact and monetary reward in weight loss, lipid change, and blood pressure reduction with adolescents. *Behavior Therapy*, 1982, 13(2), 175-185.

After the subjects had been seated in the laboratory for several minutes, blood pressure was measured in the right arm by a standard sphygmomameter. The subject was then left alone for five minutes, after which the nurse returned to take a second reading. The same procedure was followed 24 weeks later following the end of the subjects' participation in the weight-loss classes.

Table 7 presents pre- and post-treatment values for weight and blood pressure for all subjects. Blood pressures reported represent the average of the second reading taken at each of two assessment sessions. Pre- and post-treatment values were compared using the t-tests for paired samples.

As can be seen in Table 7, there were significant reductions in both systolic and diastolic blood pressures. In addition, the correlation between change in percent overweight and change in systolic blood pressure was significant ($r = .29$, p< .05). The correlation between change in percent overweight and change in diastolic blood pressure was not significant (r = .07, n.s.).

6.2. Modifying diet: the great sensations study

Effective nutrition education programs are in short supply. Saylor, Coates, Killen, and Slinkard (1982) reviewed available empirical evidence on the efficacy of nutrition education programs with children and adolescents. They were able to locate only 25 empirical studies published in the last twelve years. Only five of these met minimal criteria for quality investigations, including adequate follow-up, specification of treatment components, and adequate and reliable dependent measures.

We have conducted three studies demonstrating efficacy of programs based on social learning theory in producing positive nutrition behavior changes in school children. Coates. Jeffery, Slinkard (1981) developed and evaluated the Heart Health Program for elementary school students to (1) increase their consumption of complex carbohydrates and decrease their consumption of saturated fat, cholesterol, sodium. and sugar: (2) increase their level of habitual physical activity; and (3) generalize these changes to family members. The program produced substantial and maintained changes in students' eating behavior at school, knowledge about heart health food preferences, and family eating patterns.

Coates, Slinkard, Perry, and Hashimoto (1982) replicated these effects in a second study. A component analysis revealed that students receiving the instructional program alone showed positive changes in knowledge and preferences but no changes in behavior. Only those students and their families receiving the full program (containing both instructional and motivational components) showed positive and sustained (across summer vacation) change in diet.

The Great Sensations Program was a nutrition education project developed for high school students (Coates et al., 1983). It was designed to (1) decrease students' consumption of salty snacks and (2) increase students' consumption of fresh fruit snacks. The overall programs were designed following principles of social learning: informative instruction, participatory classroom activities, personal goal setting, feedback, and reinforcement. The class program was delivered in six lessons during regular health education classes. A parent involvement program consisted of mailers and telephone calls to parents to teach parents to encourage changes in student snacking habits. A school-wide media program was designed to provide out-of-class peer support for student modification in salty snack foods. The program was evaluated in one high school using a 2 by 2 design. A second high school served as a no-treatment control. Assessments were taken at baseline, post-program, at the end of the school year, and at the beginning of the school year following summer vacation. The school-wide program was effective in decreasing consumption of salty snack foods and of increasing consumption of target snack foods. However, only those students receiving class instruction maintained those changes until the end of the school year. No changes were maintained across summer vacation. These outcomes suggest that school programs developed using principles of social learning can be effective in facilitating important behavior changes at home and at school. It would be useful to follow this research with (1) studies of methods for enhancing the immediate and long-term reduction of sodium intake and (2) studies of the impact of these reductions on blood pressure.

6.3. The Minnesota blood pressure study

Gillum, Elmer, and Prineas (1981) randomized 80 school children with blood pressures above the ninety-fifth percentile for age and sex but below 130/90 mm Hg to a family intervention program or to a control group. Twenty children aged six to nine years and their families began a program to modify the family diet toward a goal of 70 mEg sodium per person per day. Families in the treatment group attended four, biweekly, intensive 90 minute lecture-demonstrations followed by 90-minute maintenance sessions occurring at bimonthly intervals over the remainder of the year. Children and parents attended separate 60-minute sessions, followed by a joint 60-minute, low-sodium refreshment period with discussion. Adherence was assessed by three-day food records and urine collections in children and adults. Effects were assessed one year following randomization. Results are difficult to interpret, as only 17 of the 41 randomized to treatment completed the full intervention. Twenty-four hour urinary sodium excretion data were available for only 11 intervention families. There was a significant decrease in urinary sodium excretion from the beginning to 6 months of 34 mmol/24 hrs. Poor parent compliance hampered collection and analysis of their data. At one year, urinary sodium excretion data were available on 32 control and 32 intervention families. Both groups showed slight increases in overnight urinary sodium excretion ($+3.7$ and $+2.9$ mmol 10 hrs, respectively). There were no significant differences in changes in blood pressure between the two groups. Thus, this study demonstrates the difficulty of conducting this kind of intervention and research and suggests that effective programs may require better access to subjects.

6.4. Modifying potassium and sodium

Parfey et al., (1981) selected students whose parents did (PHT n=12) or did not (PNT n=11) have hypertension. During a two-week period of no added sodium, the PHT group had higher blood pressure and plasma noradrenaline levels than the PNT group. At the end of the four weeks of the high sodium diet, the blood pressure levels of both groups was significantly higher than baseline. In contrast, when the low-sodium diet was supplemented with potassium, the blood pressures of the PHT group fell significantly, while the blood pressure in the PNT group rose slightly. Thus, it is possible that a combination high potassium/low sodium diet may be needed to lower blood pressure in those predisposed to hypertension.

6.5. Diet and medication combined: ADAPT

A Dietary/Exercise Alteration Program, in combination with low dose medication, was developed by the Bogalusa Heart Study as a model pediatric hypertension program (Frank et al., 1982). ADAPT was designed to accomplish a daily 1 to 2 gm. sodium intake. Six core components included private consultation, education classes, self-administered training, tool development, school lunch modification, food procurement, and physical activity classes. The objective was to empower children to make judicious choices using class instruction (12 lessons), regular consultations, and tools (sodium counters, recipe books). A supportive environment was fostered by grocery store, restaurant, and school lunch programs. Medication consisted of low-dosage diuretic (Chlorthiadone) and B-blocker (Propranolol) prescribed at 25 percent of usual therapeutic dose for a given weight. Participants in this study were 100 children, 8 to 18 years of age, who tracked at or above the ninetieth percentile based on 36 blood pressure readings. These students were randomly assigned to treatment and control groups. Fifty children in the midrange (fiftieth to sixtieth percentile) served as an additional comparison group. In the first six months of intervention, blood pressures of treatment subjects decreased of mm Hg systolic and diastolic and remained 5/3 mm Hg lower than the risk comparison group. They reached the fiftieth to sixtieth percentile for the population. Sodium intake was not decreased significantly. Obese treatment subjects grained less weight than their counterparts in the high obese group.

This program is important because it demonstrates the potential of blood pressure lowering in a segment of the population. It must be noted, however, that the intervention required three full-time nutritionists, a part-time physical activity coordinator, and supporting staff from the Bogalusa Heart Study. Given the effort, it is questionable whether any effect beyond that attributable to the medication can be demonstrated. Moreover, many would question the desirability of administering medications to adolescents and children if other approaches have not been evaluated.

6.6. A comprehensive program: know your body

Botvin and Eng (in press) conducted school-based studies to reduce elevated blood pressure in high school students as part of the "Know Your Body" program. The focus of the program was on nutrition education, weight control, stress management and cardiovasacular function. The program involved only those students in the top decile of blood pressure determined during two school-wide screenings. Blood pressures were decreased for students in the treatment school (129.71/82.25 mm Hg to 122.04/77.92). Systolic pressures in the matched control school remained the same, while diastolic pressures increased. This study demonstrates the potential of school-based programs to treat blood pressure. The authors stressed, however, the need to conduct treatment during class times and preferably as part of the regular school program.

6.7. A school-based relaxation and biofeedback program

Ewart, Coates, and Simon (1983) are conducting studies in Baltimore City Schools with the following specific aims:

1.   to conduct blood pressure screenings in selected high schools in Baltimore City to develop distributions of blood pressure for this population and to identify individuals above the ninetieth centile systolic and/or diastolic
2.   to investigate the short and long-term effects of relaxation on blood pressure when it is taught and practiced regularly in the school classroom (Study 1)
3.   to study behavioral correlates of response and non-response to treatment (Studies 1 and 2)

School-based programs offer several advantages, both scientific and clinical, over more traditional clinic or laboratory based investigations in studying the efficacy of relaxation in lowering blood pressure in adolescents.

1.   *Compliance can be enhanced because relaxation will be taught and practiced in the classroom during a regular class period.* This has scientific significance (for an adequate test of hypotheses) and clinical significance (to insure the efficacious use of the strategies). We have attempted previously to conduct clinic-based treatments for elevated blood pressure with adolescents (see Coates and Masek, 1982). This was less than optimal. While students reported enjoying relaxation and understand its importance for controlling blood pressure, returning to the hospital for repeated treatments was burdensome. Conducting programs in schools during regular class sessions will enhance correct and consistent application of relaxation.

2.   *Maintenance will be enhanced because school and family-involvement programs can be developed and implemented in a cost-effective manner. Maintenance is one of the key issues in health behavior change.* Treatments in schools will facilitate maintenance of change. Parents and peers can be encouraged to participate in programs designed to promote maintenance. By working in the schools, we can develop peer support for use of relaxation and reach parents through existing school-based networks.

During the fall of 1982, we conducted blood pressure screenings for all students in the ninth and tenth grades at a selected high school. Although 1172 students were listed on school enrollment records, only 866 could be reached by mail or phone to obtain parent consents. Parental refusals were few (9%) and 86 percent of the available pool of 866 subjects were screened during the five-week program.

Subjects with a systolic or diastolic blood pressure above the eighty-fifth percentile of the screening distribution (i.e. > 120 mm Hg systolic or > 73 mm Hg diastolic) were screened a second time two months later. Those with either blood pressure above the 120/73 criterion were considered eligible for the PMR intervention study. A total of 79 students were randomized to the treatment (class) or no-treatment control group.

The class, offered as an elective in the Health Education program entitled "High Blood Pressure," was taught for full credit by an accredited teacher. Students attended during regular class hours every day and received credit for participation for the semester-long course.

The class itself covered a full range of topics related to high blood pressure including basic physiology, factors related to and the consequences of hypertension, and diet, exercise, and medication. The unique features of this class were systematic teaching in relaxation and temperature biofeedback. Students were introduced first to progressive muscle relaxation in the following steps:

1. The teacher explained the purpose and goals of the program.
2. The teacher then guided the students through the relaxation procedure.
3. Data were collected on student's ability to attend and complete the procedures with 90 percent accuracy.

Following this, students were introduced to differential relaxation. Differential relaxation teaches discrimination of "tense" versus "relaxed" muscles. It also teaches students to relax on command.

The following steps were used:

1. Teacher read background information, reviewed essential terms, and demonstrated the exercise.
2. Students were grouped in twos. One student was on the floor and relaxed his/her body with three complete breaths. The student then followed the teachers' instructions. The partner checked to make sure the student could tense specific muscles while the others remain relaxed.
3. Students changed roles.
4. Data were collected on student's ability to tense and relax muscles.

Finally, students learned to use a finger thermometer as a biofeedback system to assist in monitoring his/her blood pressure during tense or calm situations. The following steps were used:

1. The teacher explained the purpose of the finger thermometer.
2. The teacher explained and demonstrated how to use the finger thermometer.
3. The teacher then guided the students as they worked with the finger thermometers.
4. Baseline data were taken during the first few days of the program.
5. Data were taken on students' progress.
6. There was a question and answer period.

Of the 40 students randomized to PMR treatment, 30 attended regularly. PMR sessions were conducted on an average of 4 days per week over a 12 week period: in all, PMR was practiced 45 times. Attendance data showed that subjects participated in an average of 38.4 PMR sessions. Systematic observation showed that, during these sessions, compliance (as judged from absence of visible movement in nine muscle groups averaged 80 percent (seven out of nine groups immobile)).

To evaluate generalization of PMR and finger temperature mastering, behavioral observation and finger temperature data were collected from both treatment and control subjects during the pretest and posttest phases of the study. Subjects were asked to "sit and relax" for five minutes in an unfamiliar school office while temperature and behavioral data were recorded by a trained observer who was unaware of the PMR treatment and was blind to subjects' experimental status. Treatment subjects' breathing rates dropped from a pretest baseline of 17.2 to 11.2 cycles per minute ($t = 5.24$, $p < .001$), where as breathing of controls increased from 16 to 17.2 per minute. PMR students' finger temperature rose from 91.4 to 94.0 ($t = 1.66$, $p < .06$) while controls dropped from 93.3 to 92.3 (n.s.). PMR students' muscle movement remained stable (60.3% to 63% motionless) while controls became less quiet at posttest (55.5% to 49.4% motionless $t = 2.56$, $p < .02$). Analyses of covariance (using pretest as the covariate) revealed statistically significant differences between treatment and controls in breathing rate ($t = 5.75$, $p < .001$) and muscle movement ($t = 3.62$, $p < .001$), but not finger temperature ($t = 1.28$, n.s.).

Having shown that students mastered PMR, we next investigated possible BP effects. Fourteen treatment subjects and 17 controls had been eligible for the study on the basis of their systolic pressure. Group pre-post comparisons showed that treatment subjects' systolic pressures dropped from a mean of 127.8 ± 7.8 mm Hg to 122.8 ± 7.1 mm Hg, ($t = 2.33$; $p < .03$), and that corresponding pre and post values for the controls were 127 ± 5.8 mm Hg and 127.3 ± 9.7 (n.s.). An analysis of covariance with posttest BP as the dependent variable and the pretest BP as the covariate revealed the difference in systolic pressure to be statistically significant ($t = 1.85$; $p < .04$), suggesting that treated subjects had lowered their BP relative to the assessment-only controls.

Among the treatment subjects, 20 had been eligible on the basis of diastolic pressure, as had 23 controls. Treatment subjects diastolic pressures dropped from a pretest mean of 78.3 ± 3.7 to 68.7 ± 10.7 ($t = 3.56$; $p < .003$) and corresponding values for controls were 79.0 ± 6.1 and 67.8 ± 8.6 ($t = 6.00$; $p < .001$). Differences between groups were not significant.

Data from behavioral observation and finger temperature measurement demonstrate that adolescent subjects will comply with PMR training offered as a routine exercise in health education classes, that they enjoy this activity, that they can master the PMR technique, and that they attain a greater degree of relaxation in a generalization setting than do untrained controls as judged from ratings by a naive observer. We conclude that our PMR training curriculum is effective in a school classroom context and is ready to be replicated in research with other adolescent school populations during the coming year.

In view of the above considerations, our research in Year II will have the following objectives:

1. to conduct follow-up BP measures with subjects given PMR training in Year I to assess long-term effects

2. to train Year I controls in PMR to replicate skill mastery and BP findings

3. to conduct BP screening in another Baltimore City Public High School with a student population that is predominantly black

4. to select all 9th and 10th grade students above the 85th percentile of systolic or diastolic BP to participate in further studies

5. to evaluate BP reactivity in all of these selected subjects using a computer video game we have developed and to collect overnight urine specimens from these individuals to derive an estimate of dietary sodium levels

6. to replicate the PMR study design used in Year I with black adolescents who have been characterized on BP reactivity to stress and on dietary sodium

7. to measure changes in BP level and BP reactivity occurring at posttest and relate these to subjects pretest reactivity and dietary and dietary characteristics.

Conclusion

We believe that we have made a convincing case of the need to help programs for the primary prevention of hypertension among children and adolescents. The best means for accomplishing that objective and even whether or not it can be accomplished remains highly questionable.

We place priority on the following research questions.

1. The significance of cardiovascular reactivity remains to be established. It has been hypothesized that reactivity may be a risk factor for the development of hypertension or atherosclerosis. Until the link is established, intervening on this variable is of questionable utility.

2. Nonetheless, it remains an interesting question to determine the most efficacious methods for reducing reactivity. Relaxation may be an effective mode, but our research suggests that increasing competence to perform stressful actions may be equally effective in moderating blood pressure reactivity.

323

3.   Research should be done to examine the relative efficacy of relaxation, biofeedback, weight loss, sodium restriction, and physical activity on blood pressure lowering among adolescents. The effects of these treatments should be examined singly and in combination.

4.   Along with this, we need studies of variables which moderate the impact of treatments and variables which predict differential effectiveness of specific techniques for individual patients.

References

Baer, P. E., Vincent, J. P., William, B. J., Bourianoff, G. G. and Bartlett, P. C. Behavioral response to induced conflict in families with a hypertensive father. *Hypertension* 2, 1 (1980) 70-71.

Barnett, P. H., Hines, E. A., Schirger, A., and Gage, R. P. Blood pressure and vascular reactivity to the cold pressor test. *Journal of the American Medical Association* 183 (1963) 143-146.

Berenson, G. S. *Cardiovascular risk factors in children: The early natural history of atherosclerosis.* New York, Oxford, 1980.

Berenson, G. S. et al. A model of intervention for prevention of early essential hypertension in the 1980's. *Hypertension* 5 (1983) 41-53.

Bergman, L. R. and Magnusson, D. Overachievement and cathecholamine excretion in achievement-demanding situations. *Psychosomatic Medicine* 41 (1979) 181-188.

Blumenthal, J. A. et al. Type A behavior patterns and coronary atherosclerosis. *Circulation* 58 (1978) 634-639.

Brand, R. J., Rosenman, R. H., Sholtz, R. I., and Friedman, M. Multivariate prediction of coronary heart disease in the Western Collaboration Group Study compared to the Framingham study. *Circulation* 53 (1976) 348-355.

Chenoweth, A. C. High blood pressure: A national concern. *Journal of School Health* 43 (1973) 307-308.

Chiang, B. N., Perlman, I. V., and Epstein, F. H. Overweight and hypertension. *Circulation* 39 (1969) 403-421.

Coates, T. J. et al. The great sensations study: Modifying snack food preference of high school students. (unpublished manuscript) University of California, San Francisco, 1983.

Coates, T. J., Jeffery, R. W., and Slinkard, L. A. The heart health program: Introducing and maintaining nutrition changes among elementary school children. *American Journal of Public Health* 71 (1981) 15-23.

Coates, T. J., Jeffery, R. W., and Slinkard, L. A., Killen, J. D., and Danaher, B. G. Frequency of contact and contingent reward in weight loss, lipid change, and blood pressure reduction in adolescents. *Behavior Therapy* 13 (1982) 175-185.

Coates, T. J. et al. Monetary incentives and frequency of contact in weight loss, blood pressure reduction, and lipid change in adolescents. *Behavior Therapy*, in press.

Coates, T. J., Parker, E., and Kolodner, K. Stress and cardiovascular disease. Does blood pressure reactivity offer a link? In Coates, T. J., Petersen, R. C., and Perry, C. P. eds. *Promoting adolescent health: A dialogue on research and practice*. New York: Academic Press, 1982.

Coates, T. J., Slinkard, L. A., Perry, C. P., and Hashimoto, G. Heart health eating and exercise: A replication and component analysis. (unpublished manuscript) University of California, San Francisco, 1982.

Comstock, G. W. An epidemiologic study of blood pressure levels in a biracial community in the Southern United States. *American Journal of Hygiene* 65 (1957) 271.

Corcoran, A. C., Taylor, R. D., and Page, I. H. Controlled observations on the effect of low sodium diet therapy in essential hypertension. *Circulation* 3 (1951) 1.

Court, J. M., Hill, G. H., Dunlop, M., and Boulton, T. J. C. Hypertension in childhood obesity. *Australian Pediatric Journal* 10 (1974) 296-300.

Cutilletta, A. F., Benjamin, M., Culpepper, W. S., and Oparil, S. Myocardial hypertrophy and ventricular performance in the absence of hypertension in spontaneously hypertensive rats. *Journal of Molecular Cardiology* 10 (1978) 689-703.

Dahl, L. K. Salt and hypertension. *American Journal of Clinical Nutrition* 25 (1972) 231.

Davignon, A. et al. Hemodynamic studies of labile essential hypertension. In New, M. I. and Levine, L. S., (eds.) *Juvenile hypertension*. New York, Raven Press, 1977.

de Castro, R. J. et al. Hypertension in adolescents. *Clinical Pediatrics* 15 (1976) 24-26.

Demobroski, T. M. et al. Components of the type A coronary-prone behavior pattern and cardiovascular responses to psycho-motor performance challenge 1 (1978) 159-176.

Dole, V. P. et al. Dietary treatment of hypertension: Clinical and metabolic studies of patients on the rice-fruit diet. *Journal of Clinical Investigation* 29 (1950) 1189.

Dube, S. K., Kapoor, S., Ratner, H., and Turnick, E. L. Blood pressure studies in black children. *American Journal of Diseases of Children* 129 (1975) 1177-1180.

Emurian, H. H., Emurian, C. S., and Brady, I. V. Effects of a pairing contingency on behavior in a three person programmed environment. *Journal of the Experimental Analysis of Behavior* 29 (1978) 319-329.

Ewart, C. K., Coates, T. J. C., and Simons, B. School-based relaxation to lower blood pressure. NHLBI Grant # R01-HL39431, 1981-1984.

Falkner, B. et al. Cardiovascular response to mental stress in normal adolescents with hypertensive parents. *Hypertension* 1 (1979) 23-30.

Farquhar, J. et al. The Stanford Three Community Study. *Lancet* (1977)

Feinleib, M. et al. Studies of hypertension in twins. In Paul, O. (ed.) *Epidemiology and control of hypertension. New York: Grune and Stratton, 1975.*

Finnerty, F. A., Shaw, L. W., and Himmelsback, C. Hypertension in the inner city II. Detection and follow-up. Circulation 47 (1973) 76-78.

Frank, G. C. et al. Infant feeding patterns and their relationship to cardiovascular risk factor variables in the first year of life. (unpublished manuscript) Louisiana State University, 1982.

Friedman, M. et al. The relationship of behavior pattern A to the state of the coronary vasculature: A study of 51 autopsied subjects. *American Journal of Medicine* 44 (1968) 525.

Haynes, S. G., Feinleib, M., and Kannel, W. B. The relationship of psychosocial factors to coronary heart disease in the Framingham study. *American Journal of Epidemiology* 111 (1980) 37-38.

Haynes, S. G. et al. The relationship of psychosocial factors to coronary heart disease in the Framingham Study II. Prevalence of coronary heart disease. *American Journal of Epidemiology* 107 (1978) 384-402.

Heyden, S., de Maria, W., Barbee, S., and Morris, M. Weight reduction in adolescents. *Nutrition and Metabolism* 15 (1973) 295-304.

Hines, E. A. Reaction of blood pressure of 400 children of standard stimulus. *Journal of the American Medical Association* 108 (1937) 1249-1250.

Holland, W. W. and Beresford, S. A. A. Factors influencing blood pressure in children. In Paul, O. (ed.) *Epidemiology and control of hypertension.* New York: Grune and Stratton, 1975.

Horan, M. J., Kennedy, H. L., and Padgett, N. E. Do borderline hypertensive patients have labile blood pressure? *Annals of Internal Medicine* 94 (1981) 466-468.

Insel, P. M. and Chadwick, J. H. Conceptual barriers to the treatment of chronic disease: Using pediatric hypertension as an example. In Coates, T. J., Peterson, A. C., and Perry, C. (eds.) *Promoting adolescent health: A dialog on research and practice.* New York: Academic Press, 1982.

Kass, E. H. et al. Familial aggregation of blood pressure and urinary Kallikrein in early childhood. In Paul, O. (ed.) *Epidemiology and control of hypertension.* New York: Grune and Stratton, 1975.

Kilcoyne, M. Natural history of hypertension in adolescence. *Pediatric Clinics of North America* 25 (1978) 47-53.

Klein, B. E. et al. Longitudinal studies of blood pressure in offspring of hypertensive mothers. In Paul, O. (ed.) *Epidemiology and control of hypertension.* New York: Grune and Stratton, 1975.

Kotchen, J. M., Kotchen, T. A., Schwertman, N. L., and Kuller, L. H. Blood pressure distributions of urban adolescents. *American Journal of Epidemiology* 99 (1974) 315-324.

Krantz, D. S. Investigations of the extent of coronary atherosclerosis, Type A behavioral and cardiovascular response to social interaction. *Psychophysiology* 18 (1981) 654-664.

Langford, H. G. and Watson, R. L. Electrolytes and hypertension. In Paul. O. (ed.) *Epidemiology and control of hypertension* New York: Grune and Stratton, 1975.

Langford, H. G. and Watson, R. L. Electrolytes, environment, and blood pressure. *Clinical Science and Molecular Medicine* 45 (1973) 111s-113s.

Lawler, K. A., Allen, M. T., Critcher, E. G., and Standard, B. A. The relationship of physiological response to coronary-prone behavior pattern in children. *Journal of Behavioral Medicine* 4 (1981) 203-216.

Londe, S., Bourgoigne, J. J., Robson, A. M., and Goldring, D. Hypertension in apparently normal children. *Journal of Pediatrics* 78 (1971) 569-577.

Miall, W. E. and Lovell, H. G. Relation between change in blood pressure and age. *British Medical Journal* 2 (1967) 600-664.

Miller, R. A. and Shekelle, R. B. Blood pressure in tenth grade students. *Circulation* 54 (1976) 993-1000.

National Center for Health Statistics. Blood pressure of youths 12-17 years. Publication (HRA) 77-1645, Vital and Health Statistics, Series II, No. 163. Washington, D.C.: U.S. Government Printing Office, 1977.

National Center for Health Statistics. Blood pressure of youths 6-11 years. Publication (HRA) 74-1617, Vital and Health Statistics, Series II, No. 135. Washington, D.C.: U.S. Government Printing Office, 1973.

Paffenbarger, R.S., Thorne, M. C., and Wing, A. L. Chronic disease in former college students: VIII. Characteristics of youth predisposing to hypertension in later years. *American Journal of Epidemiology* 88 (1968) 25-32.

Reisen, E. et al. Effect of weight loss without salt restriction on the reduction of blood pressure. *New England Journal of Medicine* 198 (1978) 1-6.

Sannerstedt, R., Bjure, J., and Vannaukas, E. Correlation between echocardiographic changes and systemic hemodynamics in human arterial hypertension. *American Journal of Cardiology* 26 (1970) 117-120.

Saylor, K., Coates, T. J., Killen, J. D., and Slinkard, L. A. Nutrition education research: Fast or famine. In Coates, T. J., Petersen, A. C., and Perry, C. (eds.) *Promoting adolescent health: A dialog on research and practice.* New York: Academic Press, 1982.

Siegel, J. M. and Leitch, C. J. Assessment of the Type A behavior pattern in adolescents. *Psychosomatic Medicine* 43 (1981) 45-46.

Spiga, R. and Petersen, A. C. The coronary-prone behavior pattern in early adolescence. Presented at the meetings of the American Education Research Association, Boston, 1980.

Stamler, J. et al. Hypertension screening of one million Americans. *Journal of the American Medical Association* 235 (1976) 2299-2306.

Stine, O. C., Hepner, R., and Greenstreet, R. Correlation of blood pressure with skinfold thickness and protein levels. *American Journal of Diseases of Children* 129 (1975) 905-911.

Task Force on Blood Pressure Control. *Pediatrics* 59 (1977) 797-820.

Tyroler, H. A. et al. Weight and hypertension: Evans County studies of blacks and whites. In Paul, O. (ed.) *Epidemiology and control of hypertension.* New York: Grune and Stratton, 1975.

Tuthill, R. W., and Calabrese, E. J. Drinking water sodium and blood pressure in children: A second look. *American Journal of Public Health* 71 (1981) 722-729.

Tuthill, R. W., and Calabrese, E. J. Elevated sodium levels in public drinking water as a contributor to elevated blood pressure levels in the community. *Archives of Environmental Health* 34 (1979) 197-203.

Voors, A. E. et al. Cardiovascular risk factors in children and coronary-related to behavior. In Coates, T. J., Petersen, A. C., Perry, C. (eds.) *Adolescent health: Crossing the barriers.* New York: Academic Press, 1982.

Voors, A. W. et al. Studies of blood pressures in children, ages 5-14 years, in total biracial community: The Bogalusa Heart Study. *Circulation* 54 (1976) 319-327.

Voors, A. W., Webber, L. S., and Berenson, G. S. Time course studies of blood pressure in children. The Bogalusa Heart Study. *American Journal of Epidemiology* 109 (1979) 320.

Willett, W. C. Drinking water sodium and blood pressure: A cautious view of the 'second look.' *American Journal of Public Health* 71 (1981) 729-732.

Zahka, K. G. et al. Cardiac involvement in pediatric hypertension. *Hypertension* 3 (1981) 664-668.

Zinner, S. H., Levy, P. S., and Kass, E. H. Familial aggregation of blood pressure in childhood. *New England Journal of Medicine* 284 (1971) 401.

17

A FAMILY-BASED APPROACH
TO CARDIOVASCULAR RISK
REDUCTION EDUCATION

Philip R. Nader, M.D.
Tom Baranowski, Ph.D.

1. Introduction

Common sense dictates that the family command a central and continuing role in the shaping and modifying of many health-related attitudes, behaviors, and habits of individuals. The empirical evidence supports this contention. When cardiovascular health related factors are examined in a family context, a number of associations have been noted: blood pressure levels (Biron, Mongeau, and Bertrand, 1975), obesity (Garn, Cole, and Barley, 1976), eating habits and food preferences (Byran and Lowenberg, 1958), exercise (Perrier, 1979), smoking (Surgeon General, 1979), blood cholesterol levels (Garrison et al., 1979), use of alcohol (Tennant and Detels, 1976), and aspects of "coronary prone" behavior (Butensky et al., 1976). Recently, familial correlations of cardiovascular health knowledge and attitudes have also been documented (Flora et al., 1983). These findings suggest that a family-oriented approach should be considered as an important part of current community health education trials aimed to reduce cardiovascular risk (Fortmann et al., 1981; McAlister et al., 1982). Intervening with family units and capitalizing on family support systems may increase the likelihood of continued habit change (Brownell, Heckerman, and Westlake, 1978; Caplan et al., 1976; Duetscher, Epstein, and Kjelsberg, 1966; Hertzler, and Vaughan, 1979; Pratt, 1976).

This paper presents the results of a pilot project based on cognitive social learning theory which was carried out with 24 families to assist them to adopt and maintain healthful dietary and exercise behaviors.

2. Rational for Interventions

Five principles characterize this family approach to health education: (a) using educational and behavioral management techniques based upon the most recent theory for understanding how individuals learn health habits; (b) selecting as target families those with younger adults and their late elementary age school children; (c) focusing educational efforts on a limited set of interrelated health habits; (d) building upon ethnic and cultural strengths, values, habits, and preferences; and (e) utilizing systems of support within and among families to help individuals adopt and maintain positive health habits. Each principle is discussed below.

2.1 Theoretical basis for understanding health behavior

The need to integrate cognitive and behavioral change strategies with what is known about how families function and how people learn is a key reason for viewing cognitive social learning theory as a framework for health education trials (Bandura, 1977; Mischel, 1973; Parcel and Baranowski, 1981). A broad range of techniques drawn from social learning theory have been shown to be effective in assisting individuals to change their behavior over a short time period. These techniques include contracting (DeRisi and Butz, 1975), self-monitoring (Kanfer, 1975), problem solving (Heppner, 1978) and expectations change (Hoehn-Saric et al., 1966). Studies employing these techniques and adding specific family support components indicate that long-lasting health behavior changes may be obtained.

Figure 1 presents a graphic model of the social and psychological influences on health behaviors. At the core of the model is the family and relationships between and within family units. The parent models behaviors for the child, rewards (or provides incentives for) desirable child behaviors, and maintains expectations for the child. The relationship is reciprocal, that is, the child may initiate, influence, model, or request information or behavior which results in the adult engaging in a health behavior. Interactions influencing adoption or maintenance of a health behavior also occur between adults (in the same or different families) and between children (peer pressures and sibling modeling). For dietary and food choices, research has shown that one person in the family is primarily responsible for food purchasing and preparation and thus exerts influence on the family's dietary habits. This person has been called the gatekeeper (Lewin, 1943). Working with the family's food gatekeeper should be crucial in promoting family dietary changes. Many factors can influence health behaviors, other than the proposed intervention. These potential factors are identified to the left of the core factors shown in Figure 1. We have been particularly interested in measures of family structure and family process (Pratt, 1976). The educational program could not possibly have an effect on health unless family members engage in the proposed health behaviors or habits. These are the compliance variables to the right of the core variables. Behavior changes would be expected to be accompanied by changes in health attitudes, knowledge, and in increased belief in one's own ability to change specific health habits (increased self-efficacy) (Bandura, 1977). Changes in appropriate physiological indicators are measures of health status, and could provide monitors for validating alterations in health habits or behaviors.

2.2 Selecting a limited number of interrelated target behaviors

Several recent studies indicate there is no single underlying dimension in the performance of the various positive health behaviors (Langlie, 1979; Williams and Wechsler, 1972). That is, a person who regularly performs one type of health habit, for example, using a seat belt, is not necessarily also a regular exerciser, nor necessarily interested in not smoking, and so on.

However, a study by Williams and Wechsler (1972) showed that dietary and exercise behaviors were located on the same factorial dimension. This suggests that people interested in healthy diets are also likely to participate in exercise. If empirically interrelated, such behaviors should respond to similar educational programs. In addition, interrelationships exist among the physiological indicators which relate to dietary and exercise behaviors. These physiological indicators are obesity, blood pressure, total serum cholesterol, and serum triglyceride and serum lipoprotein levels (HDL and LDL).

Dietary changes (decreasing saturated fat and decreasing sodium intake) and exercise (increasing the amount of aerobic physical activity) are both behaviorally and physiologically interrelated, and may be linked to the future reduction of risk of cardiovascular disease (Byran and Lowenberg, 1958; Garn, Cole, and Barley, 1976; Weltman, Matter, and Stamford, 1980; Wood and Klein, 1974). These three and only three interrelated behaviors were targeted. Research has shown that behavior change is difficult, and the greater in magnitude or complexity the greater is the difficulty of change.

2.3 Targeting parents and their elementary school age children

Kannel (1976) suggested that greater risk reduction from life style factors was possible among 30-year-olds than among 60-year-olds because they are at an age just before the major onslaught of heart disease. Winkelstein (1981) showed the greater relative risk for ischemic heart disease from serum cholesterol and blood pressures among 35 to 54-year-olds than for older persons. Several authors have suggested preventive programs be initiated early in the family life cycle in order to prevent lifelong detrimental health habits (Butensky et al., 1976; Kannel, 1976; Winkelstein, 1981). The age range of elementary school children and their parents may be ideal for intervention. The parents of such students are generally in this highest risk age group (30s). The children (about 8 to 12 years) are cognitively and emotionally mature enough to understand and sustain interest in a program involving their family, while not yet reaching a stage of rebellion, or establishing independence, noted with adolescence (Amanat, 1979).

2.4   Selecting educational strategies based upon cultural norms and values

A project focusing on dietary and leisure time exercise behaviors must recognize the major differences among families from different ethnic and cultural backgrounds. These differences include type and structure of the family or extended family and cooking practices. For example, Black-Americans are more likely to broil and Mexican-Americans most likely to fry (Wheeler and Haider, 1979). Markedly different food preferences are obvious in Mexican diets compared to Anglo diets, even though both diets generally meet minimum daily nutritional requirements. For this reason, the educational program worked with groups of families with the same ethnic and cultural backgrounds.

2.5   Utilizing the support system to promote behavior change

There is evidence from studies of self-help groups and from a few behavior change programs in weight reduction that active family and group involvement can result in increased and longer lasting health behavior changes (Brownell, Heckerman, and Westlake, 1978; Epstein et al., 1981; Isreal and Saccone, 1979; Rosenthal, Allen, and Winter, 1980). Added to the impact of the family (Pratt, 1976) and peer group (Flora et al., 1983) on the establishment and persistence of both positive and negative health habits, the support of significant people in the environment of these individuals attempting to alter or adopt a health behavior should be a crucial variable.

Support behaviors have been classified into three types: providing material, informational, and emotional support (Baranowski, Nader, Dunn, Vanderpool, 1982; Caplan et al., 1976). Examples related to exercise might include providing a YMCA membership (material); showing a person a magazine article on fitness and weight training (informational); and complimenting an individual on his or her improved physical appearance soon after beginning a regular program of exercise (emotional).

The Family Health Project has woven these various conceptual themes into an intervention program aimed at changing salt and saturated fat consumption and level of aerobic physical activity of the whole family. The methods and results of this study are outlined below.

3. Methods

3.1 Sample and study design

The study sample consisted of 24 volunteer families, eight in each of three ethnic groups: Anglo-, Black-, and Mexican-American. The families responded to recruitment efforts utilizing print and radio media, presentations at school PTA meetings and church groups, and phone contacts by community leaders and pediatricians. All volunteer families that met project criteria were admitted to the project. These criteria were: (a) having at least one child in the third to sixth grade and (b) being free from known cardiovascular or other severe chronic illness. Families were randomly assigned to treatment or comparison groups, within ethnic categories.

3.2 Measures

Questionnaires determining demographic characteristics, cardiovascular risk knowledge (Farquhar, ongoing study), self-efficacy and importance of performing specific health behaviors (Farquhar, 1982), measures of social support for dietary or exercise change (Baranowski et al., 1982), and frequency estimates of consumption of various food items which are high in: sodium, potassium, saturated fats, and unsaturated fats, were employed prior to and following intervention (Baranowski, Doria, and Evans, 1982). Both experimental and control groups participated in pre and post physiological measurements including height, weight, blood pressure, a blood sample for serum lipoprotein analyses, and a submaximal exercise test on the bicycle ergometer.

The dietary measures deserve additional comment. The dietary data reported here were assessed at one point in time to reflect the consumption behavior for the previous two weeks. Food categories (e.g., "pickled foods") were employed with a brief listing of specific items included in the category (for example, "pickles, relish, sauerkraut, or pickled corn"). A six category response scale was employed, including "never," "rarely" about once every two weeks, "about once a week," "more than once a week," "about once a day," and "more than once a day." Food categories were generated in each of four areas: high salty foods (12 categories), high saturated fat foods (13 categories), high potassium foods (22 categories), and high unsaturated fat foods (3 categories). This approach to dietary measurement has been employed in related projects (Baranowski et al., 1983).

The number of minutes engaged in aerobic exercise activity was self-reported at the posttest only in five categories of aerobic activity: walking, swimming, jogging, bicycling, and "other."

333

3.3 The educational and behavioral management program

The tone of the evening sessions was up-beat, emphasizing enjoyment and participation, rather than disease and didactic methods. The aims of the eight weekly evening sessions included having enjoyable, participative activities which the whole family would enjoy, and which would include the following social learning theory features: precisely specifying a limited number of behaviors for change at one time (behavioral specificity); assisting families to make their behavior change goals realistic and fit into their current life patterns (life-space fitting); helping families attempt change in small steps; assisting families in the self-selection of behavior change goals (self-determination); helping families identify and overcome barriers to change; and helping family members find ways they might support one another in reaching their goals (social support).

Detailed process and content protocols were developed for each session by a team including behavioral and social scientists, health educators, and exercise physiologists (Baranowski, Nader, and Vanderpool, 1981). Only an overview of selected aspects of the sessions can be presented here: (1) structure and sequences of sessions, (2) educational techniques with adults, children, and families, (3) rewards and incentives, (4) contact with control families, and (5) the "Family Olympics."

3.4 Structure and sequence of sessions

Participating families met in three ethnically homogeneous (Mexican-, Black-, and Anglo-American) groups to promote open discussion and capitalize on a common ethnic heritage of dietary behaviors and socially supportive interactions. Sessions were led by group facilitators who utilized group process techniques in order to promote interchange, discussion, and feelings of cohesiveness and support. At the first session, the facilitators made clear statements of what the project was about and what it was not about (realistic expectations). The general structure for each session is found in Figure 2. Each session began with stretching exercises (warm up and cool down) interspersed with some family activity modified to be aerobic (e.g., family kickball or dance aerobics tailored to the music and style of the ethnic group involved). These physical activities were designed to expand the repertoire available to families for participating in enjoyable exercise. After cool down, a low-fat, low-salt snack and beverage were served. For the first two sessions, facilitators brought the snacks to model desired behaviors. For the last six sessions volunteer family participants brought the snack and were encouraged to bring favorite traditional food with ingredients modified to lower salt or saturated fat content, (for example, corn bread with no salt, potassium baking powder, and vegetable oil rather than bacon drippings. The children brought the snack for the last session. During each session, time was allotted for educational discussions for adults and children separately. This part of the session is discussed in the next section. Families reconvened as units for the behavior management

334

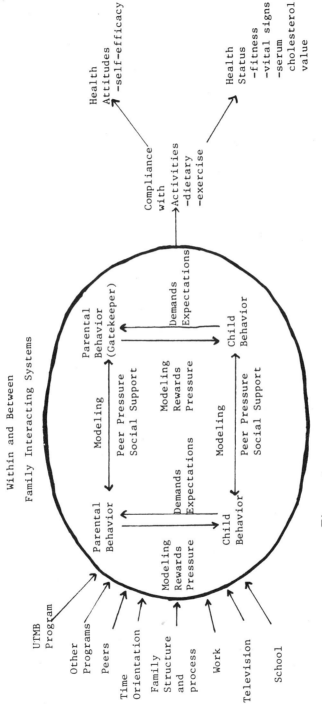

Figure 1. Graphic model of family health project intervention.

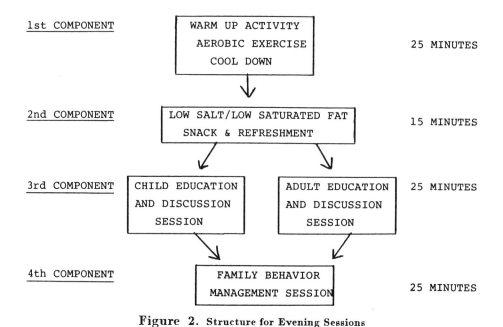

1st COMPONENT   WARM UP ACTIVITY / AEROBIC EXERCISE / COOL DOWN   25 MINUTES

2nd COMPONENT   LOW SALT/LOW SATURATED FAT / SNACK & REFRESHMENT   15 MINUTES

3rd COMPONENT   CHILD EDUCATION AND DISCUSSION SESSION   ADULT EDUCATION AND DISCUSSION SESSION   25 MINUTES

4th COMPONENT   FAMILY BEHAVIOR MANAGEMENT SESSION   25 MINUTES

Figure 2. Structure for Evening Sessions

sessions. Printed educational materials were distributed at each session, including fact sheets and information related to diet and exercise.

The first two sessions were spent in training participants to monitor dietary and exercise behaviors., to enable identification and selection of those foods which were the biggest contributors to a participant's salt and saturated food loads, and to document their degree of activity or inactivity (self-monitoring of behavior). Two weeks were then spent on each topic: reduction of salt, reduction of saturated fats, and increasing physical activity.

3.5 Educational techniques with adults, children, and families

During the third component of each session, children met as a separate group with a high school student (attractive peer model), who led the children through previously prepared structured games and role playing activities designed to improve their ability to select low sodium and low saturated fat food items and perform aerobic activity more frequently. This was accomplished by having the children read food package labels and identify green (eat as much as you want), yellow (eat sparingly), and red (do not eat). The children were provided with colored adhesive paper dots to attach to food items in the home pantry. Role playing was designed to illustrate techniques for resisting pressures of the peer group to eat salted snack foods (social innoculation) and to develop methods for supporting family members in altering their dietary and exercise behaviors. At one session, a short cartoon film strip was employed to model low salt and low saturated fat food selection and increased exercise.

The adult educational sessions (their third component) were conceived and conducted within a personal and group problem solving framework (Kanfer, 1975). New information relating to changing diet and exercise was introduced by the group leaders, outside sources, and the participants themselves. Participants' experiences in attempting changes were shared. The groups formulated solutions for overcoming participant-identified problems. Social reinforcement and support for attempted changes were provided on an individual basis.

The families were reunited for the fourth component of each evening session. This component also employed a problem solving approach. Behavioral self-management (Kanfer, 1975) was an important technique in this component. Participants (adults and children) individually self-monitored on a daily basis dietary and aerobic exercise behaviors. For example, at the end of each day in the project the person would record how frequently she or he consumed bread and bread products and high salt sauces and canned vegetables (among a variety of other targeted foods) and how many minutes were spent in aerobic activity that day. At the weekly session, the recording forms for the previous week were available and the person would select a specific food from among those items which were more frequently consumed than the rest (self-determination of change). For the second week of recording, the person monitored his or her social environment for consumption, including where the targeted food was eaten, at what meal/snack time, and who else was present at consumption. At the third and ensuing sessions, the person selected a behavior (food or exercise) (behavioral specificity) and stated a specific goal for change (goal setting) which was moderately less frequent than the consumption for the previous week (change in small steps to promote self-efficacy). For example, the table salt shaker may have been used in the previous week three times daily for seven days (a frequency count of 21 times). A goal for the ensuing week might be to use the salt shaker at only two meals a day (lunch and dinner) for the next seven days (a frequency count of 14 times). Individuals were also asked to specify the method they would use to achieve the change, for example, using a direct substitute (e.g., vegetable oil for lard), or some avoidance technique. Family discussions helped each member select a target food and target goal behavior. During this process, potential barriers to achieving the goal were identified. Attention was given to minimizing the number of changes necessary in the person's usual habits (life-space fitting) and how family members could assist each other in overcoming the barriers (intrafamily social support). After this discussion, each participant stated her or his goal to the larger group of families for the next week (public commitment), and received suggestions from the group for

337

ways to achieve the goal (interfamily social support). Public approval (reinforcement was provided for those discussions implying that the individuals would be able to accomplish their goal. At the next weekly meeting, the results were announced by each participant. If the goal was reached either another food or exercise was chosen or an increased goal was selected for the same food, or exercise, and the process repeated. If the goal was not reached, behavioral problem solving was undertaken to identify the reasons for the lack of goal attainment and ways to alter the goal or increase the likelihood of achieving it. Cases of failure to achieve the goal were usually due to an overly ambitious goal or the intervening of an unusual event, such as a party, holiday, or vacation.

3.6 Rewards and incentives

Rewards promoted in this project were of two kinds: intrinsic and extrinsic. Extrinsic rewards are tangible items (e.g., T-shirts, tickets to events) of perceived value, which change a person's behavior when used in a behavior management program. Intrinsic rewards are those satisfying feelings or thoughts which come from having performed some behavior. Behavior has been shown to be successfully changed through the application of extrinsic rewards, but these behaviors revert to baseline frequencies when the extrinsic rewards are withdrawn. Extrinsic rewards, therefore, were applied to those behaviors which were crucial to the intervention program, but which could be discontinued at the end of the intervention, i.e., daily self-monitoring of dietary and exercise behaviors. When participants submitted the first week of fully completed self-recording forms, they received a T-shirt with the project logo (a jogging family) and "We go for it" on the front, and the project name on the back. For each two weeks of fully completed forms, participants received coupons exchangeable for tickets to events in the community (movies, water slide, amusement parks).

The intrinsic rewards emphasized in the project were immediate; for example, improved feelings of physical well-being, better body appearance from firmer muscles, more energetic movement, pride in achievement, increased self-efficacy at obtaining behavior changes, social approval for change, and family togetherness; and long-term; for example, improved personal health, helping the family be more healthy, and possible increased longevity. To further emphasize these intrinsic rewards, participants who achieved their weekly behavior change goals received inexpensive iron-on patches (for the T-shirt) for salt achievement, saturated fat achievement, and exercise achievement. People in the family identified as being instrumental in helping the achiever reach a behavioral goal were rewarded with an "I am a social supporter" iron-on patch.

3.7 Contacts with control families

The same extrinsic reward system, data collection instruments, and printed materials were given to the control families. There were no weekly evening sessions or iron-on patches for this group.

3.8 The family olympics

At the end of the intervention sessions, but mid-way in the daily dietary and exercise recordings, experimental and control families were invited to participate in an enjoyable event designed to maintain their interest in continuing with the project. Through volunteer effort, several events were developed (e.g., family kickball, basketball relay, three-legged race); a low salt, low saturated fat picnic was held; and a drawing for prizes (including a bicycle and YMCA family membership) contributed from over thirty businesses in the community were held. The event was covered by local news media and favorable response was noted.

4. Results

4.1 Sample characteristics

Seventy-eight people were involved at some point in the project. These people were distributed by age, ethnic, and experimental groups as shown in Table 1. This included eight families in the Black-, and Anglo-American groups and seven families in the Mexican-American group. The one Mexican-American family dropped out of the project because of conflicting religious commitments.

Differences among children existed at baseline between experimental and control groups. Children in the experimental group tended to be older, and there tended to be more female children in the experimental than in the control groups. Neither difference was statistically significant.

Table 1

Participants in the Family Health Project

	EXPERIMENTAL			CONTROL		
	MA	BLACK	ANGLO	MA	BLACK	ANGLO
Child (age 18)	7	6	6	6	7	5
Adult (age 18)	5	6	6	4	5	5
	12	12	12	10	12	10

Among adults, the control group members were better educated than those in the experimental group. The Mexican-Americans had a significantly lower mean education (12.5 years) than either the Black-Americans (15.6 years) or Anglo-Americans (16.2 years). Those education differences reflect those found nationwide (U.S. Bureau of Census, 1979).

4.2 Knowledge results

The Stanford knowledge test was employed because it had been item analyzed for reliability in another study. The Stanford test did not measure all the elements of training of the current project, nor did the project teach to these knowledge items. The items in the test dealing with salt, saturated fat/cholesterol, and exercise were used to calculate risk specific knowledge scores.

Table 2 presents the posttest means (corrected for pretest scores) for the experimental and control group adults, for the total score and each of the risk specific scores. As can be seen, experimental and control groups significantly differed on the total and saturated fat scores, when corrected for preintervention scores. The results were similar when education, in addition to pretest scores, was controlled.

The children were not given the knowledge test on preassessment, to avoid their frustrations at completing a test with answers they could not be expected to know. The knowledge test was administered to children at the post intervention assessment. For the test as a whole and for each component score, the experimental group scored higher than the control group, but in no case were these differences statistically significant.

4.3 Changes in total consumption

Composite index scores were computed for the pretest and posttest consumption of the four food groupings (i.e., saturated fats, unsaturated fats, sodium, and potassium). The scores indicate how many times per day items from the particular food category were consumed. Participants with less than 75 percent of the items answered were deleted from the analysis. Change scores were computed by taking the difference (pretest minus posttest consumption). If the intervention was successful, the experimentals should have greater change scores for saturated fat and sodium consumption. Table 3 summarizes the means and standard deviations of the change scores. As can be seen, salt and saturated fat scores have differences between experimental and controls in the directions anticipated.

To test for statistical significance, a three-way analysis of variance was computed on the change scores for each of the four food groups. The independent variables included in the analysis were: treatment (experimental vs control), age (adult, child), and ethnicity (MA, BA, AA). The ANOVAs of changes in both the

Table 2

Posttest knowledge scores, corrected for pretest values, for adults only

	Control Group	Experimental Group
Total Score (33 max.)	13.0* (14)	15.9* (16)
Salt Score (4 max.)	1.3 (14)	1.1 (16)
Saturated Fat Score (11 max.)	3.4** (14)	4.7** (16)
Exercise Score (5 max.)	3.3 (14)	3.4 (16)

Legend: The number under the type of score (e.g., 33 max.) indicates the maximum possible mean score. The numbers in parentheses under the mean score indicates the number of people on whom the test was conducted.

* Indicates statistically significant differences in means between experimental and control using an F-test from an analysis of covariance with the pretest score as the covariate, $p < 0.05$.

** Indicates statistically significant differences in means between experimental and control using an F-test from an analysis of covariance with the pretest score as the covariate, $p < 0.01$.

saturated fat consumption and salt consumption had significant main effects for the treatment condition (Table 4). A main effect on ethnicity and a three-way interaction were also obtained on sodium change. The ANOVA of changes in unsaturated fat and potassium consumption both failed to produce any statistically significant F tests.

Table 3

Means and Standard Deviations
Changes (Pretest minus Posttest) in Food Consumption by
Treatment Group*

	Mean	S.D.	n
Salt			
Experimentals	1.9403	2.419	30
Controls	.351	2.206	29
Saturated Fats			
Experimentals	1.6339	1.810	31
Controls	.375	2.158	28
Unsaturated Fats			
Experimentals	.269	.863	34
Controls	.227	.616	30
Potassium			
Experimental	.700	2.174	30
Controls	2.008	3.824	26

* Sample sizes differ by measure because subjects were eliminated from analysis if 25% or more of the categories were blank.

Further analysis was conducted to clarify the main effect and three-way interaction effects in sodium consumption. Mean changes in sodium consumption by ethnic, age, and treatment conditions are found in Table 5. The Duncan procedure indicated that the differences among ethnic groups was produced by the difference between Mexican-American (high change) and Anglo-American (low change) groups ($p < 0.05$). Two priori contrasts revealed that the Mexican-American adults displayed significantly ($p < 0.05$) greater change than the rest and the Anglo-American children displayed significantly ($p < 0.05$) less change than the rest.

4.4 Changes in consumption of specific categories

Analyses were also conducted on the individual categories among the high nutrient foods to determine in which categories the changes were obtained.

Table 4

Three-Way ANOVAs of Saturated Fat and Salt Consumption Changes

Source of Variation	df	Saturated Fats		Salt	
		M.S.	F.	M.S.	F
Treatment	1	23.763	5.385**	39.076	9.371**
Age	1	.336	.076	.031	.007
Ethnicity	2	8.723	1.977	19.167	4.596**
Treatment by Age	1	.037	.008	6.049	1.451
Treatment by Ethnicity	2	4.243	.961	3.383	.811
Age by Ethnicity	2	1.632	.370	5.840	1.400
Treatment by Age by Ethnicity	2	.059	.013	11.444	2.744**
Residual	56	4.413		4.170	
Total	67	4.513		5.363	

* p< 0.10
**p< 0.05

Statistically significant changes were obtained $(x^2, p < 0.05)$ in five (out of 12) categories of high sodium items: cheeses, table salt shaker use, fast food meal, peanut butter, and bread products. These items were the categories most frequently selected by participating individuals for change. Significant change was obtained in three (out of 13) categories of high saturated fat food items: cheeses, whole milk and cakes, cookies, bread stuffing, tortillas or pie crusts made with butter, lard, animal fat, or crisco. Participants reported greater difficulty in reducing the consumption of high saturated fat than in reducing high sodium foods. No significant differences were obtained among three unsaturated fat categories.

4.5 Differences in exercise

Two types of data were collected on the exercise behaviors of the sample. On the posttest, subjects recorded the total number of minutes spent in aerobic exercise during the past week. In addition, three more subjective questions were asked on the posttest: whether there was a perceived change in exercise behavior; the perceived frequency of family activity; and perceived changes in frequency of family activity.

343

Table 5

Mean Changes in Sodium Consumption by
Ethnicity, Age, and Treatment Condition
n = 59

Ethnicity	Children		Adults		All Groups
	Treatment	Control	Treatment	Control	
Mexican-Americans	2.985	1.018	4.485	0.195	2.196
Blacks	2.600	0.685	0.612	0.696	1.126
Anglos	1.105	-1.942	0.313	0.898	0.519

The question concerning total exercise in minutes was submitted to a three-way analysis of variance using treatment condition, age, and ethnicity as independent variables. No statistically significant differences were obtained due to very large variances.

Even though no differences were obtained in minutes in exercise by treatment group, the experimental group perceived such a change, as measured by the three subjective questions. The results of chi square tests of independence along with the computed Cramer's V's are summarized in Table 6. When analyzed within age and ethnicity categories (Table 6), the relationship between treatment group and perceived changes and frequency in exercise is maintained primarily among the Mexican-Americans, and to a lesser extent, among the children.

The results of this pilot intervention provide modest support for the utility of a risk factor related behavior change effort focused at younger families. The most dramatic changes were obtained in the consumption of high salt food items. Moderate changes were achieved in saturated fat food items and no quantitative changes were demonstrated in aerobic activity.

4.6 Difficulties in exercise promotion

At least two other projects have demonstrated an inability to obtain changes in exercise patterns among healthy children (Coates, Jeffrey, and Slinkard, 1971). Several groups, however, have reported obtaining exercise changes among individuals with a variety of health problems (Atkins, Kaplan, and Timms, unpublished; Heinzelman and Bagley, 1970; Martin, 1981). From a

Table 6

Family Exercise Variables by Treatment Group
Controlling for Age and Ethnicity
Carmer's V

Variable	Total 0.27*	Age Child	Adult	Ethnicity M.A.	B.A.	A.A.
Perceived Changes in Exercise Behavior	0.27	N.S.	N.S.	0.70**	N.S.	N.S.
Subjective Frequency Family Activity	0.36**	0.44**	N.S.	0.62**	N.S.	N.S.
Perceived Changes in Family Activity	0.33**	0.36*	N.S.	0.55**	N.S.	N.S.

* chi square p< 0.10
** chi square p< 0.05

phenomenological viewpoint, exercise changes are more difficult to obtain than dietary changes. Everyone eats. Dietary change requires the substitution of some kinds of foods for others, or the elimination of certain foods altogether from an habitual eating plan. Not everyone, however, exercises. Exercise takes time as well as physical effort. Finding time in an otherwise too busy day may pose scheduling problems, in addition to the motivational problems of exerting the energy to obtain the benefits from exercise. This is a fascinating problem since many of the intrinsic rewards emphasized for project participation were exercise related: feeling better, breathing less hard after a physical exertion, trimmer physique, more strength.

Future efforts at exercise promotion will necessarily have to allocate proportionally greater time and effort to exercise promotion, assisting the person to schedule specific exercise events into the on-going life style, and developing the social support for exercise.

4.7 Social support for change

The intervention was complex, focusing on giving the family members the behavioral capability to attempt change, selecting the right aspects for change, and creating an environment supportive of change.

Data reported elsewhere indicated that the frequency of performance of specific supportive behaviors by family members for the respondent are almost nonexistent at baseline and very low in frequency after the intervention (Baranowski et al., 1982). Significantly more frequent support was obtained in the experimental than in the control group over the course of the experiment.

Methods must be employed to further enhance the frequency of these supportive behaviors to promote the desired changes. Candidate activities include role playing of supportive activities in the evening sessions and self-monitoring of the frequency of attempts at social support.

4.8 Refinements for a more effective·intervention

Many things were learned from conducting this study. As a result, we will conduct a variety of the activities a little differently in the future, besides enhancing the exercise intervention and promoting greater social support.

A major reason the adults reported continued participation in the evening sessions was that the children were having a good time, and the activity was something the family could do together. While we suspect the adults enjoyed the sessions as well, we think we could include more activities to increase their enjoyment and meet needs they see as important. An example would be to introduce games for the adults in that part of the evening meeting devoted to group discussion (the third component). Game activities might include: board games, card games, role playing games. Such games can instruct while enhancing the enjoyment of the adult participants, and thereby increase the likelihood the adults will return for future sessions.

We have also learned that paper work (e.g., self-monitoring or contracting) must be kept to a minimum among these healthy individuals who may not have the motivation of diseased individuals to perform the paperwork related methods employed in other studies.

4.9 Self-report measures of diet and exercise

The notable findings in this paper were obtained from self-report data. Some authors have criticized self-report measures on the grounds that respondents will report to an investigator what she or he wants to hear, as opposed to what actually occurred. While this phenomenon probably occurs to some limited extent in all respondents, the results obtained here lend validity to self-report measures. For example, the participants reported in the evening sessions greater difficulty in making changes in the saturated fat component than in the sodium component of their diets. This was reflected in the dietary self-report measures.

The two additional groups of categories of food items (potassium and unsaturated fats) work as a check on the validity of the self-report measures. If participants were primarily responding to the wishes of the investigators, they would have reported decreases in salt and saturated fat food items, and increases in unsaturated fat and potassium foods, and increases in exercise. The difficulty reported by others in obtaining exercise change was confirmed here using self-report measures (Butensky et al., 1976). The two categories of foods, high in particular nutrients but not specifically targeted in the evening sessions, demonstrated no change.

The primary data of concern for a behavior change intervention, therefore, should be behavioral data, not physiological data, since behavioral measures are more sensitive to behavioral changes (Petitti, Friedman, and Kahn, 1981). The rationale for physiological measures is to validate an individual's self-reported measures, or to validate the effect of an intervention on a group basis. Given the extensive intra-individual variation in many physiological variables, many repeated physiological measures may be needed to validate an individual's self-report (Liu et al., 1978; Liu, Cooper, and McKeever, 1979). This becomes very costly. Given the extensive interindividual variation in many physiological variables, many cases may be needed to group validate the effectiveness of an intervention (Liu et al., 1978; Liu, Cooper, and McKeever, 1979). Smaller sample sizes are thus possible when using self-report of behavior measures.

This does not get the health behavioral investigator off the measurement hook. Very few self-report measures have been documented to be reliable and valid for adults in general, and even fewer measures for specific ethnic group members or children.

A discussion of self-report measures of exercise in regard to children is enlightening. Most exercise self-report measures use minutes engaged in exercise as their primary unit of measurement. Children in the third to sixth grades, however, do not usually have watches, and their schedule is externally controlled, for examples, by buzzers at school, requests or commands from parents at home. It would seem unlikely that minutes are meaningful units for a child or that a child could accurately report minutes engaged in any activity. The staff of the Family Health Project are currently engaged in methodologic studies on the reliability and validity of several alternative approaches to self-reports of diet and exercise among third to sixth grade children. Such methods should enhance the ability of behavioral intervention programs to document effects.

5. Summary

An intervention was described to change behaviors related to cardiovascular risk by working with family units. The intervention is based on a cognitive social learning theory understanding of health behavior change. A low salt and low saturated fat diet and increased aerobic activity was targeted because diet and exercise have been shown to be simultaneously performed and they share physiological indicators of outcome. Families with third to sixth grade children were selected because the adults were at greatest relative risk of heart disease and the children were old enough to understand the program, yet not have entered a phase of adolescent rebellion. The program was implemented within ethnically homogeneous groups to capitalize on ethnic group strengths. Intra- and inter-family supports for change were promoted. Empirical support was demonstrated for this approach to behavior change, at least in initiating dietary change. This warrants testing this basic intervening method with larger groups of families. If further efforts prove to be successful, then the utility of including a family based approach to cardiovascular risk reduction efforts will have been demonstrated.

References

Amanat, E. Paradoxical treatment of adolescent resistance. *Adolescence* 14 (56) (1979) 851-861.

Annest, J. L., Sing, C. F., Biron, P., and Mongeau, J. G. *American Journal of Epidemiology* 110 (1979) 479-491.

Atkins, C. J., Kaplan, R. M, and Timms, R. M. Behavioral programs for exercise compliance in chronic obstructive pulmonary disease. San Diego, Calif., San Diego State University, (unpublished).

Bandura, A. *Social learning theory.* Englewood Cliffs, N.J.: Prentice Hall, 1977.

Bandura, A. Self efficacy: Toward a unifying theory of behavioral change. *Psychological Review* (1977) 191-209.

Baranowski, T., Nader, P. R, Dunn, K., and Vanderpool, N. A. Family self-help: Promoting changes in health behavior. *Journal of Communication* (1982).

Baranowski, T., Doria, J., and Evans, M. Scale for qualifying salt food consumption in large sample studies. (unpublished).

Baranowski, T., Nader, P. R, and Vanderpool, N. *Family health project: The facilitator's handbook.* Galveston: University of Texas Medical Branch, Department of Pediatrics, 1981.

Benfari, R. C. The multiple risk factor intervention train (MRFIT), III. The model for intervention. *Preventive Medicine* 10 (1981) 426-442.

Biron, P., Mongeau, J. G., and Bertrand, D. Familiar aggregation of blood pressure in adopted and natural children. In: Paul O. (ed.) *Epidemiology and control of hypertension.* New York: Stratton, 1975.

Brownell, L. D., Heckerman, C. L., and Westlake, R. J. The effect of couples training and partner cooperativeness in the behavioral treatment of obesity. *Behavioral Research and Therapy* 16 (1978) 323-333.

Butensky, A., Faralli, B., Heebner, D., and Waldron, I. Elements of the coronary prone behavior patterns in children and teenagers. *Journal of Psychosomatic Research* 20 (1976) 439-444.

Byran, M. S. and Lowenberg, M. E. The father's influence of young children's food preferences. *Journal of the American Dietetic Association* 34 (1958) 30-35.

Caplan, R. D., Robinson, E. A. R., French, J. R. P., Jr., Caldwell, J. R., and Skinn, M. *Adhering the medical regimens: Pilot experiments in patient education and social support.* Ann Arbor: University of Michigan, Institute for Social Research, 1976.

Coates, T. J., Jeffrey, R. W., and Slinkard, L. A. Heart healthy eating and exercise: Introducing and maintaining changes in health behaviors. *American Journal of Public Health* 71 (1981) 15-23.

DeRisi, W. J. and Butz, G. *Writing behavioral contracts: A case simulation practice manual.* Champaign, Ill.: Research Press, 1975.

Duetscher, S., Epstein, F. H., and Kjelsberg, M. D. Familial aggregation of factors associated with coronary heart disease. *Circulation* 33 (1966) 911-924.

Epstein, L. H., Wing, R. R., Koeske, R., Andraski, F., and Ossip, D. J. Child and parent weight loss in family-based behavior modification programs. *Journal of Consulting and Clinical Psychology* 49 (5) (1981) 674-685.

Farquhar, J., Principal Investigator: This measure was used with the permission of the Stanford Cardiovascular Risk Reduction Program at Stanford University.

Flora, J. A., Williams, P. T., Solomon, D., Fortmann, S., and Farquhar, J. W. Familiar correlations of cardiovascular health knowledge and attitudes. Abstract 68, 23rd annual conference on cardiovascular disease epidemiology, March 3-5, Heart Association, No. 33, January, 1983, p. 36.

Fortmann, S. P., Williams, P. T., Hulley, S. B., Haskell, W. L., and Farquhar, J. W. Effects of health education on dietary behavior: The Stanford three community study. *American Journal of Clinical Nutrition* 34 (1981) 2030-2038.

Garn, S. M., Cole, P. R., and Barley, S. M. Effect of parental fatness levels on the fatness of biological and adoptive children. *Ecology of Food Nutrition* 6 (1976) 1-3.

Garrison, R. J., Castelli, W. P., Feinleib, M., Kennel, W. B., Havlik, R., Padgett, S. J., and McNamara, P. M. The association of total cholesterol, triglycerides, and plasma lipoprotein cholesterol levels in first degree relatives and spouse-pairs. *American Journal of Epidemiology* 110 (3) (1979) 313.

Heinzelman, F. and Bagley, R. W. Response to physical activity programs and their effects on health behavior. *Public Health Report* 85 (1970) 905-911.

Heppner, P. O. A review of the problem-solving literature and its relationship to the counseling process. *Journal of Counseling Psychology* 25 (5) (1978) 366-375.

Hertzler, A. A. and Vaughan, C. E. The relationship of family structure and interaction to nutrition. *Journal of the American Dietetic Association* 74 (1979) 23-27.

Hoehn-Saric, R., Frank, J. D., Imber, S. D., Nash, E. H., Stone, A. R., and Battle, C. C. Systemic preparation of patients for psychotherapy - 1. Effects of therapy behavior and outcome. *Journal of Psychiatric Research* 2 (1966) 267-281.

Hopp, J. W. and Irwin, C. Nutrition and physical fitness education for families. A paper presented to the Nutrition/Food section, at the annual meeting of the American Public Health Association, Detroit, Mich., October, 1980.

Isreal, A. C. and Saconne, A. J. Follow-up of effects of choice of mediator and target of reinforcement on weight loss. *Behavior Therapy* 10 (1979) 260-265.

Kanfer, F. H. Self management methods. In F. H. Kanfer and A. P. Goldstein (eds.) *Helping people change.* New York: Pergamon, 1975.

Kannel, W. B. Prospects for prevention of atherosclerosis in the young. *Australian and New Zealand Journal of Medicine* 6 (1976) 410-419.

Kolonel, L. N. and Lee, J. Husband-wife correspondence in smoking, drinking, and dietary habits. *American Journal of Clinical Nutrition* 34 (1981) 99-104.

Kornitzer, M., Dramaix, M., Kittel, F., and DeBacker, G. The Belgian heart disease prevention project. Changes in smoking habits after two years of intervention. *Preventive Medicine* 9 (1980) 496-503.

Langlie, J. D. Interrelationships among preventive health behaviors: A test of competing hypotheses. *Public Health Reports* 94 (1979) 216-225.

Lewin, L. Forces behind food habits and methods of change. In: *Report of the committee on food habits,* Bulletin No. 108. Washington, D.C.: National Research Council, 1943. Pp. 55-64.

Liu, K., Stamler, J., Dyer, A., McKeever, J., and McKeever, P. Statistical methods to assess and minimize the role of intraindividual variability in obscuring the relationship between dietary lipids and serum cholesterol. *Journal of Chronic Disease* 31 (1978) 399-418.

Liu, K., Cooper, R., and McKeever, J. Assessment of the association between habitual salt intake and high blood pressure: Methodological problems. *American Journal of Epidemiology* 110 (1979) 219-226.

Martin, J. E. Exercise management: Shaping and maintaining physical fitness. *Behavioral Medicine Advances* 4 (1981) 1-15.

McAlister, A., Puska, P., Salonen, J. T., Tuomilehto, J., and Koskela, K. Theory and action for health promotion: Illustrations from the North Karelia project. *American Journal of Public Health* 72 (1982) 43-50.

Mischel, W. Toward a cognitive social learning reconceptualization of personality. *Psychological Review* 80 (1973) 252-267.

Parcel, G. S. and Baranowski, T. Social learning theory and health education. *Health Education* 80 (1973) 252-267.

Perrier, J. *Fitness in America, the Perrier study,* January 1979.

Petitti, D. B., Friedman, G. D., and Kahn, W. Accuracy of information on smoking habits provided on self-administered research questionnaires. *American Journal of Public Health* 71 (3) (1981) 308-311.

Pratt, L. *Family structure and effective health behavior: The energized family.* Boston: Houghton Mifflin, 1976.

Rosenthal, B., Allen, G. J., and Winter, C. Husband involvement in the behavioral treatment of overweight women: Initial effects and long-term follow-up. *International Journal of Obesity* (1980) 65-173.

Surgeon General. DHEW: Smoking health report. Publication # 79-50066, 1979.

Tennant, F. S. and Detels, R. Relationship of alcohol, cigarette and drug abuse in adulthood with alcohol, cigarette and coffee consumption in childhood. *Preventive Medicine* 5 (1) (1976) 70-71.

Tobian, L., Jr. The relationship of salt to hypertension. *American Journal of Clinical Nutrition* 32 (12) (1979) 2739-2748.

U.S. Bureau of the Census, Statistical abstracts of the United States, 1979, (199th edition) Washington, D.C.: U.S. Government Printing Office, 1979. Table 231, p. 145.

Weltman, A., Matter, S., and Stamford, B. A. Caloric restriction and/or mild exercise: Effects on serum lipids and body composition. *American Journal of Clinical Nutrition* 33 (1980) 1002-1009.

Wheeler, M. and Haider, S. Q. Buying and food preparation patterns of ghetto Blacks and Hispanics in Brooklyn. *Journal of the American Dietetic Association* 75 (5) (1979) 560-563.

Williams, A. F. and Wechsler, H. Interrelationship of preventive actions in health and other areas. *Health Services Report* 87 (10) (1972) 969-976.

Winkelstein, W., Jr. Primary prevention of ischemic heart disease: Evaluation of community interventions. *Annual Review of Public Health* 2 (1981) 253-276.

Wood, P. D. and Klein, H. et al. Plasma lipoprotein concentration in middle aged runners. *Circulation* 50 (1974) 111-115.

18

THE BEHAVIORAL MANAGEMENT OF TYPE II DIABETES MELLITUS

Robert M. Kaplan, Ph.D.[1]
Catherine J. Atkins, Ph.D.

1. Introduction

Diabetes mellitus is a major medical, personal, and public health problem. The United States National Commission on Diabetes estimates that five percent of Americans (more than ten million people) are affected by the condition. In the United States today, diabetes and its complications are the sixth most common cause of death. In 1974, 38,000 U.S. deaths were attributed directly to diabetes mellitus. As will be discussed later, diabetes is also a major cause of significant morbidity.

---
[1]Supported by Grants R01 AM 27901 and K04 00809 from the National Institutes of Health.

Many different factors may cause the onset of diabetes, and its course and prognosis vary among individuals. The term diabetes mellitus actually covers several separate and distinct conditions (Dash and Becker, 1978). In December of 1979, the National Diabetes Data Group published an extensive classification system for diabetes and glucose intolerance. In this chapter we will use the categories in this new and comprehensive system. According to a National Diabetes Data Group (1979), two of the most common forms of diabetes mellitus are Type I, Insulin Dependent Diabetes Mellitus (IDDM) and Type II, Non-Insulin Dependent Diabetes Mellitus (NIDDM). Insulin dependent diabetes (Type I) is considered to be the more severe form, and individuals in this subclass require insulin injections to preserve life. Non-insulin dependent (Type II) diabetics may use insulin to correct symptomatic hyperglycemia, but they are not ketosis prone nor do they require injected insulin in order to preserve life. In Western societies, 60 to 90 percent of non-insulin dependent diabetic patients are obese (National Diabetes Data Group, 1979). Many of these patients can control their disease through diet and exercise (Skyler, 1978; 1979).

Approximately five million Americans of all ages are being treated for diabetes, and the number of new cases is increasing by six percent every year. It is the non-insulin dependent form of the disease that is most responsible for the increased incidence of diabetes in the United States (National Diabetes Data Group, 1979). Environmental factors acting in concert with genetic susceptibility may be important precursors to the development of non-insulin dependent diabetes. The exact role of heredity in Type II diabetes is not well understood. There is little question that cases aggregate in families. Adding strength to the genetic argument is the fact that scientists observe a rate of almost 100 percent concordance in twins. Indeed, this concordance rate is much higher than the rate of 20 to 50 percent found in Type I diabetes. Geneticists have preliminary evidence that there may be a specific genetic mechanism leading to these metabolic abnormalities. Studies by a variety of groups suggest that development of Type II diabetes may be associated with the insertion of DNA polymorphism on the short arm of chromosome 11 located near (5' to) the insulin gene (Owerbach and Nerup, 1982; Rotwein et al, 1981; 1982).

Evidence for the heritable predisposition toward Type II Diabetes also comes from population studies. Some groups clearly have a strong genetic predisposition toward the condition. For example, Pima Indians have a prevelance of Type II diabetes approaching 50 percent (Bennett, 1982). The incidence appears to have increased since the 1940's coincident with changes in diet and exercise habits (West, 1974). Mexican-American populations in Texas also have a high prevelance of Type II Diabetes (as high as 14.5 percent). The frequency of cases in different Mexican-American groups is tied to the percent of the gene pool estimated to be native American (Gardner et al., 1984).

However, despite this evidence, the incidence of Type II diabetes does not clearly fit any simple Mendelian pattern. The current consensus seems to be that a substantial portion of the population is genetically predisposed toward Type II diabetes. This characteristic finds phenotypic expression in the presence of other factors such as obesity (Meyer, 1981). Excessive caloric intake and low physical activity leading to obesity are believed to contribute to the pathogenesis in genetically susceptible individuals. The clinical picture of Type II Diabetes has been made more complex by recent evidence that it is a heterogenous mix of problems. Although the majority of patients have insulin resistance, some may secrete abnormal insulin and others may have post-receptor deficits (Olefsky, 1984).

Since poor dietary habits and lack of exercise are often cited as contributory to Type II diabetes, it is possible that social behaviors and habits are partially responsible for the increased incidence of the disease (Saltin et al., 1979). Most new cases of Type I diabetes occur before age 20. For Type II diabetes, the proportion of cases identified in population surveys increases with age for both adult men and women (Barrett-Connor, 1980).

Many non-insulin dependent diabetic patients do not require medical therapy. Instead, structured dieting for reduction of caloric intake and for weight control is considered the most important aspect of the treatment (Skyler, 1978; 1979). In fact, weight loss may result in the correction of abnormal glucose tolerance in as many as 90 percent of Type II maturity-onset patients (Newburgh and Conn, 1979; Skyler, 1978).

The purpose of this paper is to review the rationale for the use of behavioral strategies for the management of Type II diabetes. In addition, we will provide evidence on the expected efficacy of these approaches. Before proceeding, it is important to note the seriousness of the diabetic condition by emphasizing some of the consequences of poor diabetes control.

2. Complications

It may be inappropriate to think of diabetes as a disease. Rather, it is an abnormality in metabolic control. Poor metabolic control over a long period of time is a health concern because it is associated with breakdowns in nearly every system within the body. In June of 1980, the National Diabetes Advisory Board convened a census conference. The conference identified some of the major complications of diabetes and offered recommendations on how they should be managed. Some of the major complications are heart disease, nephropathy, blindness, amputations of feet and toes, and parinatal mortality and morbidity (Ross, Bernstein and Rifkin, 1983). It must be noted that these are only a few of the many serious complications of the condition.

Heart Disease. Heart disease is a serious threat for patients with diabetes. A variety of studies have demonstrated that diabetic patients are at a greater risk of developing atherosclerosis than non diabetics of the same age. This condition restricts blood flow to the heart and is associated with a greater probability of heart attack and stroke. These relationships have been observed consistently for Type I diabetics (Eaton, 1979; Ganda, 1980) and recent evidence suggests that Type II diabetic patients also are at greater risk for cardiovascular problems (Stamler and Stamler, 1979; West, 1978).

Nephropathy. Nephropathy may develop in approximately half of Type I diabetics and a significant number of Type II diabetics. Diabetic patients are 17 times more likely to develop kidney disease than age matched members of the general population (National Diabetes Data Group, 1979). The most common form of diabetic nephropathy is the thickening of membranes that line the glomeruli. The glomeruli are small organs within the kidney which have an important filtering function. They sort elements that are retained by the blood and waste products that are eliminated as urine. Diseases of these organs can reduce the efficiency with which these functions are performed. Thus, the thickening of the glomeruli may result in the accumulation of waste products in the blood and elimination of essential blood components such as protein.

Eye Problems. Diabetic patients are very much at risk for serious complications of the eye. Today, diabetes is the major cause of blindness in the world. Diabetic patients are 25 times more likely to develop blindness than age matched members of the general population. The cause of vision complications in diabetic patients is usually retinopathy which is a problem with the small blood vessels in the retina. In this condition, the walls of small blood vessels become weak. As a result, the vessels can balloon out and break, leaking in blood to the vitreous humor or center area of the eye. Approximately 85 percent of diabetics develop retinopathy within 25 years after the onset of their condition. In most cases, retinopathy will cause only minor problems or temporary vision disturbances. Minor hemorrhage in the eye may block vision or produce black spots inside the eye. However, when this blood is absorbed, only small scarring remains and this scarring has little impact upon vision. In about 20 percent of the cases, there is large hemorrhage or a series of small hemorrhages within the same eye. This can cause detachment of the retina or larger scale destruction of retinal tissue and permanent blindness. Patients with this complication may beneit from photocoagulation therapy which uses a laser beam to cauterize small blood vessels or to repair detached retina. Established scar tissue in the vitreous area can be removed with a newer surgical technique known as vitrectomy. This surgery has been successful in restoring vision in many cases.

Amputations of Lower Extremities. Diabetes is the major condition necessitating amputations of lower extremities. In diabetic populations, the probability of therapeutic amputations is 15 times greater than it is in the general population. Diabetic complications account for 45 percent of all lower extremity amputations with an age-adjusted rate of 59.7/10,000 (amputations/diabetic individuals) (Most and Sinnock, 1983). The reasons these amputations are necessary is that diabetes can cause severe circulatory problems. As a result, blood flow to the extremities is often restricted. Small lesions on the foot can lead to gangrene and, in many cases, areas affected by gangrene must be removed. In order to avoid these problems, the diabetic patient must take very good care of his or her feet. This involves self-inspection, good hygiene, and the avoidance of behaviors which are associated with vascular diseases. Cigarette smoking is particularly dangerous in that it is known to cause restrictions in blood flow. Similarly, hypertension and weight should be vigorously managed.

Neuropathy. Neuropathy is disease of the nervous system. Diabetes is known to cause a variety of forms of dysfunction in the nervous system. This occurs in the peripheral nervous system as manifested by dysfunction in motor movements in the fingers, toes, arms, and legs. Patients often report tingling sensations in the extremities and difficulty with some muscular movements. Neuropathy in the autonomic nervous system often causes greater concern. As a result of these conditions, approximately 50 percent of diabetic males eventually become organically impotent and may lose partial control over important body functions such as urination. In severe cases, there is difficulty with the regulation of basic autonomic function such as blood pressure and heart rate.

Parinatal Mortality and Morbidity. Although many Type II diabetic patients are beyond child-bearing age, the existence of a diabetic condition can cause severe problems in pregnancy. Expectant diabetic mothers may find their condition is exacerbated by the pregnancy, and some forms of diabetes are only apparent during pregnancy. If not well managed during pregnancy, diabetic mothers are more likely to give birth prematurely, to have still borns, or to have deformed children. Thus, aggressive management of diabetes during pregnancy is essential. All of these problems can be exacerbated by specific behavioral patterns during pregnancy. For example, the probability of complications in the diabetic mother may be increased if she smokes cigarettes, or if she does not follow her dietary regimen.

3. Will Better Control Lead to Fewer Complications?

As we have shown, the consequences of diabetes mellitus can be severe. Naturally, diabetic patients have a high stake in avoiding complications. But will good behavior and self-control result in avoidance of complications? Unfortunately, there is no guarantee that it will. For Type I diabetic patients, there are many cases in which patients develop complications despite rigorous self-care and monitoring.

Most specialists believe that complicatons result from poor glycemic control. Poor control is usually defined as blood glucose levels above the idealized standard, either after a fast or after specified periods following glucose administration. However, the relationship between control and complications is an important epidemiological question and has been the focus of intense debate. The coverage of that debate is well beyond the scope of this paper but will be the focus of a paper by Tchobroutsky (this volume). Suffice it to say that there are at least three explanations for the development of diabetic complications. First, there is the view that diabetic complications are the result of long-term hypoglycemia produced by insulin therapy. This view has very little empirical support. The second explanation is the primary lesion theory. Advocates of this position argue that whatever caused the diabetes independently causes complications. Evidence for this view comes from a variety of clinical observations. For example, it is not uncommon for a patient to present to a neurologist with diabetic polyneuropathy prior to developing frank diabetes. Patients with cystic fibrosis may also get beta cell dysfunction and maintain very high plasma glucose levels (gt 400 mg/dl). Yet these individuals tend not to develop diabetic complications.

The majority of evidence supports the third view--that poor control is responsible for the development of complications. In the following section, some of this evidence will be reviewed. Again, we can only provide an overview in this limited space. More comprehensive reviews are available elsewhere (Tchobroutsky, 1981). It should also be noted that the U.S. National Institutes of Health has embarked on a major clinical trial in an attempt to resolve the control and complications controversy. Patients will be randomly assigned to one of two protocols that may result in differing levels of control. All patients will then be monitored for a variety of complications.

The evidence for the relationship between control and complications is different for different specific complications. First, we will consider heart disease. For Type I diabetics, chemical control of blood sugar has been known to reduce the risk of a myocardial infarction (Meinert, Knatterrud, and Prout, 1970). The development of heart disease is associated with high levels of low density lipoprotein (LDL) and low levels of high density lipoprotein (HDL) in blood serum. Epidemiological studies have shown that this pattern is more common in diabetic patients (Kannel and McGee, 1979). One study demonstrated that an insulin pump device which may lead to better diabetes control can lead to higher levels of high density lipoprotein which in turn are associated with reduced risk of heart disease (Falko; O'Dorisio and Cataland, 1982). Other studies have demonstrated that hypertension, which is a known risk factor for heart disease, is more common in diabetic patients. Further, diabetic hypertension may produce increased risks of heart disease for the diabetic patient. Some of this risk can be accounted for by the tendency toward obesity among Type II diabetic patients (Barrett-Connor et al., 1981).

On the other hand, several authors have argued that diabetic control is not clearly associated with the avoidance of macrovascular complications (see Davidson, 1978; 1981). In general, data from large scale epidemiological investigations do not show a strong benefit of diabetic control for avoidance of heart disease and stroke. A major prospective American trial on this issue (see UGDP in later section) failed to demonstrate a significant benefit of tight control (University Group Diabetes Program, 1982). However, the study has been severely attacked on methodological grounds (Kilo et al., 1980).

In summary, there is reason to suspect that tight control of Type II diabetes will result in the prevention of macrovascular complications. However, data from human population and experimental studies do not conclusively affirm this notion.

The strongest evidence for the relationship between tight control and renal complications comes from animal studies. Diabetes can be produced in animals by injecting alloxan, a chemical that destroys insulin secreting cells in the pancreas. In addition, diabetes occurs spontaneously in some animals (dogs and monkeys are good models) and can be induced by injections of hormones, such as growth hormone, that are antagonists to insulin. Thickening of the glomerular tissues occurs in monkeys with spontaneous diabetes and in dogs induced to be diabetic by either injections of growth hormone or alloxan (Davidson, 1981). More importantly, kidney problems can be retarded or reversed by good control in dogs chemically induced to be diabetic. Animals randomly assigned to tight control through multiple daily injections of insulin had fewer complications than dogs assigned to suboptimal insulin dosages (Engerman, Bloodworth and Nelson, 1977). Finally, transplantation of a diabetic rat kidney into a nondiabetic rat from the same litter results in some reversal of the vascular problems. A normal rat kidney transplanted into a diabetic animal will develop glomerulosclerosis (Davidson, 1981).

The human evidence is not so clear. Although there is strong evidence that diabetic patients develop vascular lesions in new transplanted kidneys, the evidence is often relevant only to Type I diabetes. In one well known prospective study there was a modest relationship between degree of control and diabetic nephropathy as assessed by renal biopsy. However, the mean age of onset diabetes was associated with a higher incidence of renal problems. This suggests that the problems may be more common for those who become diabetic early in life (Type I). Nevertheless, there were still Type II patients in poor control who developed renal problems (Takazakura et al., 1975).

In a major long-term clinical trial with Type II diabetics, insulin treatment was associated with improved kidney function (as measured by changes in serum creatinine) in comparison to placebo treated patients. However, elevated serum creatinine could not be attributed directly to glucose control. In other words, the insulin effect may have operated separately from the glucose effect. None of the patients were placed on dialysis, and death due to kidney failure did not appear to be common (UGDP, 1982).

In summary, animal and human data suggest that control is related to prevention of kidney problems in Type I diabetic patients. The evidence for Type II diabetics is more equivocal. Pharmicological treatments are more likely to reduce glucose levels than diet, yet the available evidence does not show a clear relationship between glucose levels and renal complications in ketosis resistant patients.

Some of the most recent evidence on the control/complications controversy comes from evaluations of a new technology known as Continuous Subcutaneous Insulin Infusion (SCII). With this new technology, an insulin pump provides continuous adjustments and infusions of insulin and may permit better control than Type I diabetics had previously enjoyed. The pump has been evaluated in Europe by the Steno Study Group. The difference in control (measured by hemoglobin Alc which is a blood test assessing control over a 120-day period) for patients using the pump was compared with the group not using the pump over a six-month period. The results suggested that those using the pump achieved better control. Other data from the study showed that the group using the pump also scored better on measures of retinal and kidney function after six months in comparison to the control group (Steno Study Group, 1982). Thus, the study indicates that good control for a six-month interval can decrease the risk indexes for some major complications.

When first published, these results created a great deal of excitement. However, within one year new evidence began to raise new questions. A study by Puklin (Yale Department of Opthalmology, unpublished) examined 27 diabetic patients with heterogeneous eye disease. All of the patients went on the insulin pump and achieved normalized blood glucose levels. Diabetic retinopathy was found in 40 of the eyes at the beginning of the treatment period. By the end of the follow-up period (mean 21.3 months after implantation of the pump) the retinopathy became more severe in 37.5 percent of the eyes. Among the 25 eyes with proliferative retinopathy, 44 percent became more severe. The only encouraging note was that no new retinopathy developed in the 14 eyes that were normal at the beginning of the study.

In summary, the relationship between control and complications is a complex and unresolved question. Nevertheless, there is significant evidence that the *probability* of complications may be reduced through effective self-care and regular medical supervision (Deckert and Larsen, 1979). For example, patients who simply visit hospitals and clinics more frequently have a greater chance of survival than those who are seen in the medical care setting less often. Yet we cannot be sure that this is the effect of control and not the result of self-selection.

After reviewing the complex literature on the relationship between metabolic control and diabetic complications, Tchobroutsky (1981) concluded that (despite some argument to the contrary) the weight of the evidence suggests microvascular and neuropathic complications in diabetes can be reduced or avoided through control of blood sugar. In the conclusion of his detailed review of control and complications, Tchobroutsky indicated that most diabetic patients must rely on behavioral methods for the avoidance of complications. He stated, "...to achieve the best control...we must use refined techniques of teaching and education, and we must convince the patient to follow his diet strictly, to divide up the daily insulin administration (at least for the majority of patients), to try to be as thin as possible, to not smoke, and to include physical exercise as a very important aid to treatment" (p. 28).

Until very recently, it was believed that diabetes should only be regulated through the use of chemical substances which alter the blood sugar. In the next section we will present the rationale for behavioral management in Type II diabetes.

4. University Group Diabetes Program (UGDP)

The largest, and clearly the most controversial, study comparing medical treatments for the treatment of Type II diabetes was the University Group Diabetes Program (1970). In this cooperative study involving twelve clinical centers, a total of 823 patients were randomly assigned to one of four treatment groups. Two groups received insulin injections, one on a variable dosage schedule and the other on a standard dosage schedule. The third group received tolbutamide. Tolbutamide is an oral hypoglycemic agent which, until the publication of the study, had been the major method of controlling Type II diabetes. The fourth or placebo group received oral inactive medication. Actually, there was originally a fifth group assigned to take phenformin. However, this treatment was discontinued early. All groups were given a special diet.

Patients in the study were carefully followed over a long period of time. After eight years and ten months, the life-death status of 818 of the original 823 patients was determined. To everyone's surprise, patients randomly assigned to receive tolbutamide had a significantly *increased* probability of death due to cardiovascular diseases in comparison to the placebo group. The two insulin groups did not differ significantly from the placebo group. In other words, none of the treatments was shown to be significantly more effective than the combination of a diet and a placebo. The University Group Diabetes Project investigators concluded that the combination of tolbutamide and diet was less effective than diet alone for the management of Type II diabetes--at least with regard to cardiovascular death. For ethical reasons, they halted the experimental trial for the tolbutamide group. Although the data with regard to the insulin treatments

361

were less clear, it was suggested that diet and exercise may be the most effective means of managing Type II diabetes (Univeristy Group Diabetes Program, 1970).

The University Group Diabetes Program was extravagant, expensive, and remains to this day one of the most controversial pieces of clinical research in the literature. Ten years passed between the initial funding of the project and its first report in the literature. The project cost more than seven million dollars and involved the participation of a large number of staff and other resources. There have been many challenges to the report as might be expected for any major and complex undertaking. Without going into the details, it is worth noting that established epidemiologists and biostatisticians found many flaws with the randomization, the dependability of the outcome measures, the uniformity across centers, and a variety of decisions which were made throughout the project (Feinstein, 1971; Kilo, Miller, and Williamson, 1980; Schor, 1971). Some of these problems were of major concern (for example, a larger proportion of the deaths in the tolbutamide groups were autopsied--raising questions about the comparability of measurement in the different groups (Schor, 1971). Further, there was an unexpected low death rate among females in the placebo group in comparison to known death rates of untreated diabetic women. A lower than expected death rate in controls may produce a bias against the treatment (Kilo et al., 1980).

Other aspects of the debate about UGDP were more affective in tone. For instance, the results of the UGDP study may have been disappointing to pharmaceutical companies because they suggest that tolbutamide should no longer be prescribed. After Feinstein (1971) published a major critique of the study, he was called a "drug-house horror," "snake-oil salesman," and was accused of engaging in activities which represented a "conflict of interest," "unbridled sensationalism," and "deliberate destruction" (p. 189). Many of the arguments were less passionate. Perhaps the most convincing was Cornfield's publication (1971). Cornfield, who had been involved with the UGDP, systematically considered many of the complaints about the experiment and the data analysis. He noted that the issues raised by others could have produced errors, but these errors are small and specifiable. In no case would the errors have caused a major change in the outcomes. For instance, it was suggested that there was poor randomization in the study, and that groups differed prior to treatment. Yet, as Cornfield notes, the groups did not differ on any of 14 baseline characteristics at the significance level of .05. Further, he presented some unpublished information supporting the original conclusion. At the end of his report, Cornfield asked other investigators not to pursue the matter further. He stated, "Although further investigation, particularly if undertaken in a non-adversarial framework, may still be useful, it seems likely that a point of diminishing returns may not be far off, and that continued analysis of the UGDP, in the hope of finding errors which alter the conclusions, will become increasingly unrewarding" (p. 1687).

The UGDP study has remained controversial. In late 1982, the group published its final report on the value of insulin therapy. The report rebutted earlier criticism and stood firmly behind the original findings. When the results were published in *Diabetes*, the official journal of the American Diabetes Association, an apologetic note from the editor stated:

"Publication in a scientific journal does not imply that the editors endorse the position taken by the authors of a piece of scientific research...

We believe that the scientific process of examination and reexamination works to sift out truth, and we are willing to rest on that" (Foster, 1982, p. ii).

At least three other prospective studies on tolbutamide have been published. None of these studies has shown that tolbutamide actually reduces the incidence of cardiovascular complications (Feldman et al., 1974; Keen, Jarrett, and Fuller, 1974; Paasikivi, and Wahlberg, 1971). One study reported that patients taking tolbutamide do not have higher HDL levels than comparable patients treated with diet alone (Barrett-Connor et al., 1983), suggesting that the toxic effect of the drug is not through an effect upon lipoproteins. Another study argued that tolbutamide in combination with diet actually prevents progression to Type II diabetes in pre-diabetic adults (Sartor et al., 1980).

Evidence from non-experimental studies tends to suggest that outcomes are best for those on diet alone. For example although the UGDP remains controversial, it clearly has changed Type II diabetes care. The reader is urged to review the conflicting reports for him or herself. At present, we find little evidence suggesting that oral hypoglycemic agents are required for the management of Type II diabetes. Yet, many will disagree with this judgment, and the debate will certainly continue. In the next section, we will review some of the evidence relevant to behavioral interventions.

5. Exercise

The value of exercise in diabetic care has been recognized for centuries (Skyler, 1979). Within the last few years, there has been increased attention to the value of exercise for diabetic persons. The Kroc Foundation held a major conference on diabetes and exercise (Vranic, Horvath, and Wahren, 1979), and there has been a very important review of exercise and diabetes by Vranic and Berger (1979). There are a variety of reasons why exercise is of value. For example, inactivity may be associated with relative resistance of insulin and a reduction in glucose tolerance (Lipman et al., 1972). Increased exercise appears to enhance sensitivity to insulin, and available insulin may be more potent (Cahill, 1971). There are certainly some complications of increased exercise. For example, for diabetic patients who are insulin dependent (Type I), increased exercise, although known to be of value, can increase the risk of hypoglycemia. Thus, in the insulin dependent diabetic, there must be a careful balance between food intake, insulin treatment, and exercise (Skyler, 1979).

These risks are greatly decreased in the Type II diabetic because these patients rarely, if ever, experience ketosis as a result of hypoglycemia. In summary, there are several advantages to increased exercise in the Type II diabetic patient. Exercise appears to:

1. increase insulin binding;

2. cause increased caloric utilization resulting in weight loss;

3. utilize free glucose;

4. reduce the risks of cardiovascular diseases which are a common complication of Type II diabetes; and

5. aid in the control of hypertension (Vranic and Berger, 1979).

Although the use of exercise in the management of Type II diabetes is encouraging, there is still a dearth of systematic studies. Most studies draw inferences from populations on non-diabetic volunteers (Sherwin and Koivisto, 1981). The preliminary evidence includes a study by Saltin and colleagues (1979) showing improved oral glucose tolerance in non-obese Type II patients after three months of physical training. Engerbretson (1965) engaged a small group of Type II patients in a six-week training program and found a lowering of fasting glucose levels in blood. However, Ruderman and colleagues (1979) did not find an improvement in oral glucose tolerance after a six-month training program. Systematic, randomized, prospective trials are virtually absent from the literature.

A variety of lines of evidence suggest that the major objective in treatment of Type II diabetes is weight loss. Several recent studies support the use of exercise in weight control. In studies designed to compare the value of diet versus exercise on weight control, it has been reported that combinations of exercise and calorie restriction produce more weight loss than either approach alone (Dahlkoetter, Callahan, and Linton, 1979; Weltman et al., 1980). Exercise may contribute significantly to weight loss because increased activity may lead to greater caloric consumption at rest (Thompson et al., 1982).

Despite the otimism about the benefits of habitual exercise, there are still very few published studies evaluating the effects of continued exercise in adult patients with diabetes. Recently, Bogandus and colleagues (1984) compared diet therapy with diet plus exercise in 18 patients with glucose intolerance. Half of the patients met the criteria for NIDDM. The addition of exercise to diet had no effect for measures of body composition or fasting plasma glucose. However, the diet plus exercise group had significantly higher carbohydrate storage rates. The results of the study are difficult to interpret because there were some important differences between groups prior to treatment. For instance, there was a significant difference in plasma glucose in response to a low fiber test meal and a remarkably strong pre-treatment difference in insulin response to intervenous glucose.

Exercise programs for obese Type II patients must be developed with great care. Indeed, there may be a significant cardiac risk for sedentary Type II patients who are given the casual advice to get more exercise. We are currently working on a randomized trial in which some patients are assigned to a supervised exercise group. Exercise begins very slowly and after careful screening including an ECG.

Although we do not have conclusive evidence for the value of exercise for the management of Type II diabetes, the available evidence looks promising. We are encouraged by a major retrospective study of long-term diabetics who did not develop complications. The vast majority of this complication-free group exercised on a regular basis (Chazan et al., 1970).

The belief that patients with Type II diabetes should increase their exercise when it is safe is gaining widespread acceptance among clinicians. However, there is a fair degree of cynicism about the likelihood that these individuals will indeed adhere to an exercise program. In other words, a behavioral change will likely result in better management of the metabolic state. What is needed is a technology for increasing the rate of adherence to the exercise program (see Martin and Dubbert, 1982, for overview of adherence to exercise).

6. Diet

Nutritional management has long been a core ingredient in diabetes care. Control of blood glucose levels requires a balance between energy expenditure and sources of energy which are obtained from food. In the normal person, a metabolic equilibrium is obtained through the secretion of insulin and other hormones. However, in the diabetic patient, insulin secretion or insulin utilization is abnormal, and therapeutic approaches are necessary to maintain normal blood glucose levels.

The relationship between diet and diabetes control is well known and discussed in virtually every publication on diabetes care. Less well known, however, are data showing that Type I and Type II diabetics differ radically in their insulin production. Although there is still some debate, evidence suggests that Type I diabetes is associated with beta cell damage caused by an antoimune response to a viral infection. Some studies have shown that diabetic children have high antibody titer to coxackie B4 virus within several months after diagnosis. Other viruses (mumps, measles, and encephalomyocarditus) have also been implicated. Certain individuals may have an inflammatory response to these infections which causes insulin to leak out of the beta cells. Then, antibodies are produced in response to excess insulin that cause beta cell destruction (Gorsuch et al., 1982). With the destruction of the beta cells (the insulin-producing portion of the pancreas), insulin secretion decreases, and the patient is required to use supplemental insulin to regulate the metabolism. The virus-auto immunity theory may explain why there is a greater number of new cases of Type I diabetes during the winter months (Gray, Duncan, and Clark, 1979). Evidence for this theory is also provided by studies showing that immunosuppressive drugs (i.e., cyclosporine) may reverse some recent onset cases (Stiller et al., 1984).

For the Type II diabetic, this picture is the opposite. Several decades ago, Yalow and Berson (1959) developed a radioimmunoassay for insulin. Until this time, it was believed that Type II diabetics had difficulties producing insulin as did Type I diabetics. However, the work of Yalow and Berson (1960) revealed that Type II diabetics actually had higher levels of insulin in blood plasma in response to oral glucose than did control or non-diabetic patients. This condition is known as hyperinsulinemia. A variety of groups have demonstrated that hyperinsulinemia is associated with obesity (Kreisberg et al., 1967). In fact, the amount of excess body weight is correlated with the degree of hyperinsulinemia (Bagdade et al., 1967b). In other words, evidence suggests that Type II diabetic patients produce too much rather than too little insulin. A variety of studies have demonstrated that obese individuals with Type II diabetes mellitus have a relatively slow response to the introduction of glucose or to meals (Genuth, 1973). Newer evidence suggests that a major problem with Type II diabetics is that they have insulin resistance which results from a modification of insulin receptors (Olefsky, 1976). The reason for this insulin resistance is not known; however, it has been suggested that it is an adaptive response to hyperinsulinemia (Ireland, Tomson, and Williamson, 1980). Other new evidence reveals that there are many forms of Type II diabetes. Some of these are associated with post-receptor deficits (Kolterman et al., 1985).

Although the data are complex, many lines of evidence suggest that dietary changes and weight loss may be an effective remedy for hyperinsulinemia and conditions associated with insulin resistance. Some studies have shown that obese patients who are placed on diets which restrict their calorie intake experience improved binding of insulin to its receptors (Archer et al., 1975). Other studies have shown that hyperinsulinemia is decreased when obese patients lose weight (Farrant et al., 1969; Pfieffer, 1974). Pfieffer's studies have shown that the correction of hyperinsulinemia with weight loss is not immediate. Instead, there appears to be a lag of a few months between weight loss and restoration of hormone regulation. Skyler (1979) observed that reduction in weight from 108.2 kgs. to 75.8 kgs. had little immediate effect upon hyperinsulinemia. However, after four more months with very little additional weight loss, hormonal response returns to normal. In summary, caloric restriction and weight loss appear to result in decreased hyperinsulinemia and increased binding of insulin to receptor sites. These findings suggest that nutritional interventions should have a beneficial impact upon blood glucose regulation for Type II diabetic patients. Since these individuals may secrete too much rather than too little insulin, these dietary interventions may have greater potential than the introduction of even further supplemental insulin.

There are some specific characteristics of diet which may be very important to the Type II diabetic patient. Some studies have shown that diets high in complex carbohydrate and fiber will increase glucose tolerance in the Type II diabetic (Brunzell et al., 1971). Crapo and associates (1976; Crapo, 1981) have found that glucose tolerance improves when Type II diabetic patients increase the ratio of complex carbohydrates to simple sugars. Jenkins (1979) reports that soluable fibers such as metamucil, pectin, and agar can reduce insulin requirements and increase glucose tolerance if taken regularly by Type II diabetics. These data suggest that, in addition to caloric restrictions, Type II diabetic patients will benefit from increased consumption of complex carbohydrates (fibers) and decreased consumption of simple carbohydrates (sugars). The most dramatic demonstrations of the value of high fiber diets have been provided by Anderson (1979). Figure 1 presents mean insulin dose required before and after diabetic patients were placed on a high carbohydrate, high fiber diet. Figure 2 shows the effect of the diet upon plasma and urine glucose, weight, and insulin dose. These data clearly suggest that high fiber diets can promote diabetic control. An excellent review of the value of sustained-release carbohydrates is provided in Jenkins (1982).

To summarize this section, the evidence is not convincing that supplemental insulin will benefit the Type II diabetic patient. Further, the use of oral hypoglycemic drugs is also controversial. These individuals tend to have hyperinsulinemia associated with the overproduction of insulin and insulin resistance. This condition is most often associated with obesity. With weight loss, insulin production appears to return to normal and there is less resistance to insulin. Diets which reduce caloric intake and those which increase the ratio of complex to simple carbohydrates may be associated with decreased risk of diabetic complications and a decreased probability of cardiovascular disease. Of course, the medical management of diabetes is complex, and any change in medicine use should be under the advisement of a qualified physician. Progression of medical therapy is usually avoidable if a trial of weight loss is unsuccessful or does not produce improved glucose control.

7. Other Behavioral Factors

Several other behavioral factors have been implicated in Type II diabetes. Some of the most widely discussed are smoking and stress. The evidence on the effects of smoking upon complications in Type II diabetes is unclear. Ireland and colleagues (1980) suggest that several studies have shown adverse effects of smoking for diabetic patients, particularly with regard to retinopathy. However, no references were given. Review of other sources suggests that these results are quite inconsistent. There is little question that smoking has damaging effects upon health status (Smoking and Health, 1979). Smoking unquestionably increases the probability of lung cancer, obstructive lung disease, heart disease, stroke, and

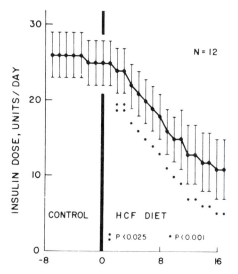

Figure 1.   Insulin doses on control and high complex fiber diets.
Source: Anderson, 1979, p. 264. Reproduced by permission.

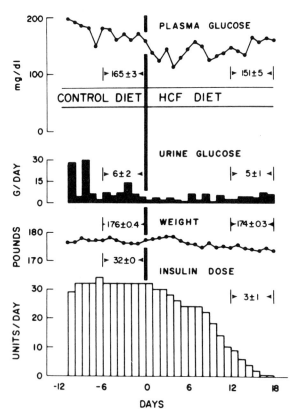

Figure 2.   Effects of High Fiber diets on glucose, weight and insulin dose.
Source: Anderson, 1979, p. 266. Reproduced by permission.

complications with pregnancy. A diabetic patient should be discouraged from smoking as should any other individual.

The real question is whether already existent risks for diabetic patients are heightened through the use of cigarettes. Several self-care manuals for diabetic patients imply that this group is at a higher risk for diseases related to smoking. Some studies confirm this. For example, smoking appears to increase insulin requirements by an average of 15 to 20 percent and may also have a significant impact upon triglyceride concentrations (Madsbad et al., 1980). However, this same study showed no effect of smoking upon retinopathy. Other studies have shown that diabetic smokers are more likely to lose bone mineral content (a complication in diabetes) than diabetic non-smokers (McNair et al., 1980). Christiansen (1978) was unable to find differences in retinopathy between smoking and non-smoking Type I diabetic patients. He did find a higher incidence of kidney disease among diabetic patients who were currently or had previously been smokers than among those who had never smoked. In contrast, West, Erdriech, and Stober (1980) found no differences between smoking and non-smoking diabetic patients for retinopathy or kidney disease. Of course, each study finds diabetic patients (independent of smoking status) at greater risk for these problems than non-diabetic individuals.

Perhaps the greatest risk for diabetic smokers is heart disease. The risks of mortality due to heart disease are 4.8 times greater than average for diabetic patients, although this risk is somewhat lesser for those in the Type II category. Cigarette smoking increases this risk further, although it is not clear that cigarette smoking interacts with the diabetic condition (Dupree and Meyer, 1980). In summary, cigarette smoking is a severe health risk and should be avoided by everyone. Data on the added risks of smoking for diabetic patients have not been consistent. However, due to increased risk of cardiovascular and lung disease and the generally shortened life expectancy for diabetic patients, this group should be discouraged from smoking. For those who have already established a smoking habit, smoking cessation programs should be recommended.

Another factor in diabetes care is the presence of psychological stress. Sarason and Sarason (1980) define stress as a feeling or reaction which occurs when faced with a situation that demands action. This may be pressure, frustration, or conflict, or it could be a positive challenge associated with demanding life circumstances. Clinical evidence suggests that stress may be associated with either hypo or hyperglycemia. In other words, stress may cause hormonal changes which result in fluctuations in blood glucose control. There is substantial evidence that stress alters hormones that affect carbohydrate metabolism. Stress can modify cortisol, epinephrine, norepinephrine, growth hormone, glucagon, and insulin (Cox et al., 1984). In addition, stress is known to be associated with changes in behavior which may also alter blood glucose control. For example, patients alter their eating patterns when under stress. A variety of authors have reported that emotional factors play an important role in the regulation of the diabetic condition (Danowsky, 1963; Kimball, 1971; Trenting, 1962; and others). Grant and his colleagues have shown that life changes as measured by the Schedule of Recent Events are associated with fluctuations in the diabetic condition (Grant et al., 1974).

In an interesting series of studies, Cox and colleagues (1984) have shown the role of patient perceived stress upon diabetic control. In one study they found that self-reported day to day hassles and frustrations correlated significantly with hemoglobin A1 among 59 adult, Type I diabetic patients. The hemoglobin A1 values were not significantly correlated with Type A behavior, social support, or self-reported compliance with the daily regimen. In another study, they found that diabetic patients typically regard stress as a very potent factor in the control of their condition.

These studies suggest that emotional factors and stress do play an important role in the regulation of the diabetic condition, and that behavioral methods for coping with stress may be an advantage in controlling the condition.

One of the most interesting examples of the value of relaxation training in clinical practice comes from a recent study by Surwit and Feinglos (1983). They hospitalized twelve obese, Type II diabetic patients. Six of the patients underwent extensive relaxation and biofeedback training while the other six were randomly assigned to a control condition which did not receive the behavioral treatment. Results from a glucose tolerance test demonstrate that those exposed to the relaxation treatment had improved glucose tolerance, as measured by incremental glucose area and two hour postprandial blood glucose. The control group did not show improved glucose tolerance. Surwit and Feinglos (1984) argue that their results are biologically plausible because relaxation is associated with an epinephrine and cortisol response and decreased sympathetic nervous system activity.

Although the Surwit and Feinglos results are very exciting, they must be interpreted with some caution. Although their patients were randomly assigned to treatment and control groups, half of the treatment group had fasting plasma glucose values less than 140 mg/dl prior to the intervention. Only one control was below the 140 mg/dl criterion. Further, the treatment group had greater glucose stimulated insulin secreation prior to treatment. The Surwit and Feinglos results are certainly very promising and deserving of continued attention.

8. Behavioral Management

In the preceeding sections, we have suggested that behavioral factors play an important role in Type II diabetes. Patients who take good care of themselves may have a higher probability of living normal lives without complications of cardiovascular disease, neuropathy, nephropathy, and retinopathy. Conversely, those who take poor care of themselves may be exposed to greater risks of these frightening conditions and might expect a shorter life of poorer quality. Diabetes patient education typically emphasizes these issues. Yet few diabetes manuals recognize the potential of behavioral interventions for the management of these conditions (Surwit, Feinglos, and Scovern, 1983).

As the evidence reviewed above suggests, Type II diabetic patients benefit from controlled dieting, exercise, and stress reduction. These are essential components of successful weight control programs. Methods that have been used and evaluated for weight loss may be of value for the management of Type II diabetes. Several authors have suggested that reduction of obesity will result in enhanced insulin utilization (Ireland et al., 1980). Since the Type II diabetic may return to normal insulin secretion and proper functioning of insulin receptors, it has been suggested that control of obesity and long-term compliance with diet and exercise programs may offer a "cure" for some patients. By 1979, the American Diabetes Association issued a policy statement suggesting that diet is the first line treatment and that oral hypoglycemic drugs should only be considered if diet therapy fails (ADA, 1979). Even ads for oral hypoglycemic drugs note that diet and exercise should be used before graduating to pharmaceutical treatments. With this in mind, let us consider the value of behavioral programs for the management of obesity.

Approximately 30 percent of American women and 15 percent of American men are obese. Obesty is defined as weighing more than 120 percent of desired weight (Abraham and Johnson, 1979). The consequences of being overweight are severe because obesity is associated with increased risk of heart disease (National Academy of Science, 1980) as well as Type II diabetes and a variety of other chronic diseases (Thorn, 1970). In addition to chronic disease problems, being overweight is associated wth many undesirable social situations, including difficulties in sexual function, inability to travel and participate in athletic activities, and greater difficulty in entering certain occupational fields (Ferguson and Birchler, 1978).

Obesity appears to be caused by a variety of factors, including a heredity predisposition (Mahoney and Mahoney, 1976), the influence of the endocrine system upon the shape a body takes, and the influence of prenatal or childhood nutrition upon the number of fat cells (Charney et al., 1976; Ravelli et al., 1976). However, despite the physiologic predisposition to gain weight, the development of obesity is a complex interaction of physiological and psychological events (Rodin, 1978). Further, there is extensive evidence that behavioral programs can be successfully used for a variety of individuals despite their physiological predisposition (Rodin, 1978; Stunkard, 1979).

In 1962, Ferster, Nurnberger, and Levitt published a detailed analysis of the behavioral control of eating. This early paper outlined the relationship between eating and reinforcement. Building upon this early effort, many behavior therapists developed programs to help individuals shed unwanted pounds. Within the last few years there has been a surge of interest in weight reduction programs and commercial programs using behavioral methods which have been quite successful. National programs such as Weight Watchers, run primarily by paraprofessionals, use a group behavioral format and see about 400,000 individuals each week (Stunkard, 1979).

The elements of at least one successful behavior modification strategy for weight loss include: 1) describing the behavior to be controlled, 2) identifying the stimuli which typically occur prior to eating, 3) employing behavioral methods for controlling eating behavior, and 4) changing the consequences of eating (Stunkard, 1979). The appropriate application of behavior therapy requires appropriate behavioral assessment. The best results are obtained when the package is tailored to fit the characteristics of a particular client (Wolpe, 1981).

A variety of studies are underway to evaluate the value of behavior modification programs specifically designed for Type II diabetic patients. Rainwater and colleagues (1983) found that Type II diabetic patients assigned to a behavior modification program lost more weight than a group given conventional treatment. Urine sugar and fasting glucose values also declined in comparison to a baseline in the treatment group. However, patients were not randomly assigned to the two groups, and the experimental subjects were more overweight prior to the treatment.

We recently completed a study evaluating the effects of various behavior modification techniques for the management of Type II diabetes. In comparison to patients in an attention control group or a no treatment control group, those experiencing behavioral programs lost more weight and significantly reduced their requirements for medication. Changes in hemoglobin A1 were nonsignificant, but in the expected direction. Wing and her colleagues in Pennsylvania are conducting a similar study, and the results are expected soon.

Recently, Rabkin, Boyko, Wilson and Streja (1983) reported a randomized trial comparing individual nutritional counseling with behavior modification for the control of Type II diabetes. Those in the nutritional counseling groups lost more weight than those in the behavior modification group. There were no differences between the groups for lipids or for fasting serum glucose. However, the study had many methodological problems. For example, those in the nutritional counseling groups were given individual attention while the behavior modification subjects were seen in groups. From the description, the nutritional counseling group appeared to be behavior modification. It included a review of eating habits, an individual meal plan, social support, and reinforcement of principles previously taught. Conversely, the behavior modification condition was offered by a nutritionist and began with a discussion of the pathophysiology of diabetes. The authors stressed that "the patient's eating habits were not assessed nor was an individualized meal plan established" (p. 51). On the basis of this study, it appears inappropriate to conclude that behavior modification is less effective than nutritional counseling.

It should be noted that the enthusiasm for behavioral interventions for weight loss has recently waned in some circles. Foreyt and colleagues (1981) have painted a rather gloomy picture of the long-term success rate after a review of all studies reporting one-year follow-up data. Brownell (1982) carefully reviewed the complexities of evaluating behavioral approaches to weight loss. Despite some skepticism, behavior modification methods, particularly some of the newer approaches (see Brownell, 1982), remain the best established techniques for achieving long-term weight control.

Although there are an enormous number of studies on the effects of diet upon weight, there are many fewer systematic studies evaluating the effects of exercise. A recent review of research on exercise and its effects on obesity does demonstrate that exercise has a very promising effect, particularly when combined with diet (Dahlkoetter et al., 1979). Thus, there is a strong rationale for using exercise as a method for controlling weight. As noted above, there is also convincing evidence that exercise has beneficial effects on the regulation of blood glucose for diabetic patients. The major problem with exercise is that adherence to exercise programs tends to be poor.

In a series of studies, Martin (1981) and his associates have evaluated several approaches to exercise management. First, they have studied factors relevant to staying in an exercise program. Smoking appears to be the single best predictor of dropping out of an exercise program. Studies have shown that 59 percent of smokers will drop out of a program. In addition, blue collar workers who are inactive during their leisure time have a greater probability of dropping out, particularly if they are smokers. Eighty percent of those characterized by this profile have dropped out of cardiac rehabilitation programs which involve exercise (Oldridge, 1979). Being overweight is also a good predictor of poor adherence to an exercise program (Andrew and Parker, 1979). Patients also drop out of exercise programs when they are in an inconvenient location (Teraslinna et al., 1969) and when the exercise programs are rigorous (Pollack et al., 1977). Yet, physical status of participants is not a strong predictor of drop-out (Martin, 1982).

There are relatively few studies that attempt to modify exercise patterns. Martin, Katell, Webster, Zegman. and Blount (1981) performed extensive studies on 85 sedentary adults and children enrolled in a community exercise program. Through this study they learned that personalized feedback and praise have a strong beneficial effect upon adherence to an exercise program. They also learned that assigning a flexible goal which is determined by the subject produces superior adherence to an exercise program than a fixed goal which is chosen by the experimenter.

Dubbert, Martin, Raczyinski, and Smith (1982) evaluated the effects of a cognitive-behavioral strategy upon the maintenance of exercise. Sixteen sedentary adults participated in a community course which required them to jog or walk briskly three times per week. Half of these adults were assigned to an experimental group in which they learned to set realistic goals and to attend to pleasant stimuli in the environment while they are exercising. They also learned to identify self-defeating thoughts, and when they occurred, to replace them with more pleasant and positive self-statements. A control group was taught to identify body sensations which are associated with athletic injuries. After three months, the eight subjects randomly assigned to the cognitive-behavioral strategy had a higher attendance rate at exercise meetings than the control group. Similarly, Atkins, Kaplan, Timms, Reinsch, and Lofback (1984) have reported that cognitive-behavioral methods are useful for increasing activity among patients with chronic obstructive pulmonary disease. In summary, we still know very little about the factors associated with maintenance of an exercise program. However, research is beginning to identify personality and behavioral factors which increase adherence to exercise programs. For the Type II diabetic patient, greater adherence to exercise programs may mean better regulation of glucose, reductions in weight, and overall better regulation of the metabolic condition.

Eating and exercise are complex behaviors, and it is naive to think that simple advice will lead to modification of these patterns. Readers interested in developing behavioral programs are advised to consult one of several references giving detailed study of cues for eating and exercise, behavioral contracts, goal settng, functional analysis of behaviors, and the like. Treatment usually involves ten sessions and completion of regular homework assignments. Although these interventions are difficult, behavior modification continues to have the best record of long-term success among the many available interventions for weight control.

It is not our purpose to review all of the behavioral approaches since that literature is voluminous. Suffice it to say that there are many behavioral strategies with proven effectiveness which are not currently in use in diabetes care. The authors are currently working on a controlled trail comparing behavioral strategies with control interventions for the management of Type II diabetes. The outcome measures for this experiment are psychological as well as biochemical. They include measures of functioning, measures of blood chemistry (glycosylated hemoglobins), and weight loss. One of the purposes of this experiment is to evaluate the cost-effectiveness of these interventions. We hope to report these results in the near future.

In a recent editorial on psychological factors in diabetes, Skyler (1982) criticized psychological studies in diabetes for using "shotgun" approaches with many outcome measures and fuzzy conceptual focus. We agree with Skyler's assessment. However, modern approaches to psychological research are theory-based. Glasgow and McCaul (1982) outlined how contemporary psychological theories, such as social learning theory, clearly specify which variables to manipulate and which variables to observe in experimental studies. The theory emphasizes the interaction between personal and environmental influences in determining behavior. Social learning theory is supported by a wide variety of empirical studies. We would hope to see more theory-based studies in future diabetes research.

9. Summary

Type II diabetes is a major public health problem affecting millions of adults in Western countries. The condition causes fear and concern because it is associated with severe complications including blindness, heart disease, kidney failure, and amputations of limbs. An experimental trial comparing oral hypoglycemic drugs for the treatment of Type II diabetes suggested that these medications may have limited value in the long term management of the condition. Instead, behavioral interventions have been recommended. Some preliminary evidence indicates that behaviors such as eating, exercise, and perhaps cigarette smoking are associated with control of the condition. In addition, exposure to stress and the ability to cope with threatening situations may also be related to diabetes control. Lacking in current management strategies for Type II diabetics is sophistication in the use of behavioral technologies including behavior modification and cognitive behavior modification. However, there are also very few systematic studies that evaluate the benefits of these interventions. We suggest this is a rich area for future research and clinical application.

References

Abraham, S. and Johnson, C. L. Overweight adults in the United States. Vital and Health Statistics of the National Center for Health Statistics, No. 51 (1979) 11.

ADA, American Diabetes Association. Policy Statement. *Diabetes Care* 2 (1979) 1-3.

Anderson, J. W. High carbohydrate, high fiber diet for patients with diabetes. In R. A. Camerini-Davalos and B. Hanover (eds.) *Treatment of early diabetes.* New York: Plenum Press, 1979.

Andrew, G. M. and Parker, J. O. Factors related to dropout of post myocardial infarction patients from exercise programs. *Medicine and Science in Sports* 11 (1979) 376-378.

Archer, J. A., Gorden, P., and Roth, J. Defect in insulin binding to receptors in obese man: Amelioration with calorie restriction. *Journal of Clinical Investigation* 55 (1975) 166-174.

Atkins, C. J. et al. Behavioral programs for exercise compliance in chronic obstructive pulmonary disease, 1983. (Unpublished paper).

Atkins, C. J., Kaplan, R. M., Timms, R. M., Reinsch, S., and Lofback, K. Behavioral exercise programs in the management of chronic obstructive pulmonary disease. *Journal of Consulting and Clinical Psychology* 52 (1984) 591-603.

Bagdade, J. D., Bierman, E. L., and Porte, D. Diabetic lipemia--A form of acquired fat-induced limpemia. *New England Journal of Medicine* 276 (1967) 427-433.

Ball, M. F., El-Khodary, A. Z., and Canary, J. J. Growth hormone response in the thinned obese. *Journal of Clinical Endocrinology and Metabolism* 34 (1972) 498-511.

Barrett-Conner, E. The prevalence of diabetes mellitus in an adult community as determined by history or fasting hyperglycemia. *American Journal of Epidemiology* 111 (1980) 705-712.

Barrett-Conner, E., Criqui, M. H., Klauber, M. R., and Holdbrook, M. Diabetes hypertension in a community of older adults. *American Journal of Epidemiology* 113 (1981) 276-284.

Barrett-Conner, E., Witzum, J., Thompson, J. K., Jarvie, G. J., Lahey, B. B., and Cureton, K. J. Exercise and obesity: Etiology, physiology, and intervention. *Psychological Bulletin* 91 (1981) 1, 55-79.

Bennett, P. H. The epidemiology of diabetes mellitus. In B. N. Brodoff and S. J. Bleicher (eds.) Diabetes mellitus and obesity. Baltimore: Williams and Wilkins, 1982. Pp. 387-399.

Bogardus, C., Ravussin, E., Robbins, D. C., Wolfe, R. R., Horton, E. S., and Sims, E. Effects of training and diet therapy on carbohydrate metabolism in patients with glucose intolerance and non-insulin-dependent diabetes mellitus. *Diabetes* 33 (1984) 311-318.

Bonar, J. R. *Diabetes: A clinical guide*. Flushing, NY: Medical Examination Publishing Company, Inc., 1977.

Brownell, K. D. Obesity: Understanding and treating a serious, prevalent, and refractory disorder. *Journal of Consulting and Clinical Psychology* 50 (1982) 820-840.

Brunzell, J. D. and Bierman, E. L. Improved glucose tolerance with high carbohydrate feeding in mild diabetics. *New England Journal of Medicine* 284 (1971) 521-524.

Cahill, G. F. The physiology of insulin in man. *Diabetes* 20 (1971) 785-799.

Campbell, D. R., Bender, C., Bennett, M., and Donnelly, J. Obesity. In J. L. Shelton and R. L. Levy (eds.) *Behavioral Assignment and Treatment Compliance.* Champaign, IL: Research Press, 1971.

Charney, E. et al. Childhood antecedents of adult obesity: Do chubby infants become obese adults? *New England Journal of Medicine* 295 (1976) 6-10.

Chazan, B. I., Balodimos, M. C., Ruan, J. R., and Marble, A. 20-5 to 40-5 years of diabetes with and without vascular complications. *Diabetologia* 6 (1970) 565-569.

Christiansen, J. S. Cigarette smoking and prevelance of microangiopathy in juvenile-onset insulin-dependent diabetes mellitus. *Diabetes Care* 1 (1978) 146-149.

Cornfield, J. The University Group Diabetes Program. A further statistical analysis of the mortality findings. *Journal of the American Medical Association* 217 (1971) 1677-1687.

Cox, D. J., Taylor, A. G., Nowacek, G., Holley-Wilcox, P. et al. The relationship between psychological stress and insulin dependent diabetic blood glucose control: *Health Psychology* 3 (1984) 63-75.

Crapo, P. A. Dietary modifications in the management of diabetes. In M. Brownlee (ed.) *Handbook of Diabetes Mellitus*, Vol. 5. New York: Garland Press, 1981.

Crapo, P. A., Reaven, G., and Olefsky, J. Plasma, glucose, and insulin response to orally administered simple and complex carbohydrates. *Diabetes* 25 (1976) 741-747.

Dahlkoetter, J., Callahan, E. J. and Linton, J. Obesity and the unbalanced energy equation: Exercise versus eating habit change. *Journal of Consulting and Clinical Psychology* 47, (1979) 5, 898-905.

Danowski, T. S. Emotional stress as a cause of diabetes mellitus. *Diabetes* 12 (1963) 183-184.

Dash, A. L. and Becker, D. Diabetes mellitus in the child. In H. M. Ketzen and R. J. Mahler (eds.) *Diabetes, Obesity, and Vascular Disease*, Vol. 2. New York: Wiley, 1978.

Davidson, M. B. *Diabetes mellitus diagnosis and treatment*, Vol. 1. New York: Wiley, 1981.

Davidson, M. B. The case for control and diabetes mellitus. *Western Journal of Medicine* 129 (1978) 193-200.

Davidson, J. K., Vander Zwagg, R., Cox, C. L. et al. The Memphis and Atlanta continuing care programs for diabetes. II Comparative analysis of demographic characteristics, treatment methods, and outcomes over a 9-10-year follow-up period. *Diabetes Care* 7 (1984) 25-31.

Deckert, T. and Larsen, M. The prognosis of insulin dependent diabetes mellitus and the importance of supervision. In R. A. Camerini-Davalos and B. Hanover (eds.) *Treatment of Early Diabetes: Advances in Experimental Medicine and Biology*, Vol. 119. New York: Plenum Press, 1979.

Dubbert, P. M., Martin, J. E., Raczynski, J. and Smith, P. O. The effects of cognitive-behavioral strategies in the maintenance of exercise. Presented at the Third Annual Meeting of the Society of Behavioral Medicine, Chicago, 1982.

Dupree, E. A. and Meyer, M. B. Role of risk factors and complications of diabetes mellitus. *American Journal of Epidemiology* 112 (1980) 100-112.

Eaton, R. P. Lipids and diabetes: The case for macrovascular disease. *Diabetes Care* 2 (1979) 46-50.

Engerbreston, D. L. The effects of exercise upon diabetic control. *Journal of the Association of Physicians in Mental Rehabilitation* 19 (1965) 74-78.

Engerman, R., Bloodworth, J. W. B., and Nelson, S. Relationship of microvascular disease in diabetes to metabolic control. *Diabetes* 26 (1977) 760-769.

Falko, J. M., O'Dorisio, D. M., and Cataland, S. Improvement of high-density lipoprotein-cholesterol levels: Ambulatory Type I diabetics treated with the subcutaneous insulin pump. *Journal of the American Medical Association* 247 (1982) 37-39.

Farrant, P. C., Neville, R. W. J., and Stewart, G. A. Insulin release in response to oral glucose in obesity: The effect of reduction of body weight. *Diabetologia* 5 (1969) 198-200.

Feinstein, A. R. Clinical Biostatistics, VIII: An analytical appraisal of the University Group Diabetes Program (UGDP) study. *Clinical Pharmacology* 12 (1971) 167-191.

Feldman, R., Crawford, D., Elashoff, R., and Glass, A. Oral hypoglycemia during the prophylaxis in asymptomatic diabetes. In proceedings of the Ninth Congress of the International Diabetes Federation. *Exerpta Medica* (1974) 574-587.

Ferguson, J. *Learning to eat: Behavior modification for weight control*. Palo Alto, Calif.: Bull Publishing Co., 1975.

Ferguson, J. M. and Birchler, G. Therapeutic packages: Tools for change. In W. S. Agras (ed.) *Behavior Modification: Principles and Clinical Applications*, 2nd Ed. Boston: Little Brown, 1978.

Ferster, C. B., Nurenberger, J. I., and Levitt, E. D. The control of eating. *Journal of Methetics* 1 (1962) 87-109.

Fisher, E. B., Delamater, A. M., Bertelson, A. D., and Kirkley, B. G. Psychological factors in diabetes and its treatment. *Journal of Consulting and Clinical Psychology* 50 (1982) 993-1003.

Foreyt, J. P., Goodrick, G. K., and Gotto, A. M. Limitations of behavioral treatment of obesity: Review and analysis. *Journal of Behavioral Medicine* 4 (1981) 159-174.

Foster, D. W. Editorial statement. *Diabetes* 31 (1982) (supplement 5) ii.

Ganda, O. P. Pathogenesis of macrovascular disease in the human diabetic. *Diabetes* 29 (1980) 931-942.

Gardner, L. I., Stern, M. P., Haffner, S. M. et al. Prevalence of diabetes in Mexican Americans. Relationship to percent of gene pool derived from native American sources. *Diabetes* 33 (1984) 86-92.

Genuth, S. Plasma insulin and glucose profiles in normal, obese, and diabetic persons. *Annals of Internal Medicine* 79 (1973) 812-822.

Glasgow, R. E. and McCaul, K. D. Psychological issues in diabetes: A different approach. *Diabetes Care* 5 (1982) 645-646.

Gorsuch, A. et al. Can future Type I diabetes be predicted? A study in families of affected children. *Diabetes* 31 (1982) 862-866.

Grant, I., Kyle, G. C., Teichman, A., and Mendels, J. Recent life of and diabetes in adults. *Psychosomatic Medicine* 36 (1974) 121-128.

Gray, R. S., Duncan, L. J. P., and Clarke, B. F. Seasonal onset of insulin dependent diabetes in relation to sex and age at onset. *Diabetologia* 17 (1979) 29-32.

Holland, D. M. The diabetes supplement of the national health survey. *Journal of the American Dietetic Association* 52 (1968) 387-390.

Ireland, J. T., Thomson, W. S. T., and Williamson, J. *Diabetes today: A handbook for the clinical team.* New York: Springer, 1980.

Jenkins, D. J. Dietary fiber, diabetes, and hyperlipidemia. *Lancet* 2 (1979) 1287-1289.

Jenkins, D. J. A. Lent carbohydrate: A newer approach to the dietary management of diabetes. *Diabetes Care* 5 (1982) 634-641.

Kannel, W. B. and McGee, D. L. Diabetes and glucose tolerance as risk factors for cardiovascular disease: Framingham study. *Diabetes Care* 2 (1979) 120-126.

Keen, H., Jarrett, R. J., and Fuller, J. H. Tolbutamide and arterial disease in borderline diabetics. In proceedings of the Ninth Congress of the International Diabetes Federation. *Exerpta Medica* (1974) 588-602.

Kilo, C., Miller, J., and Williamson, J. R. The crux of the UGDP: Spurious results and biologically inappropriate data analysis. *Diabetologia* 18 (1980) 179-185.

Kimball, C. P. Emotional and psychosocial aspects of diabetes mellitus. *Medical Clinics North America* 55 (1971) 1007-1081.

Kolterman, O. G., Revers, R. R., and Fink, R. I. Assessment of receptor and postreceptor defects in target tissue insulin action. In R. De Pirro and R. Lauro (eds.) *Handbook of Receptor Research.* Field Educational Italia Acta Medica, 1985.

Krall, L. P. *Joslin Diabetes Manual*, 11th ed. Philadelphia: Lea and Febiger, 1975.

Kriesberg, R. A., Boshell, B. R., Di Placido, J., and Roddam, R. F. Insulin secretion in obesity. *New England Journal of Medicine* 276 (1967) 314-319.

Lipman, R. L. et al. Glucose intolerance during decreased physical activity in man. *Diabetes* 21 (1972) 101-107.

Madsbad, S. et al. Influence of smoking on insulin requirement and metabolic status in diabetes mellitus. *Diabetes Care* 3 (1980) 250-252.

Mahoney, M. and Mahoney, K. *Permanent weight control.* New York: W. W. Norton, 1976.

Martin, J. E. Exercise management: Saving and maintaining physical fitness. *Behavioral Medicine Advances* 4 (1981).

Martin, J. E. The behavioral management of exercise and fitness. Presented at the meeting of the Society of Behavioral Medicine, Chicago, 1982.

Martin, J. E. and Dubbert, P. M. Exercise applications and promotion in behavioral medicine: Current status and future directions. *Journal of Consulting and Clinical Psychology* 50 (1982) 1004-1017.

Martin, J. E. et al. The effects of feedback, reinforcement, and goal selection on exercise adherence. Presented at the 15th Annual Convention of the Association for the Advancement of Behavior Therapy, Toronto, 1981.

McNair, P. et al. Bone loss in patients with diabetes mellitus: Effects of smoking. *Mineral and Electrolyte Metabolism* 3 (1980) 94-97.

Meinert, C. L. et al. A study of the effects of hypoglycemia agents on vascular complications in patients with adult-onset diabetes. *Diabetes* 19 (1970) (supplement), 789-830.

Most, R. S. and Sinnock, P. The epidemiology of lower extremity amputations in diabetic individuals. *Diabetes Care* 6 (1983) 87-91.

National Academy of Sciences. *Toward healthful diets.* Food and Nutrition Board, Division of Biological Sciences, Assembly of Life Sciences, National Research Council of the National Academy of Sciences. Washington, D.C., 1980.

National Diabetes Data Group. Classification and diagnosis of diabetes mellitus and other categories of glucose intolerance. *Diabetes* 28 (1979) 1039-1057.

Newburgh, L. H. and Conn, J. W. A new interpretation of hyperglycemia in obese, middle-aged persons. *Journal of the American Medical Association* 112 (1979) 7-11.

Oldridge, N. B. Compliance of post myocardial infarction patients to exercise programs. *Medicine and Science in Sports* 11 (1979) 373-375.

Olefsky, J. M. The insulin receptor: Its role in insulin resistance of obesity and diabetes. *Diabetes* 25 (1976) 1154-1164.

Owerbach, D. and Nerup, J. Restriction fragment length polymorphism of the insulin gene in diabetes mellitus. *Diabetes* 31 (1982) 275-277.

Paasikivi, J. and Wahlberg, F. Preventive tolbutamide treatment and arterial disease in mild hyperglycaemia. *Diabetologia* 7 (1971) 323-327.

Pfeiffer, E. F. Obesity, islet function and diabetes mellitus. *Norm., Metab. Research Supplement* 4 (1974) 143-152.

Pirart, J. Diabetes mellitus and its degenerative complications: A prospective study of 4,400 patients observed between 1947 and 1973. *Diabetes Care* 1 (1978) 168-188.

Platt, W. G. and Sudovar, S. G. The social and economic costs of diabetes: An estimate of 1979. Ames Division, Miles Laboratories, 1979.

Pollack, M. L. et al. Effects of frequency and duration of training on attrition and incidence of injury. *Medicine and Science in Sports*, 1 (1977) 31-36.

Rainwater, N., Ayllon, T., Frederiksen, L. W., Moore, E. J., and Bonar, J. R. Teaching self-management skills to increase diet compliance in diabetics. In R. B. Stuart (ed.) *Adherence Compliance and Generalization in Behavioral Medicine*. New York: Brunner/Mazel, 1983.

Raskin, S. W., Boyko, E., Wilson, A., and Streja, D. A. A randomized clinical trial comparing behavior modification and individual counseling in nutritional therapy of non-insulin-dependent diabetes mellitus: Comparison of the effect on blood sugar, body weight, and serum lipids. *Diabetes Care* 6 (1983) 50-56.

Ravelli, G., Stein, Z. A., and Susser, M. W. Obesity in young men after famine exposure in utero and early infancy. *New England Journal of Medicine* (1976) 295-349.

Report of the National Committee on Diabetes to the Congress of the United States, Vol. 3. Department of Health, Education, and Welfare, Washington, D.C., 1975.

Rodin, J. Somatopsychics and Attribution. *Personality and Social Psychology Bulletin* 4 (1978) 4, 531-540.

Ross, H., Bernstein, G. and Rifkin, H. Relationship of diabetes mellitus to long-term complications. In M. Ellenberg and H. Rifkin (eds.) *Diabetes Mellitus Theory and Practice* (3rd ed.) New Hyde Park: Medical Examination Publishing Co., 1983. Pp. 907-926.

Rotwein, P. et al. Insulin gene polymorphism and diabetes. *Diabetes* 31 (1982) 185a.

Ruderman, N. B., Ganda, O. P., and Johansen, K. The effect of physical training on glucose tolerance and plasma lipids in maturity onset diabetes. *Diabetes* 28 (1979) (supplement 1) 89-92.

Sacket, D. L. and Haynes, R. B. *Compliance with therapeutic regimens.* Baltimore: Johns Hopkins University Press, 1976.

Saltin, B. et al. Physical training and glucose tolerance in middle-aged men with chemical diabetes. *Diabetes* 28 (1979) (supplement 1) 30-32.

Saltin, B. et al. Physical training and glucose tolerance in middle-aged men with chemical diabetes. *Diabetes* 23 (1979) 30.

Sarason, I. G. and Sarason, B. R. *Abnormal Psychology*, 3rd Ed. Englewood Cliffs, N.J.: Prentice Hall, 1980.

Sartor, G. et al. Ten-year follow-up of subjects with impaired glucose tolerance. Prevention of diabetes by tolbutamide and diet regulation. *Diabetes* 29 (1980) 41-49.

Schor, S. The University Group Diabetes Program. A statistician looks at the mortality rate. *Journal of the American Medical Association* 217 (1971) 1671-1675.

Sherwin, R. S. and Koivisto, V. Keeping in step: Does exercise benefit the diabetic? *Diabetologia* 20 (1981) 84-86.

Skyler, J. S. Psychological issues in diabetes. *Diabetes Care* 4 (1982) 656-657.

Skyler, J. S. Diabetes and exercise: Clinical implications. *Diabetes Care* 2 (1979) 307-311.

Skyler, J. S. Nutritional management of diabetes mellitus. In H. M. Katzen and R. J. Mahler (eds.) *Diabetes, Obesity, and Vascular Disease*, Vol. 2. New York: Wiley, 1978.

Smoking and Health: A Report of the Surgeon General. Washington, D.C.: U.S. Department of Health, Education, and Welfare, 1979.

Stamler, R. and Stamler, J. A symtomatic hyperglycemia and coronary heart disease. *Journal of Chronic Disease* 32 (1979) 683-691.

Steno Study Group. Effect of six months of strict metabolic control and eye and kidney function in insulin-dependent diabetics with background retinopathy. *Lancet* (1982) 121-123.

Stiller, C. R., Dupré, J., Gent, M. et al. Effects of cyclosporine immunosuppression in insulin-dependent diabetes of recent onset. *Science* 223 (1984) 1362-1366.

Stuart, R. B. A three dimensional program for treatment of obesity. *Behavior Research and Therapy* 9 (1971) 177-186.

Stunkard, A. J. Behavioral medicine and beyond: The example of obesity. In O. F. Pomerleau and J. P. Brady (eds.) *Behavioral Medicine: Theory and Practice.* Baltimore: Williams and Wilkins, 1979.

Surwit, R. S. and Feinglos, M. N. Relaxation-induced improvement in glucose tolerance is associated with decreased plasma cortisol. *Diabetes Care* 7 (1984) 203.

Surwit, R. S. and Feinglos, M. N. The effects of relaxation on glucose tolerance in non-insulin-dependent diabetes. *Diabetes Care* 6 (1983a) 176-179.

Surwit, R. S. and Feinglos, M. N. Diabetes and behavior a paradigm for health psychology. *American Psychologist* 38 (1983) 255-262.

Takazakura, E. et al. Onset and progression of diabetic glomerulosclerosis. A perspective study based on serial renal biopsies. *Diabetes* 124 (1975) 1-9.

Tchobroutsky, G. Metabolic control and diabetic complications. In M. Brownlee (ed.) *Handbook of Diabetes Mellitus*, Vol. 5. New York: Garland Press, 1981.

Teraslinna, P., Partanen, T., Koskela, A., and Oja, P. Characteristics effecting willingness of executives to participate in an activity program aimed at coronary heart disease prevention. *Journal of Sports Medicine and Physical Fitness* 9 (1969) 224-229.

Thorn, G. W. Alterations in body weight. In E. Wintrobe (ed.) *Harrison's Principles of Internal Medicine*. New York: McGraw-Hill, 1970.

Treuting, T. F. The role of emotional factors in the etiology and course of diabetes mellitus: A review of the recent literature. *American Journal of Medical Science* 244 (1962) 93-109.

University Group Diabetes Program: Affects of hypoglycemic agents on vascular complications in patients with adult-onset diabetes. VIII: Evaluation of insulin therapy: Final report, *Diabetes* 31 (1982) (supplement 5).

University Group Diabetes Program: A study of the effects of hypoglycemic agents on vascular complications in patients with adult-onset diabetes. *Diabetes* 19 (1970) (supplement 2) 747-830.

Vranic, M. and Berger, M. Exercise and diabetes mellitus. *Diabetes* 28 (1979) 147-167.

Vranic, M., Horvath, S., and Wahren, J. (eds.) Proceedings on a conference on diabetes and exercise. *Diabetes* (1979) (supplement 1) 1-113.

Weltman, A., Matter, S., and Stanford, B. Caloric restriction and/or mild exercise: Effects on serum lipids and body composition. *The American Journal of Clinical Nutrition* 10 (1980) 1002-1009.

West, K. M. *Epidemiology of diabetes and its vascular lesions*. New York: Elsevier, 1978.

West, K. M. Diabetes in American Indians and other native populations in the new world. *Diabetes* 23 (1974) 841-847.

West, K. M., Erdreich, L. S., and Stober, K. A. Absence of a relationship between smoking and diabetic microangiopathy. *Diabetes Care* 3 (1980) 250-252.

Wolpe, J. Behavior therapy versus psychoanalysis. *American Psychologist* 36 (1981) 159-164.

Yalow, R. S. and Berson, S. A. Assay of plasma insulin in human subjects by immunological methods. *Nature*. London. 184 (1959) 1648-1649.

19

THE COMMUNITY STUDIES
OF THE STANFORD HEART DISEASE
PREVENTION PROGRAM

Nathan Maccoby, Ph.D.
John W. Farquhar, M.D.
Stephen P. Fortmann, M.D.

In 1971, the Stanford University Heart Disease Prevention Program undertook a three-community study to discover a method for risk reduction that would be generally applicable. (Farquhar et al., 1977; Maccoby et al., 1977). Why did we pick a community as the unit of education for reducing risk of cardiovascular disease? (1) If an individual therapist-instructor is used with either one person at a time or even with a small group of persons, the problem of general risk reduction is just too large to manage. Such an undertaking would be very expensive, and the number of therapist-instructors needed would be astronomically high. (2) Risk reduction involves behavior changes that have a very long-term-- even a lifetime--duration. Cessation of smoking or not beginning to smoke cigarettes is behavior that has to last if it is going to make a contribution to the reduction of risk of disease. Similarly, changes in diet involving reductions in dietary cholesterol, saturated fats, salt, and calories generally call for permanent changes in eating habits. (3) These changes need to take place not in the clinic but in the context of people's environments. Furthermore, the community nexus can contribute greatly to the maintenance of changes in life style. The home, the school, the work place (Meyer and Henderson, 1974) and other community settings are the environments in which such behavior occurs, and therefore it must be practiced there. Furthermore, these institutions are potential sources of support for new behavior, or they can constitute obstacles to such changes (Farquhar, 1978).

The use of communities enables the research to trace the processes by which the effects are either achieved or not achieved and permits one to study the process of mass adoption of health innovations.

Maintenance of change has a much better chance of being successfully achieved, and the role of social support in this process can be examined.

Generalizability of both methods and results for achieving improved health and reduced risk can be discovered.

Even some of the alternative methods can become more effective when done in the context of a community-wide health education effort.

Diffusion of changed behavior and social support are potentially important sources of strength for the information and maintenance of new habits. Peers can play a very important role in the process. For example, school children can be influenced not to begin cigarette smoking if peer models occupy visible roles as nonsmokers (McAlister, Perry, and Maccoby, 1979).

We were interested in discovering a method for helping people to change their life styles so as to reduce their risk of suffering a cardiovascular event such as a myocardial infarction - a heart attack - or a stroke. We were searching for a method that would accomplish this objective, would not require overwhelming numbers of therapists, and would be relatively inexpensive per person assisted.

One research finding which greatly influenced the project design was a formulation of Cartwright's. Analyzing household sample survey data on the purchasing of United States Savings Bonds, Cartwright (1949) found data suggesting that mass media alone were not effective in achieving sales, but that personal solicitation, in the context of the campaign, was. To account for this finding, he posited that in order to influence behavior, it was necessary to stimulate action or behavioral structures in addition to cognitive and motivational ones. Thus, from this perspective, it was seen that the intervention might be enhanced by the addition of an interpersonal element designed to stimulate or "trigger" specific behavior changes linked to cardiovascular health, through interpersonal influence and group process.

386

Another of the primary theoretical orientations guiding intervention planning was derived from Bandura's (1969, 1977) social learning theory. In a later formulation, Bandura (1978) describes a reciprocal interaction model. According to this conceptual approach, human behavior is regulated by immediate situational influences and by the person's performance skills and his anticipations of the consequences for different courses of action, rather than by such global constructs as personality traits. Bandura views the interplay between environment and behavior as a reciprocal influence process in which the environment shapes the individual's behavior, but the person also shapes his environment. Thus, the person's relationship with the environment is an open system, always modifiable by providing the individual with appropriate skills for self-management and motivating him or her to make use of them.

Social learning theory and research--particularly with the introduction of cognitive training for self-control--offers a promising conceptual framework for stimulating community-wide behavior change. We noted, however, that the techniques employed in behavior change typically have involved face-to-face training either one on one or, at most, one instructor to a class. Although often effective, these training methods are extremely costly when one considers the numbers of people served. In a group situation, members tend to conform to the behavioral patterns of the other group members. Data have generally supported this generalization. Lumsdaine and Janis (1953), Festinger and Maccoby (1964), McGuire (1964) and Roberts and Maccoby (1973) have formulated and tested some hypotheses on counterarguing and persuasion aimed at discovering bases for stable change.

In 1972 our group (Farquhar et al., 1977)--investigators at the School of Medicine and the Institute for Communication Research at Stanford University-- began a field experiment in three northern California communities in order to study the modification of risk factors in cardiovascular disease through community education. The major tactical choices for such a campaign are mass media, face-to-face instruction, or combinations of the two. Study of previous mass-media campaigns directed at large, open populations established the potential effectiveness of the media in transmitting information, altering some attitudes, and producing small shifts in behavior, by means of choices among consumer products, but failed to demonstrate that the media alone substantially influence more complex behavior (Bauer, 1964; Cartwright, 1949; Robertson et al., 1974; Star and Hughes, 1950). Yet the habits influencing cardiovascular risk factors are very complex and of long standing, are often reinforced by culture, custom, and continual commercial advertising, and are unlikely to be very strongly influenced by mass media alone. Face-to-face instruction and exhortation also have a long history of failure, particularly with respect to recidivism, as noted above, in efforts to influence diet (Stunkard, 1975) and smoking (Bernstein and McAlister, 1976).

387

After considering the powerful forces which reinforce and maintain the health habits that we wished to change, and in view of past failure of health education campaigns, we designed a heretofore untested combination of extensive mass media with a considerable amount of face-to-face instruction. We chose this method not so much because it was potentially widely applicable, but because it was a method that we judged most likely to succeed (Mendelsohn, 1973). We could then compare a more generally applicable treatment, though one not quite so promising of results, with the more sure one. Therefore, another community was selected in which we administered treatments via mass media alone. We also chose to include three elements typically ignored in health campaigns: (1) the mass media materials were devised to teach specific behavioral skills, as well as to perform the more usual tasks of offering information and affecting attitude and motivation; (2) both the mass media and, in particular, the face-to-face instruction were designed to embody many previously validated methods of achieving changes in behavior and self-control training principles; and (3) the campaign was designed on the basis of careful analysis of the specific needs and the media consumption patterns of the intended audience. Our overall goal was to create and evaluate methods for effecting changes in smoking, exercise, blood pressure and diet which would be both cost-effective and applicable to large population groups.

1. Elements of Media Design

Any campaign aimed at inducing specified techniques, knowledge, and behavior changes involves a number of processes. We conceptualized the interface between behavior change and mass media management in terms of the number of steps that are required to move the target population from initial awareness of, and interest in, the problem to the adoption and maintenance of the advocated attitudes or behavior (McGuire, 1984; Rogers, 1982). While, for any individual, depending on his or her precampaign state, some steps may be more or less important than others, and although for some individuals the sequence in which the processes are engaged may vary across a total population, it is possible to characterize the general processes to be engaged. By doing so, we specify the major elements of the campaign design.

(1) *Agenda-setting.* By agenda-setting, we mean the process by which an individual's attention is obtained and focused on specified issues and problems. It has been shown that while the mass media may be limited in their ability to persuade, they can be effective in setting agendas, that is, for bringing specified matters to the public's attention (McCombs and Shaw, 1972) for discussion or debate.

(2) *Informing.* Once a particular topic or subject matter is on the public agenda and is perceived as a salient issue, a campaign must present the logical set of propositions, in layperson's terms, that brings the issue home and sets the stage

for individual action. This self-appraisal through information and analysis is a crucial stage in furthering long-range objectives.

(3) *Motivation.* The population must be given positive incentives to change their behavior and must be given support and encouragement in maintaining their new behavior.

(4) *Training.* Once members of the population understand their personal relationship to the problem, they must be taught how to modify risk-related behaviors including how to handle barriers and personal costs.

(5) *Self-Maintenance.* If newly acquired habits are to be maintained, they must be under self-control. Prompting or self-cueing at the appropriate time and place can be critically important in this maintenance process.

These five campaign elements are by no means definitive or mutually exclusive. No doubt messages may serve different functions for different receivers, and messages may serve different functions for the same receiver. For example, a message may simultaneously persuade, inform, and motivate a single receiver (Cartwright, 1949).

Changes in overt human behavior are known to occur through the basic series of intermediary processes outlined above; however, a number of other contributing variables such as age, sex, socioeconomic status, current health beliefs, and previous education are important influences on the accomplishment of these basic processes (Dervin, 1982). These influences are complex; often they are contradictory or negative. Our design of a community-wide intervention strategy attempted to take into account the full range of potential influences operating at the individual level, the primary group level, and at the social or macro-environmental levels (DeFleur and Ball-Rokeach, 1974).

2. The Five City Study

The next logical step in the research was to build on the Three Community Study in order to investigate further means of achieving community-wide reduction in cardiovascular disease. We sought, therefore, to study larger communities, when long-term sustained changes in life style could occur so that reductions in morbidity and as well as risk could be achieved.

While intensive face-to-face instruction under an umbrella of mass media appeared likely to meet the above criteria, the number of such skilled trainers required would be prohibitive. We therefore looked for alternative, more exportable means for education, and community organization suggested itself as the means. Organizations in many instances have long histories of existence in most communities and if properly stimulated and led could well serve as the vehicle of long-term education for health.

Table 1

Three-Community Study Design

	1972	1973		1974		1975	
Watsonville (W)	Baseline Survey (S1)	•Media campaign •Intensive instruction (II) (2/3 of high risk participants)	Second Survey (S2)	•Media campaign •Intensive instruction (II) Summer Followup	Third Survey (S3)	•Maintenance (low-level) Media campaign	Fourth Survey (S4)
Gilroy (G)	Baseline Survey (S1)	•Media campaign	Second Survey (S2)	•Media campaign	Third Survey (S3)	•Maintenance (low-level) Media campaign	Fourth Survey (S4)
Tracy (T)	Baseline Survey (S1)		Second Survey (S2)		Third Survey (S3)		Fourth Survey (S4)

The design and plans for the Five City Study grew out of our analysis of the strengths and weaknesses of the Three Community Study. The current project differs, in its broad features, from the Three Community Study in several important ways:

(1) The two communities selected for education are larger and more socially complex than those in the previous study, and the health education program is designed to benefit the entire population.

(2) Three cities, rather than the one town of the Three Community Study, were selected as controls, resulting in a total population sized of 330,000 in the five cities as compared to 43,000 in the Three Community Study.

(3) The project is running for nine years, and a community organization method has been devised in order to provide a cost-effective and lasting program of community health promotion.

(4) People selected from a broader age range are taking part in the surveys. Repeated independent samples are being drawn every two years to monitor community-wide changes independent of survey effects, in addition to the cohort design of the Three Community Study.

(5) With the cooperation of local health officials, the Five City Program is monitoring the annual rates of fatal and non-fatal cardiovascular events in the five cities.

Overall, the Five City Project represents an ambitious new chapter in experimental epidemiology of potential relevance both to etiologic hypothesis testing and to a field application of cardiovascular disease control methods. It also presents a significant opportunity for testing generalizable behavioral health education methods.

3. Overall Goals and Design

The major aim of the FCP is to test the hypothesis that a significant decrease in the multiple logistic of risk for the educated communities will lead to a decline in morbidity and mortality from cardiovascular disease beyond that attributable to the secular trend. A six-year education program is designed to stimulate and maintain the changes in life style that should result in a community-wide reduction in risk of cardiovascular disease. Population surveys, epidemiological surveillance, and other assessment methods are being combined to evaluate the effects of the education program. The overall design is illustrated in Figure 1.

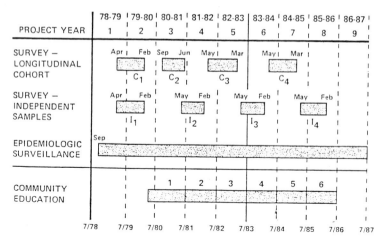

Figure 1.   Study Design and Timeline for Three Community Study.

3.1 Community education program

There are three goals of the education program. The first and broadest goal is to achieve a transformation in knowledge and skills of individuals and in the educational practices of organizations such that risk factor reduction and decreased morbidity and mortality are achieved.

The education program is to continue for a total of six years in order to create changes of approximately this magnitude, which are needed to detect changes in morbidity and mortality.

A second goal is to carry out the education program in a fashion that creates a self-sustaining health promotion structure, embedded within the organizational fabric of the communities, that continues to function after the project ends.

The third goal is to derive a model for cost-effective community health promotion from the experience and data accumulated in the Three Community Study and in this study, the broad features of which would have general applicability in many other American communities and we expect to the prevention of other chronic diseases.

3.2 Field trial

The Five City Project is also a field trail that will evaluate the effectiveness of the overall educational program in achieving the following:

(1) Changes in cardiovascular risk factors assessed through repeated longitudinal and cross-sectional population of surveys and by unobtrusive assessment of selected aspects of population behavior in the two treatment and two of the three reference communities.

(2) Changes in cardiovascular morbidity and mortality, assessed through the continuous community epidemiologic surveillance of fatal and non-fatal heart attack and stroke in two treatment and three reference communities.

Community-wide changes in knowledge, behavior, and risk, independent of survey effects, will be determined through comparison of the four biennial cross-sectional sample surveys conducted in the two education cities and two of the three reference communities. Studies of the process of change in individuals will be possible by comparison of the longitudinal cohort surveys. This cohort was drawn from the first independent sample.

4. Education Program

4.1 Risk factors and health behaviors

The major cardiovascular risk factors selected for attention are cigarette smoking, arterial blood pressure (BP), and plasma cholesterol. These risk factors have been linked with several "health behaviors" that, therefore, become major areas for the intervention. These are smoking, nutrition, exercise, hypertension (i.e., treatment of high BP), and obesity. Nutrition behavior will affect BP (salt, weight control), cholesterol (dietary saturated fat, cholesterol, and fiber; weight control), exercise will affect BP (weight control), and hypertension treatment will affect' that proportion of the BP distribution appropriate for drug therapy. The other factors are identified to enable specific plans when appropriate.

While it is convenient in this presentation, and in planning, to consider each risk factor and health behavior separately, there is a danger of perceiving a fragmentation that is not present. In fact, the educational program in use in the Five City Project is highly unified. It is most difficult, in fact, for it to be otherwise. People who attend a smoking cessation class, for example, are encouraged to begin exercise, substitute healthy foods for the smoking habit, and so forth. We are presenting a single life style that is most likely to be healthy. It involves being vigorous, active, self-confident, eating a wide variety of enjoyable foods, and not smoking. It is this basically healthy and happy image that binds together the various elements of the intervention and makes the educational programs and materials coherent and, to an extent, indivisible.

4.2 Theoretical perspective

A variety of perspectives and theoretical formulations need to be blended to successfully design and carry out the educational program described above. In addition to the clearly relevant field of community organization, which creates a receptive environment for our education materials and programs, we have also found it necessary to borrow from an additional perspective to create the blend needed for success. The communication-behavior change framework is based on a social psychological perspective relevant to the individual and group learning that is needed within the overall Community Organization method. This perspective is particularly germane to the content of educational materials produced.

The communication-behavior change framework offers a perspective on how individuals and groups change knowledge, attitudes, and behavior. Our picture of the change process draws on the work of others: the social learning model of Bandura (1969, 1977, 1982); the hierarchy of learning model of Ray and others (1973); the communication-persuasion model of McGuire (1969); the attitude change model of Fishbein and Ajzen (1975); and the adoption-diffusion model of Rogers (1982). The communication-behavior (CBC) model emphasizes the features that are relevant to the community-based education and health promotion. There are several underlying assumptions implicit in these portrayals of the change process:

- o A need for change exists and room for change exists.
- o Initial and final states are measurable.
- o Education forces have adequate social legitimacy.
- o Adequate time exists within the design for the change to occur.

This approach proposes a series of steps that people go through as they gradually adopt the advocated behavior. The concept of behavior change as an orderly sequence of steps is admittedly an idealized version of real life and may not be an accurate portrayal of the process of change for all people in all areas. For some people, on some topics, the sequence of steps may vary. On some issues, one or two steps will be much more important than others. One message may perform more than one function. However, this conceptualization does help us to develop a clearer picture of how to devise a course of action. As we proceed, we can also use it as an evaluation framework to observe the shift of population groups over time in the direction of the intended project goals. These steps are listed below and briefly discussed. The corresponding communication function is identified for each step.

Become Aware (Gain Attention). Agenda-setting is to gain the public's attention and focus it on certain specific issues and problems. In the public's mind, the existence of the problem must be established, and an awareness of potential solutions promoted. Generally, in our society, the mass media play an important role in agenda-setting.

Increase Knowledge (Provide Information). Once a particular topic or subject matter is on the public agenda, an educational program must present information, in layman's terms, that makes the issue interesting and understandable. Messages need to be designed that make the issue personally meaningful and that set the stage for action. The messages must be retained in a way that assist the person to act in a different way in the future.

Increase Motivation (Provide Incentives). Change is more likely when individuals perceive clearly the personal and social benefits of change, which can be enhanced by appropriate communications.

Learn Skills (Provide Training). Where changes in complex habits of long standing are involved, it may be necessary to provide skills training in how to start making changes, both by providing step-by-step instruction and by promoting the availability of self-help and professional resources.

Take Action (Provide Cues). Ideally, this phase of an overall strategy would provide educational inputs that act as cues to trigger specific actions. Messages would indicate clear action paths to stimulate the trial adoption of new behaviors.

Maintenance (Provide Support). At this stage, inputs are required to provide a sense of social support and approval, and as a reminder of both the short-term and long-term personal and social benefits of the changes undertaken. Both gaining self-efficacy and learning self-management methods are important aspects of the maintenance phase.

4.3 Community organization

Community organization has been desiged to perform a significant role both in the initial success and in the durability of the program. As a result the educational program is being conducted in a manner that encourages involvement from the outset by local community groups, which we anticipate will lead to local ownership and control.

The following assumptions have been made: (1) mass media education alone are powerful, but its effects may be augmented by community organization; (2) interpersonal influence can be relatively cheaply enhanced through community organization and can thus allow a multiplier effect to occur that should increase behavior change; (3) organizations can expand the educational program's delivery system in ways important to achieving community-wide health education; (4) organizations can help the process of community adoption of risk reduction programs as their own and thus increase the likelihood of long term continuing health education programs and behavior change in their communities; (5) formation of new organizations can be catalyzed by our external efforts to increase the array of groups concerned with health education and health promotion.

4.4 Formative evaluation

Formative research, as distinct from summative research, is intended to provide data for use in designing educational strategies, to design particular programs and materials to meet specific objectives for the target audiences, and to monitor the progress of the educational program. A general criticism in the literature of health communication campaigns is their lack of formative research and, therefore, the resulting design of materials and programs which are not able to meet the objectives of the overall effort. A core element of the education program is our ability to conduct and utilize formative research.

4.5 Broadcast media programs

The overriding goal of broadcast media in the Five City Project is the encouragement of lasting behavior change that will result in risk factor modification and ultimately in reductions in morbidity and mortality. Underlying this major goal are two sub-goals: broadcast media serve the function of encouraging direct behavior change as well as encouraging indirect change through support of community events. Some products are designed to support community programs (particularly by encouraging recruitment), while others are designed to create direct changes in knowledge, attitudes, and behavior on the part of the target audience. Some products are hybrids. For example, a smoking cessation television show could encourage cessation as well as recruit individuals into smoking cessation programs available in the community.

There are a variety of factors which determine the format, content, and time sequence of broadcast media products. Certain products are essential to support programs as requested by community groups. Some media products are requested by the mass media outlets themselves. However, the decision-making process for most media productions is highly related to our particular risk factor goals over a certain period of time. Such planning is based on a careful understanding of risk factor goals, the knowledge, attitudes, behaviors, and desires for change among the majority of adult smokers, and estimates of the efficacy of various approaches to cessation and maintenance.

4.6 Print media programs

Print media are able to provide higher information-density messages on a particular topic than broadcast media, which are most capable of presenting relatively low information-density messages. Print can be read and re-read at a user's own pace. Therefore, it provides a large amount of information in a user-oriented format. It is an especially important vehicle for skills training. With this in mind, the Five City Project has invested a substantial amount of effort in the use of print media, particularly in topic areas which require more than the superficial amount of information that can be provided in typical broadcast media. To this end, formative research efforts of a variety of kinds have been used to design, modify, and distribute printed material.

5. Evaluation

5.1 Health surveys

Each health survey includes both physiological and behavioral measures. The array of physiological measures are:
- o Body height and weight.
- o Blood pressure by two methods (mercury manometer and a semi-automated machine).
- o Non-fasting venous blood sample analyzed for plasma thiocyanate (as a measure of smoking rate), total cholesterol, triglycerides, and cholesterol content of lipoprotein subfractions.
- o Expired air carbon monoxide.
- o Urinary sodium, potassium, and creatinine (as an index of prior sodium chloride intake).
- o A low-level bicycle exercise test (as a measure of fitness).

Lipid analyses follow long-established methods of the Lipid Research Clinics Program (U.S.D.H.E.W., 1974). Plasmathiocyanate determination followed the procedure of Butts, Kuehneman, and Widdowson (1974) as used in the Multiple Risk Factor Intervention Trial. Expired air carbon monoxide is measured on the Ecolyzer apparatus.

Blood pressure is obtained as indirect brachial artery pressure (systolic and fifth phase diastolic) on participants sitting at rest for at least two minutes before the first measurement. Pressures are obtained twice using a Sphygmetrics Infrasonic Automatic Blood Pressure Recorder (SR-2), and this is followed by dual measurements of pressure using a standard mercury sphygmomanometer and auscultation. Urine samples are frozen after collection and shipped to Stanford twice monthly. Standard laboratory procedures are used for determination of urinary sodium and potassium (flame photometry), and creatinine.

The low-level exercise test, using a Schwinn electric brake stationary cycle ergometer is performed following blood pressure measurement and blood sampling. Measurements are made of pulse rate using a Quinton Instrument Cardio-tachometer. A small proportion of participants are excluded from the test according to very conservative criteria. The test is designed to obtain a pulse rate index of relative fitness after a standard work load that is estimated to be 70 percent of maximum aerobic capacity.

Behavioral measures include a broad range of attitude and knowledge assessments, behavioral intention measures, self-reported behavior, and dietary and physical activity recalls. In addition, questions are asked for use in formative evaluation, such as attitudes toward different types of educational materials. Of course, standard demographic and medication use data are collected.

5.2 Epidemiologic surveillance

The purpose of community epidemiologic surveillance is to allow the calculation of comparable, city-specific rates for total mortality, cardiovascular mortality, fatal myocardial infarction, nonfatal myocardial infarction, fatal stroke, and nonfatal stroke. The mortality rates mentioned are obtainable from vital statistics, but these rely on the unaided interpretation of death certificate diagnoses. A common method for obtaining the morbidity rates is to identify a cohort of individuals and follow them through time with repeated, thorough examinations which discover the occurrence of new events. Such cohort studies are large and expensive. Thus the needed mortality statistics are available inexpensively, but their accuracy is suspect. The morbidity data are unobtainable except at great expense. Community surveillance is designed to obtain accurate mortality statistics and to obtain morbidity statistics at an acceptable cost.

Potential fatal events are identified from death certificates and non-fatal events from hospital discharge records. Non-clinical or "silent" infarctions are not identified. All potential events are investigated by hospital chart review of family interview. The resulting data are reviewed at Stanford by trained analysts using standard criteria for each type of event. The analysts are unaware of the community of origin and review the cases independently. A final endpoint is assigned by a computer algorithm applied to the analysts' digest of each case. If the analysts disagree on a case, it is reviewed by a physician.

6. Summary

A considerable proportion of the chronic disease in the United States is amenable to prevention or delay. The most important causes of death in the 55 through 74 year-old age group--cardiovascular disease, cancer, alcoholism, and trauma--have significant environmental and behavioral components. The Stanford Five City Project is an application of a community-wide approach to the control of cardiovascular disease through health changes in behavior. This approach may be generalizable to other disease control efforts as we learn to unite the medical, behavioral, communication, and social sciences to solve problems.

References

Ajzen, I. and Fishbein, M. *Understanding attitudes and predicting social behavior.* 1980.

Bandura, A. Self-efficacy mechanism in human agency. *American Psychologist* 37 (1982) 122-147.

Bandura, A. The self system in reciprocal determinims. *American Psychologist* 33 (1978) 344-358.

Bandura, A. *Social learning theory.* Englewood Cliffs: Prentice Hall, 1977.

Bandura, A. *Principles of behavior modification.* New York: Holt, Rinehart and Winston, 1969.

Bernstein, D. A. and McAlister, A. The modification of smoking behavior: Progress and problems. *Addictive Behavior* 1 (1976) 195-236.

Bauer, R. The obstinate audience: The influence process from the point of view of social communications. *American Psychologist* 19 (1964) 319-328.

Butts, W. C., Kuehneman, M., and Widdowson, G. M. Automated method for determining serum thiocyanate to distinguish smokers from non-smokers. *Clinical Chemistry* 20 (1974) 1344-1348.

Cartwright, D. Some principles of mass persuasion. *Human Relations* 2 (1949) 253-267.

DeFleur, M. L. and Ball-Rokeach, S. *Theories of mass communication.* New York: David MacKay, 1974.

Dervin, B. Mass communications: Changing conceptions of the audience. In R. Rice and W. Paisley (eds.), *Public communication campaigns,* Beverly Hills, Calif.: Sage Publications, 1982.

Farquhar, J. W. The community-based model of life-style intervention trials. *American Journal of Epidemiology* 108 (1978) 103-111.

Farquhar, J. W., Maccoby, N., Wood, P. D., Alexander, J. K., Breitrose, H., Brown, B. W., Jr., Haskell, W. L., McAlister, A. L., Meyer, A. J., Nash, J. D., and Stern, M. P. Community education for cardiovascular health. *Lancet* (1977) 1192-1195.

Festinger, L. and Maccoby, N. On resistance to persuasive communications. *Journal of Social and Abnormal Psychology* 68 (1964).

Fishbein, M. and Ajzen, I. Belief, attitude, intention and behavior. In *Introduction to theory and research.* Reading, Mass.: Addison-Wesley, 1975.

Lumsdaine, A. A. and Janis, I. L. Resistance to counterpropaganda produced by one-sided and two-sided communication. *Public Opinion Quarterly* 17 (1953) 311-318.

Maccoby, N., Farquhar, J. W., Wood, P. D., and Alexander, J. K. Reducing the risk of cardiovascular disease: effects of a community-based campaign on knowledge and behavior. *Journal of Community Health* 3 (1977) 100-114.

Mendelsohn, H. Some reasons why information campaigns can succeed. *Public Opinion Quarterly* 37 (1973) 50-61.

Meyer, A. J. and Henderson, J. B. Multiple risk factor reduction in the prevention of cardiovascular disease. *Preventive Medicine* 3 (1974) 225-236.

McAlister, A. L., Perry, C., and Maccoby, N. Adolescent smoking: Onset and prevention. *Pediatrics* 63 (1979) 650-658.

McCombs, M. and Shaw, D. The agenda setting function of the mass media. *Public Opinion Quarterly* 36 (1972) 176-187.

McGuire, W. J. The nature of attitudes and attitude change. In G. Lindzey and E. (eds.) *The handbook of social psychology* Aronson, 3rd Edition. Reading, Mass.: Addison Wesley, 1984.

McGuire, W. J. The nature of attitude change. In G. Lindzey and E. Aranson (eds.) *The handbook of social psychology*. Reading, Mass.: Addison Wesley, 1969.

McGuire, W. J. Reducing resistance to persuasion: Some contemporary approaches. In L. Berkowitz (ed.) *Advances in experimental social psychology, Vol. 1.* New York: Academic Press, 1964.

Ray, M. L., Sawyer, A. G., Rothschild, M. L., Heelers, R. M., Strong, E. C., and Reed, J. B. Marketing communication and the hierarchy of effects. In P. Clarke (ed.) *New models for mass communication research.* Beverly Hills, Calif.: Sage Publications, 1973.

Roberts, D. F. and Maccoby, N. *Information processing and persuasion:* Counterarguing behavior. In P. Clarke (ed.) *New models for mass communication research.* Beverly Hills, Calif.: Sage Publications, 1973.

Robertson, L. S., Kelley, A. B., O'Neill, B., Wixom, C., Eisworth, R., and Haddon, W. A controlled study of the effort of television messages on safety belt use. *American Journal of Public Health* 64 (1974) 1071-1080.

Rogers, E. M. *Diffusion of innovations.* New York: Free Press, 1982.

Star, S. and Hughes, H. M. Report of an educational campaign: The Cincinnati plan for the United Nations. *American Journal of Sociology* 55 (1950) 826-833.

Stunkard, A. J. From explanation to action in psychosomatic medicine: The case of obesity. *Psychosomatic Medicine* 37 (1975) 195-236.

United States Department of Health Education, and Welfare. *Lipid Research Clinics manual of laboratory operations, Vol. 1. Lipid and lipoprotein analysis.* Washington, D.C.: Government Printing Office, 1974. (DHEW Publication No. (NIH) 75-628).

20

EXERCISE AND DISEASE PREVENTION

Roy J. Shephard, M.D., Ph.D.

1. Introduction

One of the exciting conclusions reached by the current generation of health educators, and reiterated by speakers at this conference, is that much of the tremendous social and economic burden of poor health could be corrected by a change of life style. In the United States, the annual cost of all illnesses has been estimated at $473 billion (for convenience, costs are here recalculated to their 1982 US equivalent):

Direct costs	213.1 B
Indirect costs	
Morbidity	103.4 B
Premature death	156.9 B

As much as a quarter of this enormous expense has been traced to the abuse of alcohol and tobacco (Berry and Boland, 1977; Fletcher, 1959; Luce and

Table 1

Possible health effects of regular physical exercise

Acute Illness

Infections/Immune response
Perceived health
Absence from duty

Chronic Illness

Risk profile
Ischaemic heart disease
Alcohol and cigarette abuse
Other diseases

Physical Injury

Geriatric Care

Schweitzer, 1978). How much more of the total can be attributed to other problems of life style, particularly sedentary habits?

The present paper will examine critically the potential contribution of an increase of physical activity to both health and health care economics. It will not be restricted to the NATO military population, partly because the amount of information on this group is quite limited (Allen, Brown, and O'Hara, 1978; Bardsley, 1978; Jung, 1978; Löllgen and Pleines, 1978; Shephard, 1978a), and partly because containment of the ever-increasing costs of medical and geriatric care have major implications for all aspects of national budgets. Topics to be considered include the impact of exercise upon acute and chronic illnesses, physical injury, and geriatric care (Table 1).

2. Acute Illness

2.1. Infections

Occasional animal experiments have suggested some interaction between immune reactions and vigorous physical activity. The pattern of change described was reminiscent of the general adaptation syndrome (Selye, 1974) and it was thus unclear whether the increased energy expenditure itself or the stressful nature of the task (treadmill running with repeated electric shocks, or swimming to exhaustion in cold water) was responsible for the immunological change.

Only a limited number of human studies have been completed to date; these suggest that exercise has little influence upon immune function in men and women (Jokl, 1931). Nevertheless, in some instances the redistribution of blood flow and/or the cardiac work-load induced by vigorous exercise have had an adverse effect upon the clinical course of a disease (Bourne and Wedgwood, 1959; Krikler and Zilberg, 1966; Russell, 1947; Weinstein, 1952).

2.2. Perceived health

Of greater practical importance is the possibility of changing an individual's perceived health (Table 2). Behavioral scientists increasingly recognize the absence of any sharp boundary between health and disease. Many if not most absences from duty, and many if not most medical consultations lack any clear organic cause.

An increase of physical activity thus has potential to induce a substantial decrease in the proportion of those reporting sick through its influence upon such variables as mood state, self image, and life satisfaction.

A controlled experiment with primary school students (Lavallée and Shephard - in preparation) showed that 5 hours of additional physical activity per week *increased* the absenteeism of six-year-old children by three days (from 10 to 13 days per year), but in those aged 7 to 12 years absenteeism was similar for experimental and control samples (each group missing only four days of school per year). Our interpretation of these data is that parents were unwilling to expose very young children to a combination of school and some minor infection, and that this reluctance was increased by our introduction of a required physical activity program. However, both immune function and the child's perception of his or her own health were unchanged by the daily hour of vigorous activity, so that attendance became comparable in experimental and control groups once the age of over-protection was past.

Russian industry has a militaristic approach to illness. Some authors describe the sick worker as a saboteur! Reports by Dodov and colleagues (1975) and Pravosudov (1978) suggest that health is better in physically active employees than in those who are inactive. Worker-athletes consulted doctors four times less often than their sedentary counterparts, and only 22.5 percent of consultations led to absence from work, compared with 50 to 60 percent in the case of consultations by non-athletes. Worker-athletes also suffered fewer episodes of acute respiratory disease, and these episodes were of shorter duration than in control subjects. Both "catarrhal" and "non-catarrhal" diseases were less frequent in the athletes. Benefit was also seen in neuroses, "hypertonia," chronic cholecystitis, and other morbid conditions. Unfortunately, at least a part of the benefit that was attributed to exercise could really have arisen from other factors, including self-selection of sport by healthy individuals and an absence of cigarette addiction in endurance competitors.

Table 2

Changes of perceived health from participation
in regular physical activity

Population	Response	Author
Schoolchildren Age 6	Increased absenteeism	Shephard and
		Lavallée (unpublished)
Age 7-12	No effect	
Russian worker-athletes	Four-times fewer medical consultations	Pravosudov (1978)
	Less respiratory disease neuroses Hypertonia cholecystitis	
	Less respiratory disease neuroses Hypertonia cholecystitis	
Ontario insured population	Reduced health care costs	Quasar (1976)
Purdue University staff	Fewer health insurance claims	Corrigan et al. (1980)
Life Assurance Company	Fewer hospital days	Shephard et al. (1981, 1982)
	Fewer medical consultations	
	Fewer drug purchases	

Similar difficulties of distinguishing between cause and effect have marred some North American studies. Thus a cross-sectional analysis (Quasar, 1976) suggested that medical payments under the Ontario Health Insurance Plan would be reduced by 5.5 percent if all participants aged 20 to 69 years developed at least an average level of cardio-respiratory fitness. An extrapolation of coronary risk factors from the same data base indicated the potential for an annual saving of $45 million, due to a reduction in the incidence of clinical ischemic heart disease.

However, it remains arguable that absence of disease allowed high fitness levels, rather than the converse. Likewise, Corrigan (1980) found that continuing participants in Purdue University's Adult Fitness Program had lower medical insurance claims than program drop-outs; unfortunately, there were differences of personality between the two groups of subjects, and a process of self-selection rather than exercise could have been responsible for the discrepancy in perceived health.

Shephard, Cox, and Corey (1981) carried out a controlled study of perceived health in the context of employee fitness. After introduction of a new program, employees at the experimental company reported fewer visits to the industrial nurse or physician, fewer out-patient visits, less hospitalization, and fewer purchases of prescription drugs. Relative to the behavior of workers at a closely matched control company, Ontario Health Insurance Plan records also showed a lesser hospital usage and fewer physician visits, economic savings from these two items alone being equivalent to $233 per employed worker per year (Shephard, et al., 1982). Interestingly, the increase of physical activity generated no increase of claims for either electrocardiographic or orthopedic services in the experimental sample.

2.3 Absence from duty

Perceived health is but one factor influencing absenteeism. Nevertheless, it is interesting that several investigations have shown a reduction of absenteeism subsequent to the introduction of employee fitness programs (Table 3).

Pravosudov (1978) reported beneficial effects of sports participation among factory, office, and professional workers, absenteeism being reduced by 3 to 5 days per worker-year. Romanian studies noted a similar order of benefit (Barhad, 1979; Pafnote, Voida, and Luchian, 1979). Erwin (1978) observed a 23 percent drop in unpaid absenteeism when an employee fitness program was introduced by the People's Credit Jewellers. A similar uncontrolled initiative among employees of the New York State Department of Education (Bjurstrom and Alexiou, 1978) yielded a small but useful reduction of absenteeism amounting to 4.7 hours per employee-year.

A controlled trial in the Dallas police force (Byrd, 1976; Mealy, 1976) cut the sick-leave of the experimental group by 34 percent relative to controls (a saving of 8.5 days per employee-year). At the Metropolitan Life Insurance Company, experimental subjects decreased their absences from 6.3 to 4.9 days per year, while controls worsened their absenteeism from 5.6 to 7.0 days per year (Garson, 1977). Likewise, Johnson and Johnson saw a 9 percent decrease of sick days among participants in a life style program, relative to a 13 percent increase for control subjects within the same organization (C. Wilbur - personal communication).

The Toronto Life Assurance Study (Cox, Shephard, and Corey, 1981) had very similar findings. High adherents to an employee fitness program showed a

Table 3

Effects of regular physical activity upon absence from duty

Population	Response	Author
Russian Worker-athletes	3-5 days reduction per year	Pravosudov (1978)
Romanian workers	Decreased absenteeism	Barhad (1979)
		Pafnote et al. (1979)
U.S. jewelry company	23% decrease of unpaid absenteeism	Erwin (1978)
U.S. civil servants	4.7 hours per year decrease	Bjurstrom and Alexiou (1978)
U.S. police	34% reduction of sick-leave	Byrd (1976)
		Mealy (1976)
Candadian office workers	2.8 days per year reduction	Garson (1977)
U.S. corporation	22% reduction of absenteeism	Wilbur (unpublished)
Candaian office workers	22% reduction of absenteeism in high adherents	Cox et al. (1981)

41.8 percent decrease in absenteeism over the first year of the program, this being a substantial advantage relative to the 20.1 percent decrease seen at the control company and the 23.2 percent decrease noted in non-participants at the experimental company.

The economic gain from the decreased absenteeism is undeniable. Our calculations are based upon 20 percent participation in the fitness program, which is realistic in industry, although a better attendance might be possible at a military installation. We have also assumed a replacement charge of 1.75 times the number of absentee days (although sickness of highly specialized personnel could have a much greater impact). The saving estimated on this basis is about $30 per year worker-year. However, it remains unproven that exercise is uniquely responsible for the less frequent absences from duty. Most employee fitness programs offer facilities for modifying other adverse life styles (including specialized clinics for cigarette, alcohol, and drug abuse, weight reduction, stress relaxation, and so on). It is well-documented that the smoker, for example, takes sick-leave more frequently and for longer periods than a non-smoker (British Medical Journal,

1974). Moreover, even if exercise is the only treatment offered to the worker, it remains arguable that any impact upon absenteeism is simply a "Hawthorne" type effect, the individual reacting favorably to the greater interest of command or management in his or her welfare.

3. Chronic Illness

3.1 Risk profile

Cross-sectional comparisons (Allen, Brown, and O'Hara, 1978; Cooper et al., 1977; Hickey et al., 1975; Montoye, 1975; Quasar, 1976) have noted that physical inactivity is associated with an adverse cardiovascular risk profile in both civilian and military populations. Items such as cigarette smoking, obesity, a high serum cholesterol, a low HDL/LDL cholesterol ratio, and hypertension are all over-represented in the inactive segment of the population. Quasar (1976) further assumed a causal relationship between aerobic power and several risk factors, estimating the reduction of cardiovascular disease that would result if all of the Ontario population were brought to at least an average level of cardiorespiratory fitness. Certainly, there is some justification for such an approach, since prospective studies have demonstrated that an increase of physical activity can correct such adverse findings as obesity, moderate hypertension, and a low HDL/LDL ratio. However, it is less clear whether the sudden correction of risk factors by vigorous exercise can fully reverse the hazards accumulated by a lifetime with an adverse risk profile.

One convenient method of summarizing an individual's risk-taking behavior is to complete the Canadian Health Hazard Appraisal test. Self-reports on matters such as smoking, drinking, and the use of car seatbelts are translated into a "composite risk score," an "appraised age" (the age of an individual who would have the same chance of dying over the next ten years), and a "compliance age" (the appraised age of the individual if all possible corrections were made to risk profile). Löllgen and Pleines (1978) have applied a similar concept of "age equivalent" in their studies of the West German forces. We used the Health Hazard Appraisal instrument to test the behavior of a large group of white-collar workers before and six months after introduction of an employee fitness program (Shephard et al., 1982). Although the population were nine months older at the final examination, male adherents to the program had reduced their appraised age by as much as two years, correcting much of the discrepancy between apraised and compliance age. The altered risk score reflected not only the enhanced physical activity, but also a reduced consumption of alcohol and cigarettes, plus some reduction of systemic blood pressure. Again, there is the criticism that a sudden change of "appraised age" may not indicate any immediate increase of longevity. Nevertheless, there is little doubt that the long-term changes of behavior that we observed had a favorable impact upon health.

A two-year reduction of appraised age has some economic value in that the younger individual functions more efficiently. However, the full benefit will not be realized in the military context of a fixed retirement age. Likewise, the full value to industry will accrue only if company and union policies allow the individual to work two years longer than sedentary colleagues. The "added value" of two years' labor is rated at 25 percent of salary. Given an annual income of $14,500, with 20 percent participation in the exercise classes, and distribution of the benefit over a working span for 40 years, the saving averages $36 per employed worker-year.

3.2 Ischemic heart disease

While much suggestive evidence has accumulated linking vigorous physical activity with protection against ischemic heart disease (Table 4), it remains virtually impossible to devise the key experiment that surmounts all technical criticism. The most important problem is that sustained vigorous activity is almost always self-selected.

3.3 Cross-cultural comparisons

Certain populations such as Somali camel-herders (Lapiccirella et al., 1962), the Masai (Mann et al., 1965) and traditional Inuit (Milan, 1980) have a low incidence of ischemic heart disease, and this has been attributed to the high level of habitual activity in such communities. However, it is extremely difficult to obtain accurate figures on the intensity, frequency, and duration of physical activity among nomadic populations, and differences of diet, environment, and genetic background preclude direct comparisons of disease incidence with the "white" citizens of developed countries.

3.4. Athletes

Most athletes sustain a high level of physical activity for a number of years, but again it is difficult to draw conclusions about the preventive value of such exercise from comparisons with more sedentary members of the population. There is intense selection for most sports by body build (Polednak, 1972), and this factor undoubtedly influences the liability to ischemic heart disease and other chronic disorders. Further, an abnormally high proportion of athletes die of violence (accidents, suicides, and murders), distorting the mortality rates for middle age. Many athletes abandon all voluntary physical activity before they reach the coronary prone years, and in middle age some former sportsmen smoke more, drink more, and take less exercise than their supposed sedentary counterparts (Montoye et al., 1956). While one study of champion cross-country skiers described a 4.3 year advantage of longevity relative to controls (Karvonen et al., 1974), the majority of the athletes were also life-long-non-smokers. Abstinence from cigarettes rather than a life-time of endurance exercise may thus have been responsible for their long lifespan. It is also arguable that some of the controls did not become involved in endurance sport because of pre-existing ill-health.

Table 4

Sources of evidence linking physical activity and prevention of ischemic heart disease

Cross-cultural comparisons

Somali camel herders (Lapiccirella et al., 1965)
Masai (Mann et al., 1965)
Inuit (Milan, 1980)

Athletes

Michigan State study (Montoye et al., 1956)
Cross-country skiers (Karvonen et al., 1974)

Occupational comparisons

Bus drivers/conductors (Morris et al., 1966)
Railway clerks/switch workers (Taylor et al., 1962)
Kibbutzim workers (Brunner and Manelis, 1971)
Longshoremen (Paffenbarger, 1977)

Leisure activities

British civil servants (Morris et al., 1973)
Harvard alumni (Paffenbarger et al., 1977)
Framingham Study (Dawber, 1980)
Western Collaborative Study (Rosenman et al., 1977)
Health Insurance Plan of New York (Shapiro et al., 1977)
Gothenburg study (Tibblin et al., 1975)

Randomized trials

U.S. pilot trial (Remington and Schork, 1967)
World Health Organization (1977)

3.5 Occupational comparisons

Comparisons of the incidence of disease between supposedly active and inactive occupations have encountered many problems. Often, the mental stress has differed substantially between active and sedentary tasks (Morris et al., 1966), although it has been argued that this variable has little impact upon the risk of ischemic heart disease. In only a few instances (bus drivers versus bus conductors, Kibbutzim workers, mail carriers versus mail clerks, railway clerks versus railway switch workers) has there been a reasonable socio-economic match between active and inactive samples (Shephard, 1981). Difficulties have also arisen from the imprecision of job classifications, inadequate differences of activity between supposed heavy and light work, and a selective transfer of symptomatic employees from active to sedentary occupations (Taylor et al., 1962). Nevertheless, the most serious criticism of investigations has been an initial self-selection of employment. Several studies (Morris, Heady, and Raffle, 1956; Paffenbarger, 1977; Rosenman, Bawol, and Oscherwitz, 1977) have shown significant differences of coronary risk factors between active and inactive employees even when they are first recruited. The one study where it was claimed that workers had no choice over their type of employment was the Kibbutzim investigation of Brünner and Manelis (1971). This, incidentally, showed a large protective effect among the active subjects. Nevertheless, it is hard to believe that the physique of recruits did not have some influence upon job allocation by the management committee of the Kibbutz.

Perhaps the most complete of the occupational studies is an examination of Californian longshoremen carried out by Paffenbarger (1977). In contrast to most other investigations, Paffenbarger used modern multivariate techniques to adjust for the influence of other coronary risk factors. After allowance for the effects of age, smoking, and hypertension, a low occupational energy expenditure led to a 3.3 fold increase in the risk of sudden death and a 1.6 fold increase in the risk of delayed death. The overall impact of a sedentary job seemed a doubling of cardiovascular risk. Paffenbarger (1977) noted several advantages of his study relative to previous investigations: (i) job categories were clearly defined by the union concerned, (ii) the "active" part of the work-force had a high rate of energy expenditure while on the job (22-31 kJ min^{-1}, although 45 percent of paid time was allowed for "recovery pauses"), and (iii) there was little possibility that the differential mortality could have arisen by job transfer, since the largest benefit was seen in workers who died suddenly, without warning symptoms. Nevertheless, there were fewer heavy smokers and fewer subjects above the average weight for height in the group who were undertaking heavy work; to this extent, they were self-selected. Paffenbarger used his multiple regression function to estimate that if the longshoremen increased their energy expenditure, reduced their systolic blood pressure below the population mean, and smoked less than a pack of cigarettes per day, the risk of fatal heart attacks would decrease by 88 percent.

3.6 Leisure activities

With automation, there is an ever-decreasing likelihood that occupational activity will be sufficiently vigorous to have therapeutic value. Attention is thus being directed towards leisure pursuits as sources of physical activity.

Most attempts to relate the choice of leisure activities to the incidence of ischemic heart disease have encountered the difficulty that such pursuits are by definition self-selected. The individual who chooses a vigorous form of relaxation often has other positive health attributes or a socio-economic advantage which is responsible for much of any difference in disease incidence that is observed (Rosenman, Bawol, and Oscherwitz, 1977; Young and Willmot, 1973). Morris and colleagues (1973) carried out a prospective study of leisure habits in a large and socially homogenous group of British civil servants. Over the survey period, 238 subjects developed first clinical attacks of myocardial infarction. By choosing 476 case-matched controls, they demonstrated that those who developed clinical disease were significantly less likely to have engaged in either 5 minutes per day of near maximum effort (vigorous recreation, keeping fit, running) or 30 minutes per day of heavy leisure activity (> 31 kJ min^{-1}). The development of electrocardiographic abnormalities was likewise correlated with an inactive leisure (Epstein et al., 1976). The overall difference of energy expenditure between controls and those who developed infarction was only 750 kJ day^{-1}, substantially less than had been reported in many occupational studies. Crucial issues in the Morris investigation remain (i) the success of case-matching and (ii) the possibility that an interest in physical activity may reflect a generally more favorable life style. Height, body mass, and skinfolds were identical for the two groups, but the active sample included fewer subjects with a serum cholesterol > 6.4 mmol l^{-1}, more subjects with a blood pressure $> 150/90$ mm Hg, and slightly fewer smokers (26 versus 32 percent. These differences could have made some contribution to the favorable prognosis.

Paffenbarger and colleagues (1977) accumulated 117,680 person years of experience in a long-term follow-up of the leisure pursuits of Harvard alumni. Substantial protection against ischemic heart disease (an attack rate of 1.00 versus 1.64) was seen in those subjects who accumulated an additional leisure expenditure of at least 8,000 kJ per week, an energy differential reminiscent of that reported from many of the occupational comparisons. The age-adjusted cardiac fatality rate of his sample showed a steep downward gradient as leisure expenditures increased from 2,000 to 10,000 kJ wk^{-1}. Protection against heart disease was observed independent of smoking habits, blood presure, body build (Quetelet index), or family history. Moreover, the prognosis was as good in those who had first become active after leaving the university as in varsity athletes, weakening the potential criticism that the active group was "self-selected." While some reviewers have suggested that the energy expenditure needed for protection against cardiac disease is unrealistically heavy in the context of community fitness, they overlook the fact that Paffenbarger and colleagues found benefit from walking one mile per day (attack ratio 1.00 : 1.26) and from climbing 50 stairs per day (attack ratio 1.00 : 1.25).

The Framingham study (Dawber, 1980; Kannel, 1979) noted both the incidence of coronary heart disease and mortality were higher in those members of

411

the community judged to be inactive by a simple 24-hour history. However, in this investigation, other risk factors were judged more important than the exercise history. Rosenman (1970) reported that the impact of exercise varied with personality, protection being seen in "type A" but not "type B" subjects; activity was also without benefit if subjects had a high diastolic pressure, a high triglyceride level, or a low HDL/LDL cholesterol ratio. Shapiro and colleagues (1969) summed occupational and leisure activity in interpreting data from the Health Insurance Plan of New York. A low level of activity was associated with an increased incidence of heart attacks, but was unrelated to the incidence of angina. Tibblin, Wilhelmsen, and Werkö (1975) examined the activity patterns of all male citizens of Gothenburg born in 1913; there were some problems of socio-economic matching, but nevertheless a trend towards inactive leisure was seen in those who subsequently developed coronary disease.

3.7 Randomized trials

Considerations of cost and scale have so far deterred most investigators from attempting randomized, controlled trials of the preventive value of exercise (Remington and Schork, 1967; Taylor et al., 1966). One exception is a World Health Organization (1977) study. This has matched entire factories for size, geographic location, and nature of the manufactured product. At experimental factories, advice has been given on diet, smoking cessation, reduction of body mass and hypertension, and an increase of physical activity has been encouraged. Over a two-year period of observation, several coronary risk factors (serum cholesterol, systolic blood pressure, and the number of continuing smokers) have shown small but favorable changes relative to workers in control factories.

3.8 Economic considerations

Klarman (1964) set the costs of cardiovascular disease in the United States at $31 billion, measured in 1962 dollars. Allowance has already been made for the largest of the costs he identified (premature death) through the "change of appraised age," above. The residue ($11.3 B in 1962) is now worth about $35.8 B. Assuming the epidemiologists are correct in suggesting that vigorous activity could halve this item, and that 20 percent of the population would participate in the necessary exercise classes, the saving would amount to $36 per employed worker-year.

3.9 Alcohol and cigarette abuse

An excessive consumption of alcohol and cigarettes has been a particular problem in military populations, due to preferential rates of duty on these commodities. Involvement in regular exercise may play some role in curtailing the burden of disease due to cigarettes and alcohol abuse (for instance, chronic obstructive lung disease and cirrhosis). The maximum likely benefit is a doubling of successful cigarette withdrawal, with some reduction of alcohol consumption. In the civilian population, the end result might thus be 10 percent rather than 35 percent regular smokers (Morgan, Gildiner, and Wright, 1976), 25 percent of the population reversing the cardiac and respiratory penalties of their addiction over a period of ten years.

Some of the potential health savings have already been considered, particularly a reduction of cardiac disease. Subtracting the cardiac component, other direct costs of illness and indirect costs of morbidity attributable to smoking and alcohol abuse rob the general U.S. economy of $70.2 B per year. Reasonable assumptions concerning an exercise program in an industrial population are (i) 20 percent participation in the exercise classes, (ii) 20 percent of the exercise volunteers are initially smokers and/or heavy drinkers, and (iii) 50 percent of the exercisers are cured of their addiction. Thus, at best 2 percent of the sample (2/35 addicts) will stop smoking or excessive drinking. On this basis, the benefit to the U.S. economy drops to $4.01 B per year, or $37 per employed worker-year. Assigning $20 of the saving to alcoholism and $17 to smoking, and allowing maximization of the smoking withdrawal benefit over ten years from a residual career-span of 20 years, the effective saving drops to $33 per employed worker-year. Part even of this figure may have been included already under allowances for absenteeism and health insurance claims.

3.10 Other diseases

Relatively little attention has been given to the value of regular exercise in the prevention of other chronic diseases (Table 5).

It is known that physical activity can induce a small but therapeutically important reduction of systemic blood pressure (presumably mainly through a reduction of obesity). It seems likely by analogy that regular exercise may have some value in preventing the development of hypertension. This is supported by observations of lower than average resting blood pressures among competitors in Masters' Athletic competitions (Kavanagh and Shephard, 1977).

Likewise, it is known that obesity predisposes to maturity-onset diabetes and regular exercise decreases the demand of such patients for insulin. It is thus logical to infer that exercise may also be helpful in the prophylaxis of this condition.

Gross obesity increases the risk of developing several disorders such as cholecystitis. There is also a greater likelihood of complications following abdominal surgery (wound rupture, local infection, respiratory disease) and (particularly in older people) an increased probability of bone fractures and degenerative conditions of the knee and hip. To the extent that physical activity controls obesity, it may have preventive value in all of these conditions.

Exercise also has a powerful effect in elevating mood--possibly through the release of endorphins (Farrell et al., 1982). There is thus a vast and unexploited

413

Table 5

Conditions other than ischemic heart disease where
exercise may have preventive value

Cigarette and alcohol abuse
Obesity
Hypertension
Maturity onset diabetes
Cholecystitis
Surgical complications
Bone fractures
Degenerative conditions of joints
Anxiety/depression
Osteoporosis/osteomalacia

---

potential for the treatment of that low-level anxiety and depression which makes up so much of general practice.

Lastly, there is some evidence that regular weight-bearing activity (whether by the astronaut or the senior citizen) is important in checking osteoporosis, osteomalacia, and associated renal complications (Sidney, Shephard, and Harrison, 1977; Smith and Babcock, 1973).

4. Physical Injury

Injudicious exercise can be a major cause of physical injury, and indeed exercise programs for middle aged and older adults are sometimes marred by disabling injuries in up to 50 percent of participants (Kilböm et al., 1969; Mann et al., 1969). Nevertheless, the Toronto Life Assurance Study illustrates that it is possible to conduct a gently progressive fitness program for sedentary adults without any increase of medical insurance claims for musculo-skeletal problems (Shephard et al., 1982).

In some industries, lifting injuries are more frequent among supervisors than among manual workers (Guthrie, 1963; Shephard, 1974), suggesting the prophylactic value of good muscle condition. Pravosudov (1978) claimed that "professional traumatism" was 2 to 10 times less common in workers who took up physical culture than in those who did not. However, as in many cross-sectional comparisons, interpretation of his results is made difficult by the question of self-selection. Undoubtedly, incipient back-pain would have deterred some Russian workers from becoming athletes. Mealy (1976) carried out a longitudinal experiment on Dallas police officers. Initially, the group was affected by a high incidence of low-back problems, pulled muscles, strained ankles, and strained wrists. However, there was a striking reduction in the incidence of such conditions among a group who participated in a six-month employee fitness program.

The economic impact of a reduction in industrial injuries could be substantial. In Canada, about one worker in nine is injured every year, with a compensable injury rate of 11.5 per 100 worker-years; an average of 1,400 hospital beds per day are also occupied by the injured, and one worker in 6,000 is killed every year. Injury-related absenteeism amounts to 1.3 days per worker-year with average claims of Canadian $500 per injury. If 20 percent of employees participated in a fitness program, and their injury rate was halved, the economic saving would amount to at least U.S. $40 per employed worker-year.

5. Geriatric Care

Another major burden upon the health-care system is the need to provide acute, chronic, and extended residential care to the elderly (Shephard, 1978). In the United States, the cost amounts to U.S. $1,954 per annum for every citizen over the age of 65 years and Canadian figures (equivalent to U.S. $1,720 per annum) are of the same order.

How far could an increase of physical activity reduce such ill-health and the associated costs? Cynics have argued that the person who becomes physically active and thus avoids a heart attack at 45 or 50 years of age merely lives to collect a substantial pension and demand expensive geriatric care! Certainly, in strict economic terms, benefit would be maximal if physical activity merely postponed death to the year of retirement from active service.

Nevertheless, a fiscal as well as a humanitarian argument can be advanced for an increased physical activity in the retired segment of the population. Firstly, managers of pension funds may be reassured that the active senior citizen does not seem to enjoy any great increase of longevity. Rather, because of greater strength, flexibility, and aerobic power, many of the group remain capable of self care for a further eight to nine years. During this period, about two-thirds die of various causes, so that the requirement for extended residential care is potentially reduced by as much as two-thirds in an active population (Shephard, 1978).

Since fitness cannot be stored, this presupposes providing facilities to ensure that adequate exercise is taken by the elderly population. It also presupposes that the major cause of institutionalization is a limitation of physical condition, whereas in some individuals factors such as a deterioration of sight or a loss of recent memory may be the major reason for residential care.

415

Many of the mental conditions that affect the elderly are unlikely to be helped by regular physical activity. However, involvement in a vigorous exercise program could counter anxiety and depression in at least a proportion of senior citizens. Likewise, enhanced social contacts could counter the "disengagement from society" that commonly threatens the retiree. The combined influence of these two factors could readily reduce the demand for mental care by at least 10 percent.

Lastly, there is an impact upon more general medical conditions. For some disorders such as cardiovascular and respiratory disease, the benefit may be smaller than in a younger person. However, the effects of any debilitating disease and of any piece of major surgery are more severe for the person who is in poor physical condition. Taking account of both the relative incidence of various acute and chronic disease, and their susceptibility to exercise, a one-third reduction of acute and chronic disorders might be anticipated among senior citizens who sustain regular and vigorous physical activity.

The specialized exercise programs needed by the elderly are sufficiently expensive to include in the cost/benefit calculation. Allowing U.S. $185 per participant-year, with a 20 percent exercise participation rate, the net annual benefit from reductions in various types of geriatric care would amount to some $35 per employed worker-year.

6. Recapitulation

The health benefits of regular exercise, as discussed in this paper, can conveniently be recapitulated in terms of the estimated savings to the general economy (Table 6). Almost three-quarters of the fiscal benefit apparently relates to acute rather than chronic health care costs, although it is arguable that a proportion of the hospital charges reflect acute exacerbations of chronic disease.

The military economist and the industrialist may note other fiscal benefits, including an improvement in the quality and quantity of work, and a reduction of turnover. Specialized workers are very expensive to train, and a reduction of turnover is a major argument for fitness programs.

How likely is it that the various economic savings will be realized in practice? We have assumed throughout an exercise participation rate of 20 percent. This is undoubtedly optimistic for a community fitness installation, but may be an under-estimate for military populations. It seems realizable in the general work environment, where barriers to participation are fewer than in the community (no travel costs, convenient facilities available through payroll deduction, supportive atmosphere from management and peers). On the other hand, the available class time (commonly 2 or 3 thirty minute sessions per week) is small relative to the activity undertaken in many experiments where health benefit has been reported, and it seems likely that an employee program will only achieve its intended health

Table 6

Health benefits of regular exercise, expressed as estimated economic savings

	Saving per employed worker-year
Acute illness	
Medical consultations and hospital usage+	$233*
Absenteeism/perceived health	30
Industrial injury	40
Chronic illness	
Risk profile/appraised age	36
Cardiovascular disease (other than appraised age)	36
Cigarette and alcohol related diseases	33
Geriatric care	35
	$443

Note: In an industrial setting, an employee fitness/life style program may reap roughly equal dividends from an increase of worker satisfaction, gains of productivity, and decreased turnover.

+Some of the hospital costs may reflect acute hospitalization for chronic disease.

*All savings are expressed as equivalent 1982 U.S. Dollars.

objectives if it is viewed as a groundwork for a more comprehensive personal exercise prescription. Finally, there is the ever-present obstacle of selection. While the participation rate is only 20 percent, it is quite probable that those recruited will be the healthy 20 percent of the working population; for maximum effectiveness, we must continue to search for methods of reaching and influencing the other 80 percent of employees, including those who have adverse attitudes to health and society.

References

Allen, C. L., Brown, T. W., and O'Hara, W. J. Aerobic fitness and coronary risk factors. In C. Allen (ed.) *Proceedings of the first RSG4 Physical Fitness Symposium with special reference to military forces.* Downsview, Ont.: Defence and Civil Institute of Environmental Medicine, 1978.

Bardsley, J. Canadian Forces Life Quality Improvement Programme, pp. 264-270. In C. Allen (ed.) *Proceedings of the first RSG4 Physical Fitness Symposium with special reference to military forces.* Downsview, Ont.: Defence and Civil Institute of Environmental Medicine, 1978.

Barhad, B. Physical activity in modern history. *Physiologie* 16 (1979) 117-122.

Berry, R. E. and Boland, J. P. The work-related costs of alcohol abuse. In C. J. Schramm (ed.) Alcoholism and its treatment in industry. Baltimore: Johns Hopkins University Press, 1977.

Bjurstrom, K. A. and Alexiou, N. G. A program of heart disease prevention for public employees. *Journal of Occupational Medicine* 20 (1978) 521-531.

Bourne, G. and Wedgwood, J. Heart disease and influenza. *Lancet* 1 (1959) 1226-1228.

British Medical Journal. Smoking and colds. *British Medical Journal* 3 (1974) 594.

Brünner, D. and Manelis, G. Physical activity at work and ischemic heart disease. In O. A. Larsen and R. O. Malmborg (eds.) *Coronary heart disease and physical fitness.* Baltimore, Md.: University Park Press, 1971.

Byrd, R. Impact of physical fitness on police performance. *Police Chief.* (December, 1976) 30-32.

Cooper, K. H., Meyer, B. U., Blide, R., Pollock, M., and Gibbons, L. The important role of fitness determination and stress testing in predicting coronary incidence. *Annals of the New York Academy of Science* 301 (1977) 642-652.

Corrigan, D. L. Effect of habitual exercise on total health as reflected by non-accidental insurance claims. *Action: American Association of Fitness Directors in Business and Industry Journal* 3 (1980) 7-8.

Cox, M., Shephard, R. J., and Corey, P. Influence of an employee fitness programme upon fitness, productivity and absenteeism. *Ergonomics* 24 (1981) 795-806.

Dawber, T. R. *The Framingham study.* Cambridge, Mass.: Harvard University Press, 1980.

Dodov, N., Ploshtakov, P., Patcharazov, V., and Nikolova, L. Active sport as a factor for diseases with temporary working incapability among industrial workers. In A. H. Toyne (ed.) *Proceedings of 20th world congress in sports medicine.* Melbourne: Australian Sports Medicine Federation, 1975.

Epstein, L., Miller, G. J., Stitt, F. W., and Morris, J. N. Vigorous exercise in leisure time, coronary risk factors, and resting electro-cardiogram in middle-aged male civil servants. *British Heart Journal* 38 (1976) 4093-409.

Erwin, J. People's Jewelry Company: where a new personnel director introduced recreation and employee services and watched absenteeism fall dramatically. *Recreation Management* 21 (1978) 18-19.

Farrell, P. A., Gates, W. K., Maskeid, M. G., and Morgan, W. P. Increases in plasma β endorphin/β lipoprotein immune - reactivity after treadmill exercise in humans. *Journal of Applied Physiology* 52 (1982) 1245-1249.

Fletcher, C. M. Chronic bronchitis. Its prevalence, nature and pathogenesis. *American Review of Respiratory Disease* 80 (1959) 483-494.

Garson, R. D. Pilot project on Metropolitan Life Fitness Program. Unpublished report. Toronto: Metropolitan Life Assurance Co., 1977.

Guthrie, D. I. A new approach to handling in industry. A rational approach to the prevention of low back pain. *South African Medical Journal* 37 (1973) 654-656.

Hickey, N., Mulcahy, R., Bourke, G. J., Graham, I., and Wilson-Davis, K. Study of coronary risk factors related to physical activity in 15,171 men. *British Medical Journal* 3 (1975) 507-509.

Jokl, E. Serologische Untersuchungen an Sportslevten. *Journal of Experimental Medicine* 77 (1931) 65-101.

Jokl, E. *The clinical physiology of physical fitness and rehabilitation.* Springfield, Ill.: Charles C. Thomas, 1958.

Jokl, E., Jokl-Ball, M., Jokl, P., and Frankel, L. Notation of exercise. In D. Brunner and E. Jokl (eds.) *Medicine and sport, Vol. 4. Physical Activity and Aging.* Basel: Karger, 1970.

Jung, K. Incidence of coronary risk factors in pilots of the German Air Force and possibilities to prevent them. In C. Allen (ed.) *Proceedings of the first RSG4 Physical Fitness Symposium with special reference to military forces.* Downsview, Ont.: Defence and Civil Institute of Environmental Medicine, 1978.

Kannel, W. B. Cardiovascular disease: A multifactorial problem (insights from the Framingham Study). In M. L. Pollock and D. H. Schmidt *Heart disease and rehabilitation.* Boston: Houghton Mifflin, 1979.

Karvonen, M. J., Klemola, H., Virkajärvi, J., and Kekkonen, A. Longevity of endurance skiers. *Medical Science Sports* 6 (1974) 49-51.

Kavanagh, T. and Shephard, R. J. The effects of continued training on the aging process. *Annals of the New York Academy of Science* 301 (1977) 656-670.

Kilböm, A., Hartley, L. H., Saltin, B., Bjüre, J., Grimby, G., and Astrand, I. Physical training in sedentary middle-aged and older men. I. Medical evaluation. *Scandinavian Journal of Clinical Laboratory Investigation* 24 (1969) 315-328.

Klarman, H. E. Socio-economic impact of heart disease. In E. C. Andrus (ed.) *The heart and circulation. Second national conference on cardiovascular disease, Vol. 2. Community services and education.* Washington, D.C. U.S. Public Health Service, 1964.

Krikler, D. M. and Zilberg, B. Activity and hepatitis. *Lancet* 2 (1966) 1046-1047.

Lapiccirella, V., Lapiccirella, R., Abboni, F., and Liotta, S. Enquête clinque, biologie et cardiographique parmi les tribus nomades de le Somalie qui se nourissent seulement de lait. *Bulletin of WHO* 27 (1962) 681-697.

Löllgen, H. and Pleines, J. Estimation of cardio-pulmonary function by means of the age equivalent. In C. Allen (ed.) *Proceedings of the first RSG4 Physical Fitness Symposium with special reference to military forces.* Downsview, Ont.: Defence and Civil Institute of Environmental Medicine, 1978.

Luce, B. R. and Schweitzer, S. O. Smoking and alcohol abuse: A comparison of their economic consequences. *New England Journal of Medicine* 298 (1978) 569-571.

Mann, G. V., Garrett, H. L., Farhi, A., et al. Exercise to prevent coronary heart disease. *American Journal of Medicine* 46 (1969) 12-27.

Mann, G. V., Schaffer, R. D., Anderson, R. S., et al. Cardiovascular disease in the Masai. *Journal of Atherosclerosis Research* 4 (1965) 289-312.

Mealy, M. New fitness for police and firefighters. *Physician and Sports Medicine* 7 (1976) 96-100.

Milan, F. *The human biology of circumpolar populations.* London: Cambridge University Press, 1980.

Montoye, H. J. *Physical activity and health. An epidemiologic study of a entire community.* Englewood Cliffs, N.J.: Prentice Hall, 1975.

Montoye, H. J., Van Huss, W. D., Olson, H., Hudec, A., and Mahoney, E. Study of the longevity and morbidity of college athletes. *Journal of the American Medical Association* 162 (1956) 1132-1134.

Morgan, P., Gildiner, M., and Wright, G. R. Smoking reduction in adults who take up exercise: A survey of a running club for adults. *CAHPER Journal* 42 (1976) 39-43.

Morris, J. N., Adams, C., Chave, S. P. N., Sirey, C., Epstein, L., and Sheehan, D. J. Vigorous exercise in leisure time and the incidence of coronary heart disease. *Lancet* 1 (1973) 333-339.

Morris, J. N., Heady, J. A., and Raffle, P. A. Physique of London busmen: Epidemiology of uniforms. *Lancet* 2 (1956) 569-570

Morris, J. N., Kagan, A., Pattison, D. C., Gardner, M. J., and Raffle, P. A. B. Incidence and prediction of ischaemic heart disease in London busmen. *Lancet* 2 (1966) 553-559.

Paffenbarger, R. S. Physical activity and fatal heart attack: Protection or selection? In E. A. Amsterdam, J. H. Wilmore, and A. N. deMaria (eds.) *Exercise in cardiovascular health and disease.* New York: Yorke Medical Books, 1977.

Paffenbarger, R. S., Hale, W. E., Brand, R. J., and Hyde, R. T. Work-energy level, personal characteristics and fatal heart attack: A birth cohort effect. *American Journal of Epidemiology* 105 (1977) 200-213.

Pafnote, M., Voida, I., and Luchian, O. Physical fitness in different groups of industrial workers. *Physiologie* 16 (1979) 129-131.

Polednak, A. P. Longevity and cardiovascular mortality among former college athletes. *Circulation* 46 (1972) 649-654.

Pravosudov, V. P. Effects of physical exercises on health and economic efficiency. In F. Landry and W. A. R. Orban (eds.) *Physical activity and human well-being.* Miami, Fla.: Symposia Specialists, 1978.

Quasar. *The relationships between physical fitness and the cost of health care.* Toronto: Quasar Systems, Ltd., 1976.

Remington, R. D. and Schork, M. A. Determination of number of subjects needed for experimental epidemiologic studies of the effect of increased physical activity on incidence of coronary heart disease. Preliminary considerations. In M. J. Karvonen and A. J. Barry (eds.) *Physical activity and the heart.* Springfield, Ill.: Charles C. Thomas, 1967.

Rosenman, R. H. The influence of different exercise patterns on the incidence of coronary heart disease in the Western Collaborative group study. In D. Brunner and E. Jokl (eds.) *Physical activity and aging.* Baltimore, Md.: University Park Press, 1970.

Rosenman, R. H., Bawol, R. D., and Oscherwitz, M. A 4-year prospective study of the relationship of different habitual vocational physical activity to risk and incidence of ischemic heart disease in volunteer male federal employees. *Annals of the New York Academy of Science* 301 (1977) 627-641.

Russell, W. R. Poliomyelitis, the pre-paralytic state, and the effect of physical activity on the severity of paralysis. *British Medical Journal* 2 (1947) 1023-1038.

Selye, H. Stress and nation's health. In W. A. R. Orban (ed.) *Proceedings of the National Conference on Fitness and Health.* Ottawa: Information Canada, 1974. Pp. 65-74.

Shapiro, S., Weinblatt, E., Frank, C. W., and Sager, R. V. Incidence of coronary heart disease in a population insured for medical care (HIP). *American Journal of Public Health* 59 Supplement 1-101 (1969).

Shephard, R. J. *Men at Work. Applications of ergonomics to performance and design.* Springfield, Ill.: Charles C. Thomas, 1974.

Shephard, R. J. Summary of Session 3. In C. Allen (ed.) *Proceedings of the first RSG4 Physical Fitness Symposium with special reference to military forces.* Downsview, Ont.: Defence and Civil Institute of Environmental Medicine, 1978a.

Shephard, R. J. *Physical activity and aging.* London: Croom Helm, 1978b.

Shephard, R. J. *Ischemic heart disease and physical activity.* London: Croom Helm, 1981.

Shephard, R. J., Corey, P., and Cox, M. Health hazard appraisal - the influence of an employee fitness programme. *Canadian Journal of Public Health.* 73 (1982) 183-187.

Shephard, R. J., Corey, P., Renzland, P., and Cox, M. The influence of an industrial fitness programme upon medical care costs. *Canadian Journal of Public Health.* 73 (1982) 259-263.

Shephard, R. J., Cox, M., and Corey, P., Fitness program participation: Its effects on worker performance. *Journal of Occupational Medicine* 23 (1981) 359-363

Sidney, K. H., Shephard, R. J., and Harrison, J. Endurance training and body composition of the elderly. *American Journal of Clinical Nutrition* 30 (1977) 326-333.

Smith, E. L. and Babcock, S. W. Effects of physical activity on bone loss in the aged. *Medical Science Sports* 5 (1973) 68.

Taylor, H. L., Klepetar, E., Keys, A., et al. Death rates among physically active and sedentary employees of the railroad industry. *American Journal of Public Health* 52 (1962) 1697-1707.

Taylor, H. L., Parlin, R. W., Blackburn, H., and Keys, A. Problems in the analysis of the relationship of coronary heart disease to physical activity or its lack, with special reference to sample size and occupational withdrawal. In K. Evang and K. L. Andersen (eds.) *Physical activity in health and disease.* Baltimore: Williams and Wilkins, 1966.

Tibblin, G., Wilhelmsen, L., and Werkö, L. Risk factors for myocardial infarction and death due to ischemic disease and other causes. *American Journal of Cardiology* 35 (1975) 514-522.

Weinstein, L. Physical activity and poliomyelitis. *Boston Medical Quarterly* 3 (1952) 11-16.

World Health Organization. The Prevention of Coronary Heart Disease, WHO Report ICP/CVD 002(10). (Copenhagen, 1977).

Young, M. and Willmot, P. *The symmetrical family.* London: Routledge and Kegan Paul, 1973.

CONTRIBUTORS

Catherine J. Atkins, Ph.D., *Chapter 18*, Center for Behavioral Medicine, San Diego State University, San Diego, California 92182

Pietro Avogaro, M.D., *Chapter 4*, Unit for Atherosclerosis, Hyperlipaemias and Diabetes, Ospedale Regionale, 30100, Venice, Italy

Massimo Biondi, M.D., *Chapter 15* Istituti di Psichiatria, Universita di Roma, Viale dell 'Universita 30, Roma, Italy

Denis Burkitt, M.B., *Chapter 6*, The Old House, Bussage, Stroud, Glos. G16 8AX, England

V. Cairns, Ph.D., *Chapter 12*, Research Center GSF, Institute for Medical Informatics and Health Services Research (MEDIS), Department of Epidemiology, Ingolstädter Landstrasse 1, D-8042 München-Neuherberg, Federal Republic of Germany

Thomas J. Coates, Ph.D. *Chapter 16*, Division of General Internal Medicine, University of California, San Francisco, School of Medicine, San Francisco, California 94143

Michael H. Criqui, M.D., M.P.H., *Chapters 2 and 5*, Division of Epidemiology, University of California, San Diego, School of Medicine, M-007, La Jolla, California 92093

Fabienne Elgrably, M.D., *Chapter 11*, University Pierre et Marie Curie, U.E.R. Broussais-Hôtel-Dieu, Service de Diabétologie, Hôtel-Dieu Hospital, 1 place du Parvis Notre-Dame, 75181 Paris Cedex 04, France

Craig K. Ewart, Ph.D., *Chapter 16*, John Hopkins School of Hygiene and Public Health, Baltimore, Maryland

John W. Farquhar, M.D. *Chapter 19*, Stanford Heart Disease Prevention Program, Stanford University, Stanford, California 94305

Stephen P. Fortmann, M.D., *Chapter 19*, Stanford Heart Disease Prevention Program, Stanford University, Stanford, California 94305

J. Geboers, M.Sc., *Chapter 7*, Akademisch Ziekenhuis St.-Rafael, Capucijnenvoer 33, B-3000, Leuven, Belgium

Andrew P. Haines, MB, MRCP, MRCGP, *Chapter 10*, MRC Epidemiology and Medical Care Unit, Northwick Park Hospital, Watford Road, Harrow, Middlesex, United Kingdom HA1 3UJ, England

Ursula Härtel, M.A. *Chapter 12*, Research Center GSF, Institute for Medical Informatics and Health Services Research (MEDIS), Department of Epidemiology, Ingolstädter Landstrasse 1, D-8042 München-Neuherberg, Federal Republic of Germany

423

J. V. Joossens, M.D. *Chapter 7,* Akademisch Ziekenhuis St.-Rafael, Capucijnenvoer 33, B-3000, Leuven, Belgium.

George A. Kaplan, Ph.D., *Chapter 14* Director, Human Population Laboratory, California Department of Health Services, 2151 Berkeley Way, Annex 2, Room 211, Berkeley, California 94704-9980.

Robert M. Kaplan, Ph.D., *Chapters 3 and 18,* Center for Behavioral Medicine, San Diego State University and Department of Community and Family Medicine, University of California, San Diego, School of Medicine, M-022, La Jolla, California 92093.

U. Keil, M.D., Ph.D. *Chapter 12,* Research Center GSF, Institute for Medical Informatics and Health Services Research (MEDIS), Department of Epidemiology, Ingolstädter Landstrasse 1, D-8042 München-Neuherberg, Federal Republic of Germany.

Daan Kromhout, Ph.D., M.P.H., *Chapter 8,* Institute of Social Medicine, State University of Leiden, Wassenaarseweg 62, Postbus 9605, 2300 RC Leiden, The Netherlands.

Darwin R. Labarthe, M.D., Ph.D., *Chapter 9,* University of Texas Health Science Center at Houston, School of Public Health, The EpiCenter, P. O. Box 20186, Houston, Texas 77025

Nathan Maccoby, Ph.D., *Chapter 19,* Stanford Heart Disease Prevention Program, 730 Welch Road, Stanford University, Stanford, California 94305

Philip Nader, *Chapter 17,* University of California, San Diego Medical Center, H-664-A 225 Dickinson Street San Diego, California 92103

Paolo Pancheri, M.D., *Chapter 15* Istituti di Psichiatria, Universita di Roma, Viale dell'Universita 30, Roma, Italy

Irwin G. Sarason, Ph.D. *Chapter 13,* Department of Psychology, University of Washington, NI-25, Seattle, Washington 98195

Barbara R. Sarason, Ph.D., *Chapter 13,* Department of Psychology, University of Washington, NI-25, Seattle, Washington 98195

Roy J. Shephard, M.D., Ph.D., *Chapter 20,* Director, School of Physical and Health Education, Department of Preventive Medicine and Biostatistics and Institute of Medical Sciences, University of Toronto, 320 Huron Street, Toronto, Ontario M5S 1A1, Canada

Loucas Sparos, M.D., *Chapter 1,* Associate Professor of Medicine, Department of Hygience and Epidemiology, University of Athens Medical School, GR-115 27 (Goudi) Athens, Greece.

Georges Tchobroutsky, M.D. *Chapter 11,* University Pierre et Marie Curie, U.E.R. Broussais-Hôtel-Dieu, Service de Diabétologie, Hôtel-Dieu Hospital, 1 place du Parvis Notre-Dame, 75181 Paris Cedex 04, France.

Dimitrios Trichopoulos, M.D., *Chapter 1,* University of Athens Medical School, GR-115 27 (Goudi) Athens (609), Greece.

PARTICIPANTS AT BELAGIO MEETING

Back Row: N. Criqui, N. Maccoby, T. Coates, U. Härtel, I. Sarason, B. Sarason, D. Kromhout, G. Kaplan, J. V. Joossens, P. Nader

Front Row: D. Trichopoulus, D. Burkitt, D. Labarthe, M. Criqui, R. Kaplan, C. Atkins, R. Shepard, U. Keil, M. Biondi

AUTHOR INDEX

Brandt, H., 113
Braunwald, E., 40
Breitrose, H., 385-387
Brems-Heyns, E., 105,111-112
Breslow, L., 240,242
Breslow, N., 4-6,8,11
Brewer, B.J., Jr., 59
Brightwell, D.R., 271
British Medical Journal, 406
Brook, R.H., 1,35
Bross, I.J., 165
Brown, B.W., Jr., 385-387
Brown, G.W., 227
Brown, J.H., 271
Brown, R.I.F., 272,287
Brown, T.W., 402,407
Brownell, K.D., 373,376
Brownell, L.D., 330,332,348
Brunner, D., 409-410
Brunzell, J.D., 367
Buchem, F.S.P., 122,124
Buchwald, H., 123
Buckley, J.D., 170
Buell, J.C., 75
Bunn, H.F., 194
Burch, G.E., 169,171
Burch, J.D., 168,171
Burgess, M., 292
Burke, B.S., 123,162
Burke, R., 227
Burkitt, D.P., 94-95,98,100,167
Burns, P.E., 168
Burroughs, L.F., 98
Burtel, A.G., 58
Buschard, K., 166
Bush, J.W., 32-34,36-41
Buskirk, E.R., 19
Butensky, A., 329,332,347
Butts, W.C., 397
Butz, G., 330
Byran, M.S., 329,331
Byrd, R., 405-406

Cahill, G.F., 363
Cahn, A., 123
Calabrese, E.J., 307-308
Caldwell, J.R., 330,332
Callahan, E.J., 364,373
Camacho, T., 239,243,262,264
Camerino, M., 272-273
Campbell, D.T., 9
Campeau, L., 41
Cancer Research Campaign Working
  Party Trials and Tribulations, 8,11
Caplan, R.D., 330,332
Carenza, L., 287,292
Carlier, J., 105
Carlson, L.A., 57,59
Carlsson, C., 73
Carney, S., 127
Carpenter, C.B., 273
Cartwright, D., 386,387,389
Cassel, J., 219,228,239,257,262
Castelli, W.P., 19,57-58,71-72,81,
  329
Cataland, S., 358
Cazzolato, G., 62
Celani, E., 197
Centers for Disease Control, 138
Centerwall, B.S., 69
Chadwick, J.H., 309
Chan, W.C., 172
Chandra, V., 259
Chapin, G., 145-146,148-149,151-152,
  154
Chapman, G., 166
Charney, E., 371
Chave, S.P.N., 409
Chazan, B.I., 364
Chen, M.M., 34,40
Chenoweth, A.C., 304
Chew, E., 15,19,21,29
Chiang, B.N., 315
Child, M., 171
Chirigos, M., 271

Willett, W.C., 307-308
William, B.J., 324
Williams, A.F., 331
Williams, G., 165
Williams, P.T., 329-330,332
Williams, R., 166-167
Williams, R.B., 239
Williams, R.R., 171
Williams, S.R., 166
Williamson, J., 366, 371
Williamson, J.R., 359-362
Willis, C.E., 172
Willmot, P., 411
Wilson, A., 372
Wilson, M., 127
Wilson-Davis, K., 407
Wing, A.L., 302-303
Wing, R.R., 332
Wingard, D.L., 242,245
Winkelstein, W., Jr., 331-332
Winter, C., 332
Wixom, C., 387
Wolff, C.T., 292
Wolpe, J., 372
Wood, P., 82
Wood, P.D., 331,385,387
World Health Organization, 138,409, 412

Worlever, T.M.S., 99
Worthington, R.E., 272
Wright, G.R., 413
Wuckolls, K.B., 228
Wyler, A.R., 221
Wylie-Rosett, J., 172
Wynder, E.L., 165,167,169
Yale Department of Opthalmology, 360
Yalow, R.S., 366
Yamomoto, K.J., 221
Yano, K., 68,73,77,83,171,259
Yen, S., 169-170
Ylikahri, R., 73
Young, M., 411
Young, W., 58
Yusim, S., 239
Zahka, K.G., 306
Zannis, V.I., 60
Zegman, 373
Zelen, M., 8
Zichella, L., 287,292
Ziegler, R.G., 168
Zilberg, B., 403
Zinner, S.H., 303
Zuberi, S.J., 170
Zucker, R. 227
Zukel, W.J., 125

SUBJECT INDEX